Value Engineering for Electrical Systems in Construction

Value Engineering for Electrical Systems in Construction

Hazem Hussein
Lead Electrical Engineer | MEP Consultant | Design Manager
Nijmegen, The Netherlands

WILEY

This edition first published 2026

Registered Offices
John Wiley & Sons, Inc., 111 River Street, Hoboken, NJ 07030, USA
John Wiley & Sons Ltd, New Era House, 8 Oldlands Way, Bognor Regis, West Sussex, PO22 9NQ, UK

For details of our global editorial offices, customer services, and more information about Wiley products visit us at www.wiley.com.

The manufacturer's authorized representative according to the EU General Product Safety Regulation is Wiley-VCH GmbH, Boschstr. 12, 69469 Weinheim, Germany, e-mail: Product_Safety@wiley.com.

Wiley also publishes its books in a variety of electronic formats and by print-on-demand. Some content that appears in standard print versions of this book may not be available in other formats.

Library of Congress Cataloging-in-Publication Data Applied for
Hardback ISBN: 9781394298563

Cover Design: Wiley
Cover Images: © DivVector/Getty Images, © appleuzr/Getty Images, © Bezvershenko/Getty Images

Set in 9.5/12.5pt STIXTwoText by Straive, Pondicherry, India
Printed and bound by CPI Group (UK) Ltd, Croydon, CR0 4YY
C9781394298563_050226

Contents

About the Author

Hazem Hussein is a Lead electrical engineer, MEP Consultant, and design manager with extensive international experience in the design, planning, and delivery of complex electrical works across a broad spectrum of building and infrastructure projects. His professional portfolio includes data centers, mission-critical facilities, healthcare institutions, hospitality developments, high-rise buildings, commercial and mixed-use complexes, and sustainable certified projects across Europe, the Middle East, and the Gulf region.

He has held senior engineering roles for globally recognized consultancies, including KEO International Consultants and Haskoning, contributing to the successful delivery of world-class projects.

His professional contributions have been recognized through multiple industry nominations, including being shortlisted twice for Electrical Engineer of the Year by *MEP Middle East Magazine* and *CBNME Awards*. Throughout his career, he has specialized in integrating value engineering methodologies into electrical systems design and delivery, with particular focus on optimizing functional performance, life-cycle cost efficiency, system resilience, regulatory compliance, and sustainable design outcomes. His technical expertise encompasses low- and medium-voltage power systems, lighting design, extra-low voltage (ELV) and information and communication technology (ICT) systems, emergency power infrastructure, and mechanical, electrical, and plumbing (MEP) coordination. He brings extensive experience in leading multi-disciplinary teams and managing complex design and delivery processes from early concept design through to final construction and handover across all project stages. In addition to his technical work, he has actively contributed to professional development within the engineering community through technical training, internal knowledge development programs, the development of company tools and standardization frameworks, and supporting colleagues in the practical application of electrical systems design and optimization.

He holds a Bachelor's degree in Electrical Engineering, is affiliated with various professional engineering bodies, and holds Professional certifications including PMP® and LEED® Green Associate™

Preface

Throughout my professional career, I have often observed a degree of hesitation among practitioners when it comes to engaging in value engineering (VE), particularly within the context of electrical systems. This reluctance frequently stems from a fundamental misconception—that VE is a fault-finding process intended to expose deficiencies in original design work. Such a perception can evoke defensiveness among designers, who may fear that VE challenges the sufficiency of their initial proposals. In reality, this view misrepresents both the intent and the true power of VE.

VE is not a retrospective cost-cutting exercise. Rather, it is a structured, forward-looking methodology aimed at maximizing project value through functional analysis, technical optimization, innovation, and interdisciplinary collaboration. When properly understood and implemented, VE becomes a strategic design tool that informs early stage decisions, aligns project objectives, and supports the delivery of high-performance, cost-efficient, resilient, and environmentally responsible electrical systems. In doing so, it contributes directly to broader sustainability goals by promoting resource efficiency, reducing embodied and operational carbon, and optimizing system life cycle impacts without compromising technical integrity or project functionality.

Over the years, I recognized a persistent gap in technical resources dedicated to the systematic application of VE within the mechanical, electrical, and plumbing (MEP) disciplines—particularly concerning electrical infrastructure. While VE principles have been widely documented in broader engineering contexts, their targeted application to electrical systems remained underserved. This realization prompted me to initiate a deliberate effort to document value opportunities encountered during design reviews, technical coordination, and system evaluations across numerous projects. What began as a working collection of notes has progressively evolved into this comprehensive guide.

The preparation of this book required an extensive synthesis of technical literature, industry experience, regulatory standards, and applied project practice. The goal throughout has been to translate what is often viewed as an abstract or intimidating methodology into a practical, accessible, and technically rigorous resource that supports both VE studies and general design development for professionals engaged in electrical infrastructure projects.

This book integrates both the methodological foundation of VE and its focused application within electrical works for construction projects. It first establishes a structured VE framework specifically adapted to electrical engineering, addressing functional analysis, value optimization, life cycle evaluation, and systematic decision-making. Building on this foundation, the book applies VE principles across the full spectrum of electrical systems, including load estimation, power distribution (MV/LV), earthing and lightning protection, lighting design, emergency power and uninterruptible power supply (UPS) systems, information and communication technology (ICT) and extra-low voltage (ELV) infrastructure, fire and life safety systems, and cable management. Additional chapters address regulatory compliance, financial assessment, retrofit applications, and emerging technology trends, providing a comprehensive and practice-oriented reference for professionals involved in both new construction and existing facility upgrades.

Value Engineering for Electrical Systems in Construction is intended to serve engineers, designers, consultants, project managers, and stakeholders who seek to apply VE not merely as a corrective measure, but as a proactive design philosophy. At the same time, this book also functions as a practical design guide, offering structured frameworks, system evaluation techniques, technical design considerations, discipline-specific guidance, as well as a comprehensive set of checklists, templates, and practical tools that support efficient VE workshops and real-world design application. Whether engaged in formal VE exercises or in standard design development, readers will find actionable methodologies that support high-performance, compliant, and cost-effective electrical system delivery.

As electrical systems continue to grow in complexity—driven by digitalization, sustainability targets, mission-critical performance, and evolving client demands—the application of VE becomes not only relevant but essential. This book seeks to equip professionals with both the mindset and the tools necessary to confidently apply VE in delivering optimized, compliant, and future-ready electrical systems.

Writing this book has been both a personal and professional journey. I remain deeply convinced that VE is far more than a methodology; it is a professional discipline grounded in critical thinking, technical excellence, and collaborative design. I invite readers to engage with the material, apply its principles in their own practice, and continually explore new opportunities to maximize value across the entire lifecycle of electrical infrastructure projects.

Hazem Hussein
Nijmegen, The Netherlands
June 2025

Artificial Intelligence-Assisted Visual Content Disclosure

During the preparation of this book, generative AI tools were used to assist with the development of some visual illustrations. Specifically, DALL·E 3 was employed to create draft visual compositions for selected illustrative diagrams appearing in Chapters 1, 2, 5, 7, 8, 9, 11, and 12, based on the author's technical specifications, engineering parameters, and design requirements.

All artificial intelligence (AI)-generated visual outputs were systematically reviewed, validated, and refined by the author to ensure technical accuracy, compliance with applicable engineering standards, and consistency with the book's professional content. All engineering analyses, technical judgments, and conclusions remain solely the responsibility of the author.

Acknowledgments

The development of this book reflects the culmination of years of professional practice, project experience, and ongoing professional development within the field of electrical systems and value engineering. I would like to express my appreciation to the many colleagues, industry professionals, and mentors with whom I have collaborated throughout my career. Their shared expertise, technical discussions, and practical insights have contributed to shaping the knowledge base reflected in this work.

I also wish to acknowledge the editorial and production teams at Wiley for their guidance, professionalism, and support throughout the publication process.

I am especially grateful to my family for their encouragement, support, and patience throughout the writing process.

While this book reflects my personal experience and professional perspective, I remain indebted to the broader engineering community whose collective knowledge continues to advance the practice of electrical engineering and value engineering worldwide.

Professional Use Notice

Professional Guidance and Best Practices

This publication provides general guidance on value engineering principles and practices for electrical systems design and electrical works within the construction field. The content represents established industry best practices, engineering methodologies, and professional recommendations based on the authors' experience, professional judgment, and current industry standards.

Standards and Regulatory Compliance

While this book references relevant international standards (e.g., International Electrotechnical Commission (IEC), Institute of Electrical and Electronics Engineers (IEEE), European Norm (EN), American Society of Heating, Refrigerating and Air-Conditioning Engineers (ASHRAE), and National Fire Protection Association (NFPA)) and industry guidelines, readers are required to:

- Verify the current versions of all referenced standards, as these are periodically updated and revised;
- Review and comply with applicable national, regional, and local regulations, as these may differ from international standards;
- Obtain official copies of the applicable standards, as this publication does not replace the authoritative standard documents issued by recognized standards organizations;
- Ensure full compliance with the governing codes, regulations, and statutory requirements applicable to each specific project.

Engineering Professional Responsibility

The methodologies, calculations, and recommendations contained within this publication are intended for use by qualified electrical engineers and related professionals. Readers and users of this publication remain solely responsible for:

- Exercising professional judgment in applying the content to specific project conditions and design contexts;
- Independently validating all calculations, design assumptions, and engineering decisions prior to implementation;

- Ensuring that all work fully complies with applicable codes, standards, legal requirements, and regulatory frameworks;
- Prioritizing safety, including personnel, operational, and public safety, in all design and implementation activities.

Limitations and Scope

This publication:

- Does not constitute professional engineering advice or consultation for specific projects;
- Cannot substitute for detailed, project-specific analysis performed by licensed professional engineers;
- May contain information that becomes outdated as technology, codes, and standards continue to evolve;
- Presents general principles which may not be directly applicable to all jurisdictions, project types, or unique design circumstances.

No Warranty or Liability

The authors, publisher, and affiliated organizations:

- Make no representations or warranties regarding the accuracy, completeness, or currency of the information provided;
- Accept no liability for any decisions, actions, or outcomes arising from reliance upon the content herein;
- Disclaim responsibility for any direct, indirect, incidental, or consequential damages that may result from the application or misapplication of the material;
- Recommend that qualified professional consultation be obtained for all project-specific engineering matters.

Cost and Savings Estimates

Financial estimates, cost savings examples, and economic analyses presented within this publication are:

- Provided exclusively for illustrative and educational purposes; actual project costs may vary substantially depending upon specific project conditions, geographic location, market dynamics, and execution strategies;
- Based on historical data and generalized assumptions, which may not reflect present or future pricing structures;
- Subject to multiple variables, including, but not limited to, location, scheduling, scope definition, procurement methods, and contract terms;
- Not predictive or guaranteed, as outcomes are dependent upon numerous project-specific factors and conditions.

Technology and Product References

Product names, manufacturers, and technology references cited within this publication are:

- Provided solely for informational and illustrative purposes and do not constitute endorsements or recommendations;
- Subject to change over time, as products, technologies, manufacturers, and markets evolve;
- Represent examples of commercially available solutions, not exhaustive catalogues of all viable products or suppliers;
- The responsibility of the reader to verify current specifications, availability, applicability, and suitability for intended use.

Continuing Education

Readers are strongly encouraged to:

- Maintain familiarity with evolving codes, standards, technologies, and industry best practices;
- Participate in ongoing professional development and continuing education activities relevant to their field of practice;
- Consult current manufacturer documentation, technical bulletins, and peer-reviewed publications for updated technical data;
- Engage actively with professional engineering organizations to obtain current guidance, technical resources, and professional networking opportunities.

Introduction to Value Engineering in Electrical Services in Construction Field

In today's engineering and construction industry, value engineering (VE) has emerged as a critical methodology for optimizing the balance between performance and cost. VE originated during World War II, when material shortages and procurement challenges prompted the search for functionally equivalent, cost-effective alternatives. This effort was led by Lawrence D. Miles, an industrial engineer at the General Electric Company, who formalized the foundational principles of VE to maintain quality while lowering expenditure (Miles, 1961; SAVE International, 2020).

Initially intended to substitute unavailable materials and parts without diminishing performance, VE has since matured into a structured, function-oriented methodology. It is now applied globally across industries to improve project outcomes by delivering necessary functions at the lowest total cost of ownership, without compromising safety, quality, or functionality. The evolution of this methodology—from wartime origins to global formalization—is depicted in Figure I.1.

Electrical engineering, particularly within the built environment, stands as a prime candidate for VE application. Electrical systems typically represent 10–15% of a building's total construction budget, with the proportion significantly higher in technically intensive facilities such as data centers, hospitals, and hotels (CIBSE, 2021; RSMeans, 2023). Data centers exemplify this elevated cost allocation, where electrical infrastructure—including backup generation, uninterruptible power supplies, power distribution systems, and switchgear—comprises 40–45% of total development costs (Dgtl Infra, 2024). These figures underscore the need for intelligent cost control mechanisms that do not compromise performance, resilience, or regulatory compliance.

Moreover, the nature of electrical systems—complex, interdependent, and capital-intensive—makes them uniquely suited for VE intervention. Whether designing power distribution for a mixed-use development, integrating lighting systems into architectural envelopes, or supporting automation in smart cities, electrical infrastructure demands precision and efficiency. The key challenge arises: how can we ensure electrical systems are efficient, code-compliant, and future-ready, while remaining financially viable?

VE offers a framework to address this challenge. By identifying core functions and exploring alternative methods of fulfillment, VE enables cost-effective design without degrading value. It is inherently collaborative, engaging electrical engineers, architects, cost managers, clients, and other specialists in structured, multidisciplinary analysis. This integrative process is central to the VE approach and drives the discovery of innovative, viable alternatives.

Today, the role of VE extends beyond cost rationalization. It directly supports broader industry shifts toward sustainability, energy efficiency, and life-cycle optimization. With global priorities now focused on decarbonization, renewable integration, and operational efficiency, VE has become increasingly aligned with international green building standards and long-term asset management strategies (ASHRAE, 2019; World Green Building Council, 2020).

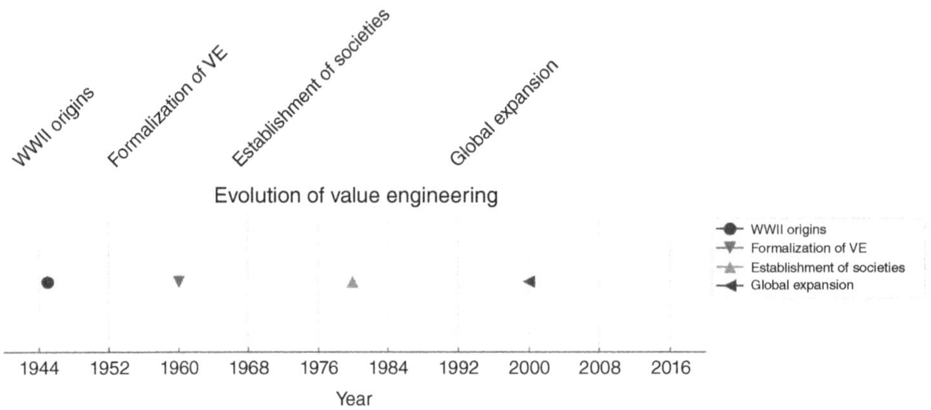

Figure I.1 Evolution of VE from WWII Origins to Global Expansion

Nevertheless, misconceptions persist. VE is often mistakenly perceived as a synonym for indiscriminate cost-cutting. In reality, effective VE requires a robust understanding of system requirements, project life-cycle goals, and technical trade-offs. It calls for a disciplined yet flexible mindset—one that balances innovation with compliance, performance with affordability, and short-term choices with long-term outcomes.

This book seeks to equip engineers, designers, project managers, and stakeholders with the tools to embed VE meaningfully into the electrical scope of construction projects. Through technical frameworks, checklists, practical examples, and case-based guidance, it presents VE not as an afterthought, but as a proactive, strategic discipline. By doing so, we aim to elevate electrical design beyond compliance—to a domain of optimized, sustainable, and resilient value.

Beyond methodology, this book also addresses the changing context in which electrical systems are designed and delivered. As the industry confronts new technological paradigms—including sustainability mandates, energy performance targets, and the integration of digital workflows such as building information modeling (BIM)—the need for proactive, value-driven thinking has never been greater. Topics such as smart grids, renewable energy integration, life-cycle costing, and financing models are no longer peripheral; they are central to modern electrical design. By positioning VE within this evolving landscape, the book aims to equip professionals with the tools to not only manage cost, but to create systems that are robust, adaptable, and aligned with long-term operational objectives.

References

ASHRAE. (2019). *ASHRAE GreenGuide: The design, construction and operation of sustainable buildings* (5th ed.). American Society of Heating, Refrigerating and Air-Conditioning Engineers.

CIBSE. (2021). *CIBSE guide K: Electricity in Buildings.* Chartered Institution of Building Services Engineers.

Dgtl Infra. (2024). How much does it cost to build a data center? Retrieved from https://dgtlinfra.com/how-much-does-it-cost-to-build-a-data-center/

Miles, L. D. (1961). *Techniques of value analysis and engineering.* McGraw-Hill.

RSMeans. (2023). *Building construction cost data 81st edn.* RSMeans Data, Gordian.

SAVE International. (2020). *Value Methodology Standard.* https://www.value-eng.org

World Green Building Council. (2020). *Bringing embodied carbon upfront.* https://www.worldgbc.org

About the Companion Website

This book is accompanied by a companion website:

www.wiley.com/go/hazem

The website includes:

- Practical Checklists in Excel sheets for Electrical Value Engineering and Design Assurance
- Practical Sheets and Templates in PDF and Excel for Effective Value Engineering of Electrical Works
- Appendices of Value Engineering for Existing Electrical Systems
- Appendices of Value Engineering for External Electrical Infrastructure Works

1

Understanding Value Engineering in Electrical Projects

1.1 Introduction to Value Engineering

Value engineering (VE) is a systematic and structured methodology designed to improve the value of a system by optimizing its functional performance while minimizing cost (SAVE International, 2020). This is achieved without compromising reliability, safety, quality, or efficiency. Within the context of electrical systems for construction, VE involves critical analysis of design choices, materials, installation techniques, and lifecycle costs to identify the most efficient and cost-effective solutions.

It is essential to emphasize that VE is not simply a means of cost. Rather, it is a disciplined process aimed at achieving maximum functionality at the lowest feasible total cost of ownership reduction (SAVE International, 2020). This involves the identification and elimination of unnecessary expenditures—whether due to overengineering, the use of inefficient technologies, reliance on outdated specifications, or redundant design features—and their replacement with technically sound and economically viable alternatives that meet the project's objectives.

VE has evolved beyond its origins as a wartime necessity into a globally recognized practice It represents a comprehensive and integrative approach to maximizing value in engineering projects, products, and processes. The central premise of VE is to ensure that each function contributes meaningfully to project objectives while maintaining quality, reliability, and safety standards (Miles, 1961).

1.1.1 VE vs. Cost Cutting

A prevalent misconception is that VE equates to cost cutting at the expense of quality or performance. In contrast, VE seeks to improve project outcomes while controlling costs through a structured, multidisciplinary approach. This distinction is outlined in Table 1.1.

In construction, VE is applied across multiple project phases to optimize cost efficiency without compromising design intent. While it is often introduced when budget constraints arise, this reactive application undermines the broader potential of VE. Ideally, VE should be integrated proactively from the earliest design stages to stimulate constructive dialogue, challenge assumptions, and identify superior alternatives.

Value Engineering for Electrical Systems in Construction, First Edition. Hazem Hussein.

Companion website: www.wiley.com/go/hazem

Table 1.1 Comparison Between VE and Cost Cutting

Aspect	Value Engineering	Cost Cutting
Primary Focus	Enhancing functionality while reducing costs	Reducing costs regardless of performance
Impact on Performance	Maintains or improves reliability and efficiency	Often results in reduced performance or quality
Methodology	Systematic analysis; multidisciplinary collaboration	Arbitrary removal of components or features
Long-Term Effect	Reduces total cost of ownership (TCO)	May increase maintenance costs and risk of system failure

For example, early VE sessions may uncover that a higher-cost switchgear is justified due to superior long-term reliability or maintenance benefits. Such insights support informed decision-making, and help avoid premature conclusions based solely on upfront costs.

VE is not only a tool for cost control but also a mechanism for improving performance across systems. Consider emergency lighting: while decentralized battery packs may offer initial savings, a centralized inverter system may be more advantageous when long-term maintenance costs and system resilience are evaluated—provided that cooling and ventilation requirements are properly addressed.

Effective VE requires a balanced assessment of both benefits and trade-offs. Take, for instance, the decision to substitute copper wiring with aluminum in healthcare or mission-critical facilities. While the latter may offer material cost savings, it demands careful evaluation of conductivity, safety, mechanical properties, and long-term reliability to avoid unintended consequences. Similarly, while commissioning processes add to initial costs, they play a vital role in uncovering design issues early, preventing costly rework or system failure during operation.

Ultimately, VE encourages the reconsideration of traditional approaches and the adoption of smarter, more sustainable alternatives. When applied proactively and strategically, VE enhances project outcomes by ensuring that every investment delivers measurable improvements in reliability, operational efficiency, and long-term value.

1.2 The Interrelation Between Initial and Running Costs

In electrical systems for construction, the relationship between initial capital expenditure (CAPEX) and operational expenditure (OPEX) is a key consideration in VE. Decisions made during the design and procurement stages directly affect not only the upfront investment but also long-term operating costs, including energy use, maintenance requirements, and system durability.

A well-implemented VE strategy seeks to achieve the optimal balance between these cost components, ensuring that any savings realized during the initial stages do not result in disproportionate increases in future operational expenses. This balance is essential for maintaining long-term value and system efficiency.

The conceptual link between initial and running costs is illustrated in Figure 1.1, which highlights how early decisions influence long-term financial and functional outcomes in electrical system planning.

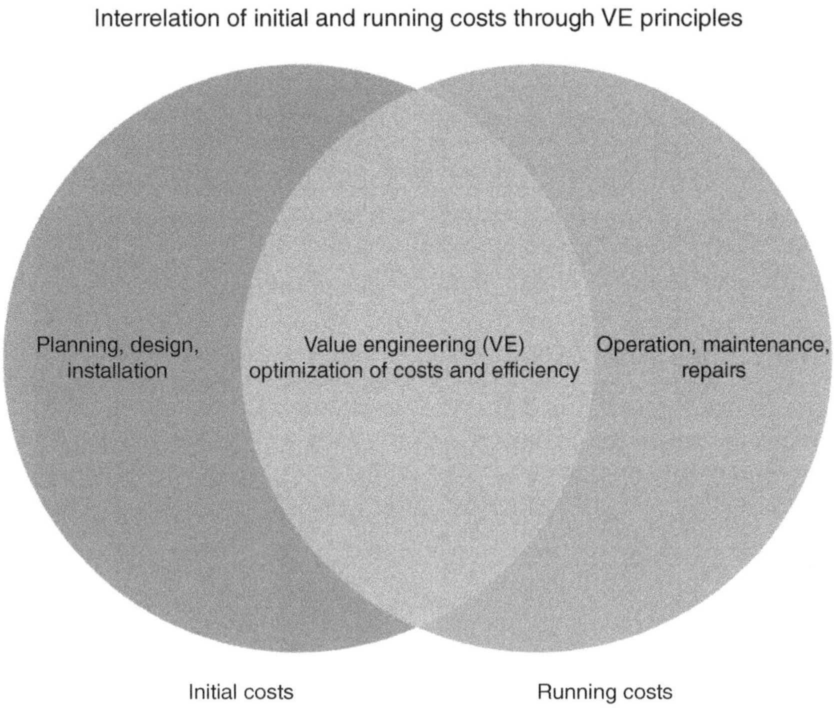

Figure 1.1 Interrelation Between Initial and Running Costs

1.3 Understanding Initial Costs (CAPEX) in Electrical Systems

Understanding and strategically managing initial CAPEX is essential in electrical services. It lays the foundation for a project's financial framework and paves the way for future operational flexibility and efficiency. Careful planning of initial costs can significantly influence the overall trajectory of a project.

Initial costs in electrical construction projects encompass all expenditures required to design, procure, and install the electrical infrastructure. These typically include:

- **Material Costs**: Cables, transformers, switchgear, circuit breakers, lighting, and other components.
- **Labor Costs**: Installation, commissioning, and testing.
- **Design and Engineering**: Planning, calculations, and compliance with applicable standards.
- **Permitting and Compliance**: Adherence to local and international electrical codes and regulations.

While traditional budgeting practices often place pressure on minimizing upfront costs, doing so without comprehensive evaluation can result in significantly higher operating costs over the system's lifecycle. These downstream expenses may outweigh any initial savings, ultimately reducing the overall value of the project.

1.4 Running Costs (OPEX) and Their Long-Term Impact

OPEX includes all costs associated with operating, maintaining, and eventually replacing electrical components throughout a system's service life. These commonly include:

- **Energy Consumption**: The efficiency of the electrical system directly influences electricity bills.
- **Maintenance and Repairs**: Component reliability affects how often maintenance or replacement is needed.
- **Downtime Costs**: Inadequate system design may result in failures, leading to service interruptions.
- **Compliance and Upgrades**: Regulatory developments may necessitate system modifications to remain compliant.

A low-cost installation may initially appear attractive, but if it results in excessive energy use, increased maintenance requirements, or a shortened lifespan, it may lead to substantial long-term costs. VE seeks to anticipate and mitigate such risks through careful analysis and planning (Dell'Isola, 1997).

1.5 Key Considerations in Balancing Initial and Running Costs

Balancing initial CAPEX with OPEX is one of the most critical aspects of VE in electrical system design. Every decision made at the design stage influences not only short-term spending but also long-term operational efficiency.

Prioritizing initial cost alone may result in higher maintenance needs, poor efficiency, or early component failure—ultimately increasing total lifecycle costs. Conversely, overinvestment in premium systems can result in excessive upfront costs without delivering proportionate operational benefits.

A well-executed VE process aims to identify the optimal balance, where investment in quality and efficiency results in cost-effectiveness over the system's lifespan. This involves a structured evaluation of several key factors.

1.5.1 The Long-Term Impact of Cost Decisions

Electrical systems are long-term assets, often with lifespans measured in decades. Early design decisions should be evaluated in terms of their future implications for performance, maintenance, and energy use. Cost cutting during early phases may provide short-term relief but create lasting issues such as:

- Higher failure rates due to low-grade materials or undersized components.
- Frequent repairs or servicing, increasing operational overhead, and downtime.
- Inefficiencies in energy distribution, contributing to higher utility costs.

Alternatively, unjustified overinvestment in high-end components without clear functional advantages may lead to unaffordable CAPEX. The key is a thorough, data-driven evaluation of lifecycle trade-offs.

1.5.2 The Role of Lifecycle Cost Analysis

Lifecycle cost analysis (LCCA) is a structured methodology for evaluating the total cost of ownership (TCO) of electrical systems (Fuller, 2010; NIST, 2022). Rather than considering only the purchase and installation costs, LCCA includes:

- Procurement and installation (materials, labor, permitting)
- Operational performance (energy consumption, system losses)
- Maintenance and service (inspections, repairs, replacements)
- System longevity (service life and upgrade cycles)
- Decommissioning (removal, disposal, recycling)

Incorporating LCCA within the VE process supports informed decision-making, ensuring that early cost savings do not result in excessive long-term expense (Whole Building Design Guide, 2023).

1.5.3 Energy Efficiency as a Cost-Balancing Factor

Energy use represents one of the most significant ongoing expenses in an electrical system's lifecycle. Apparent savings from cheaper installations may be offset by long-term inefficiencies.

Energy efficiency should therefore be a primary focus of VE assessments, considering:

- **Load Demand Optimization:** Avoiding overdesign and underdesign that lead to inefficiency or overheating.
- **Power Factor Correction**: Reducing reactive power to lower utility costs.
- **Minimization of System Losses**: Improving energy utilization through effective design.
- **Efficient Standby Power**: Ensuring backup systems operate economically during idle periods.

Efficient system design ensures long-term operational savings, and improved environmental performance.

1.5.4 Maintenance and Reliability Considerations

System longevity and maintainability are central to VE. While a low initial cost may seem beneficial, poor reliability often leads to higher maintenance and failure risks (Blanchard & Fabrycky, 2011).

Important considerations include:

- **Durability of Components**: Selection of materials and equipment suitable for extended service life.
- **Ease of Maintenance**: Systems should be accessible and designed for minimal disruption during repairs.
- **Spare Part Availability**: Readily available components reduce downtime and cost.
- **Downtime Risk**: Assessing financial consequences of outages or performance degradation.

Designs optimized for maintainability help control future service costs and improve overall system resilience.

1.5.5 Scalability and Future Adaptability

Modern electrical systems must accommodate future growth, evolving technologies, and changes in usage. Designing only for present needs can lead to expensive retrofits.

Scalable design strategies include:

- **Modular Distribution Systems**: Allowing future expansion with minimal redesign.
- **Smart Automation and Monitoring**: Enabling adaptive load and energy management.
- **Flexible Infrastructure:** Conduits and cabling that support future installations with minimal disturbance.

Initial investment in adaptability may be higher, but the reduced cost of future modifications often justifies the expenditure.

1.5.6 Conclusion: Smart Decision-Making for Cost Optimization

The relationship between CAPEX and OPEX in electrical construction is not a binary trade-off but a strategic balance, as illustrated in Figure 1.2. Both must be optimized concurrently to ensure long-term system value and performance.

A structured VE approach enables informed decisions by considering lifecycle costs, energy efficiency, maintenance, and scalability (SAVE International, 2020). The result is an electrical system that delivers optimal function at the lowest TCO—supporting sustainability, financial viability, and operational excellence throughout the building's lifecycle.

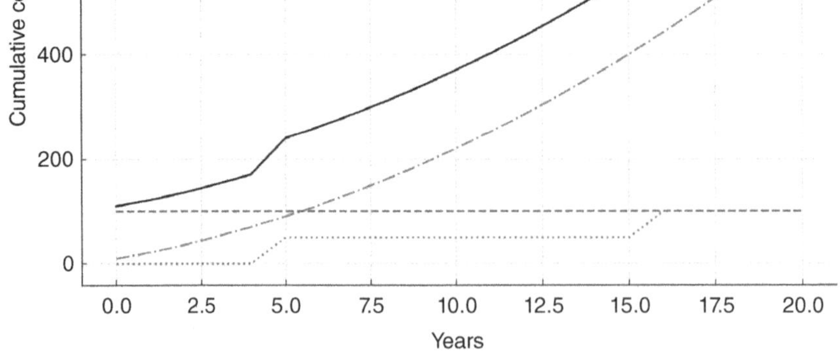

Figure 1.2 Life Cycle Cost Over Time

1.6 Collaborative Approach in Value Engineering

VE is not a solitary endeavor. Its effectiveness lies in its inherently collaborative nature, which draws upon the insights and expertise of a diverse group of professionals. This multidisciplinary approach ensures that every aspect of the electrical design is scrutinized, optimized, and aligned with the project's overarching objectives.

Key participants in the VE process include:

1) **Electrical Engineers**: These professionals bring in-depth technical expertise to ensure the design meets all functional and safety requirements while also identifying opportunities for technical and cost optimization.
2) **VE Facilitators or Specialists**: They manage the VE process, guide structured discussions, challenge assumptions, and ensure the methodology is rigorously applied throughout the team.
3) **Architects**: Architects ensure that electrical systems are well integrated with the building's spatial and aesthetic considerations.
4) **Mechanical Engineers**: Their coordination with electrical engineers ensures that systems such as Heating, Ventilation, and Air Conditioning (HVAC) and power supply are harmoniously designed.
5) **Cost Estimators**: Their cost insights support informed design decisions by evaluating the financial implications of various options.
6) **Project Managers**: They coordinate timelines, resources, and deliverables to ensure VE integration aligns with the overall project scope and schedule.
7) **Stakeholders or Owners**: Their input shapes the project's priorities, objectives, and budgetary constraints.
8) **Suppliers and Vendors**: By introducing new technologies or alternative materials, they contribute to innovation and cost-effectiveness.
9) **Environmental and Sustainability Experts**: These specialists ensure that electrical designs support broader sustainability goals, and meet environmental performance standards.
10) **Safety and Compliance Officers**: They ensure that the design and implementation align with applicable codes, standards, and statutory requirements.
11) **End-Users or Facility Managers**: Their practical experience informs decisions related to system usability, maintainability, and operational reliability.

Incorporating this range of expertise fosters a holistic approach to electrical design. It ensures the outcome is not only cost-efficient and functionally robust but also sustainable, scalable, and suited for long-term performance. The collaborative ethos of VE is illustrated in Figure 1.3, reinforcing the integrative nature of successful VE.

1.7 Navigating Collaboration in VE: Challenges and Solutions

While collaboration is essential to successful VE, it introduces a set of challenges that must be managed carefully. Below are some of the most common obstacles encountered during collaborative VE efforts, along with recommended strategies for overcoming them.

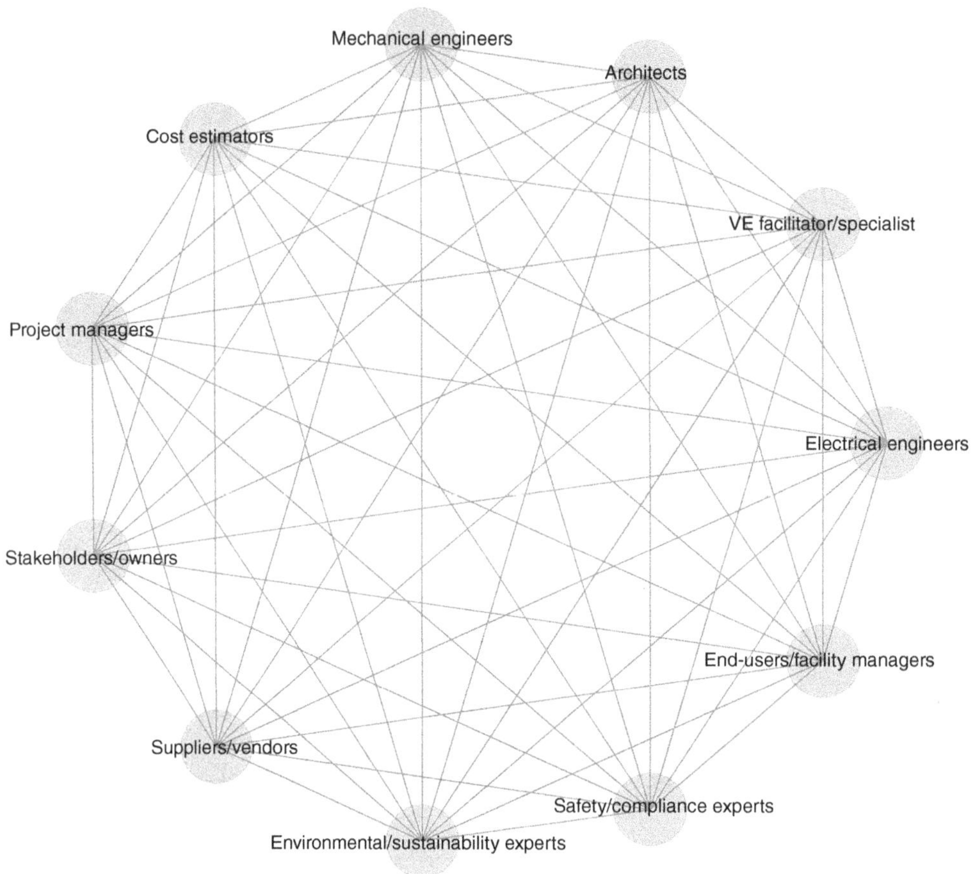

Figure 1.3 Collaborative Approach in VE

1) Aligning stakeholder interests
   - **Challenge**: Stakeholders may hold competing priorities. For example, engineers may prioritize efficiency and reliability, while clients may focus on cost reduction.
   - **Solution**: Conduct regular workshops and alignment sessions where each party can express their objectives. Structured dialogue facilitates mutual understanding and allows for consensus-driven solutions that meet technical, financial, and strategic goals.
2) Commitment to the VE process
   - **Challenge**: Not all participants may be familiar with, or committed to, the structured approach of VE, leading to resistance or superficial engagement.
   - **Solution**: Provide targeted training and orientation on VE methodology (SAVE International, 2020). Sharing successful case studies can demonstrate the tangible benefits of VE, helping to secure broader buy-in from the project team.
3) Effective communication
   - **Challenge**: A multidisciplinary team can lead to fragmented or unclear communication.
   - **Solution**: Implement standardized communication protocols and digital tools to ensure consistent, traceable information flow. Schedule regular coordination meetings to keep all participants informed and aligned.

4) Time constraints
   - **Challenge**: Collaborative VE discussions may delay decision-making, particularly when alignment is difficult.
   - **Solution**: Establish clear timelines and decision milestones from the outset. A dedicated VE facilitator should be empowered to guide discussions efficiently and refocus the team on critical project objectives.
5) Balancing innovation with practicality
   - **Challenge**: Brainstorming sessions often generate creative but impractical concepts.
   - **Solution**: Use technical feasibility studies, simulations, and pilot validations to evaluate proposed alternatives. This ensures ideas are grounded in real-world constraints, while encouraging innovation.

1.7.1 The Path Forward in VE Collaboration

Collaboration in VE thrives when diverse perspectives are unified by clear objectives, structured processes, and mutual respect. By proactively managing the above challenges, VE becomes a conduit for transformative solutions that are both visionary and implementable.

Ultimately, the success of VE lies in its ability to convert complex team dynamics into purposeful dialogue and data-driven decisions—enhancing value not only in cost terms, but in quality, performance, and sustainability (SAVE International, 2020).

References

Fuller, S. (2010). *Life-cycle cost analysis for buildings is easier than you thought.* US Forest Service Technology & Development Program. https://www.fs.usda.gov/t-d/pubs/htmlpubs/htm08732839/page01.htm

Blanchard, B. S., & Fabrycky, W. J. (2011). *Systems Engineering and Analysis* (5th ed.). Upper Saddle River, NJ: Prentice Hall.

Dell'Isola, A. (1997). *Value Engineering: Practical Applications for Design, Construction, Maintenance & Operations.* Kingston, MA: RSMeans Compan.

Miles, L. D. (1961). *Techniques of value analysis and engineering.* McGraw-Hill.

NIST. (2022). *Building life cycle cost programs.* National Institute of Standards and Technology. https://www.nist.gov/services-resources/software/building-life-cycle-cost-programs

SAVE International. (2020). Value methodology standard. https://www.value-eng.org

WBDG. (2023). *Life-cycle cost analysis (LCCA).* Whole Building Design Guide. https://www.wbdg.org/resources/life-cycle-cost-analysis-lcca

2

The Value Engineering Process in Electrical Design

2.1 The Value Engineering Process

The integration of value engineering (VE) into the design stage is a strategic approach aimed at enhancing value by optimizing costs without compromising performance or quality. This chapter outlines a systematic methodology for applying VE in electrical projects, focusing on practicality and efficiency.

The VE process in electrical design is structured around a series of key steps, each designed to identify and exploit opportunities for cost reduction while maintaining or improving system functionality. Starting with data collection, the process involves a thorough analysis of current designs, costs, and requirements. This foundational step ensures that all subsequent actions are informed and targeted.

The objective is to offer a concise yet comprehensive guide to incorporating VE into electrical design stages, equipping professionals with the knowledge and strategies needed to optimize project value efficiently. By following the outlined steps and adhering to best practices, readers can expect to achieve significant improvements in both project outcomes and overall cost savings.

This chapter sets the foundation for a deeper exploration of each step in the VE process, aiming to enhance the reader's ability to apply these principles practically and successfully in their electrical design projects.

2.2 Timing for VE Exercise

Initiating a VE exercise at the right stage in a project can significantly influence its effectiveness and the potential for value enhancement. The optimal timing for beginning a VE exercise is during the concept design stage and feasibility studies, continuing through to the detailed design phase. This timing allows for a comprehensive exploration of design alternatives without disrupting the project's progress or incurring additional costs associated with changes during later stages.

Advantages of Early VE Implementation:

- **Cost Optimization**: Early VE interventions provide the greatest opportunities for cost savings and efficiency improvements, as major design elements and systems are still flexible.

Value Engineering for Electrical Systems in Construction, First Edition. Hazem Hussein.

Companion website: www.wiley.com/go/hazem

- **Informed Decision-Making**: Engaging in VE during the schematic design phase enables decision-makers to assess various design alternatives and their implications on cost, functionality, and performance early in the project lifecycle.
- **Stakeholder Alignment**: Starting VE early ensures that all project stakeholders, including owners, designers, and contractors, are aligned with the value objectives and can contribute to the VE process effectively.

While VE is often associated with the early stages of a project, it can also be effectively applied during mid-project phases, such as preconstruction and procurement. At these stages, the project's scope and design are more defined, making it easier to identify opportunities for cost reduction and functionality improvement.

Applying VE early in a project allows for greater flexibility and creativity in finding alternative solutions. However, mid-project VE can still provide significant benefits, especially if there are changes in the project's environment or new cost-saving opportunities arise.

In conclusion, there is no single "right" time to apply VE in a project. It can be effectively used in both early and mid-project stages. The key is to make the necessary preparations and involve a dedicated VE team to ensure the best results. By incorporating VE into project management practices, teams can achieve cost savings, enhanced functionality, and improved efficiency, leading to better project outcomes (Figure 2.1).

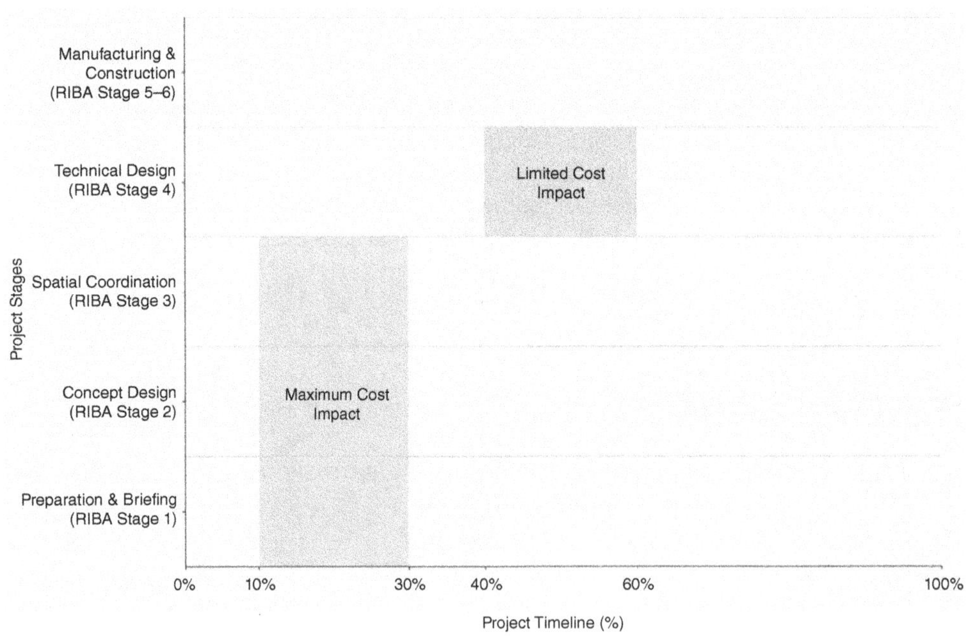

Figure 2.1 Timing for VE.

2.3 Who Should Conduct the VE Exercise?

Selecting the right team to conduct the VE exercise is crucial, as it directly impacts the depth of analysis, objectivity, and overall effectiveness of the process. The team should consist of professionals with strong technical expertise, experience in VE methodologies, and a clear understanding of the project's goals.

The choice typically falls into three categories:

1) **Third-Party VE Consultants**: Independent experts who bring an unbiased perspective and extensive industry experience.
2) **Internal Teams from the Design Consultancy**: In-house professionals who are already familiar with the project's constraints and objectives.
3) **A Hybrid Approach**: A combination of internal expertise and external consultation for a balanced evaluation.

2.3.1 Options for Conducting VE

Each approach offers distinct advantages and should be selected based on the project's specific needs, complexity, and budget constraints.

Option 1: Third-Party VE Consultants

Advantages:

- Provide an unbiased assessment, free from prior design decisions or internal influences.
- Offer specialized expertise in VE and bring a broad perspective from working on various industries and projects.
- Can identify cost-saving opportunities and efficiency improvements that the internal team may overlook.
- Introduce innovative solutions based on best practices from similar projects.

Considerations:

- Can be more expensive than using an internal team.
- May require additional time to understand project-specific constraints and details.
- Recommendations might not always align with the organization's internal processes and culture.

Option 2: Internal Teams from the Design Consultancy

Advantages:

- Have an in-depth understanding of the project's requirements and constraints.
- Can be more cost-effective, as they eliminate the need for third-party contracts.
- Ensure easier integration of VE recommendations into the ongoing design and decision-making processes.
- Facilitate better coordination with the existing project team.

Considerations:

- May lack the broader industry insights that external consultants bring.
- Internal teams could be influenced by prior design decisions, reducing objectivity.
- Might have limited experience with alternative cost-saving solutions outside their domain.

Option 3: Hybrid Approach
For complex or high-value projects, a hybrid approach can maximize the strengths of both internal and external teams.

How It Works:

1) The internal team conducts an initial review to identify areas for VE improvements.
2) Third-party consultants then provide a detailed evaluation and propose optimized solutions.
3) The final VE recommendations are integrated collaboratively, balancing practicality with innovation.

This approach allows for cost efficiency while benefiting from external expertise, leading to more effective and well-rounded value optimization.

2.3.2 Client Preferences and Project Needs

The decision on who should lead the VE exercise depends on various project-specific factors, including:

Project Complexity: Large-scale or highly technical projects may require external consultants with specialized expertise.
Budget Constraints: Organizations with limited funds may find internal teams to be a more viable option.
Client Preferences: Some clients may have prior experiences—positive or negative—with external consultants and may prefer an in-house approach.

Regardless of the chosen approach, the key is to begin the VE process early, engage all stakeholders, and maintain a structured and collaborative approach. This ensures that the project achieves maximum value, efficiency, and cost savings (Figure 2.2).

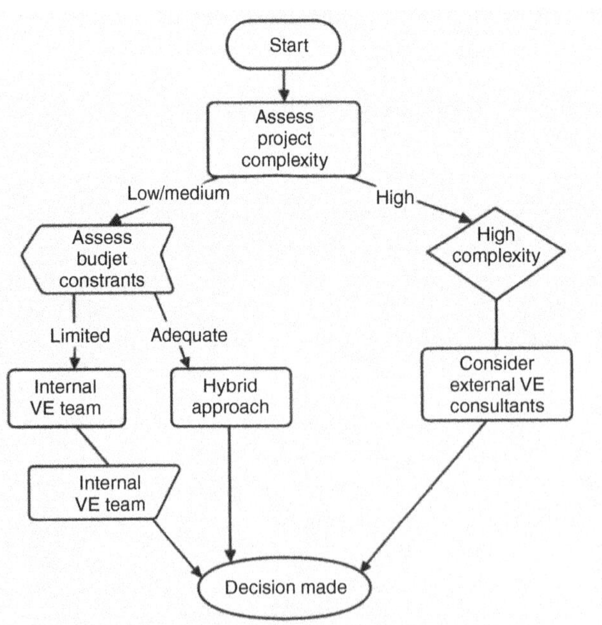

Figure 2.2 Decision Making Flowchart for Selecting VE Team.

2.4 The VE Process Steps

The VE process in electrical design follows a structured methodology aimed at enhancing system efficiency while reducing costs. By breaking down the design process into systematic steps, engineers and project managers can evaluate each component's necessity, functionality, and cost-effectiveness.

This structured approach ensures that all design decisions are made based on data-driven insights, technical feasibility, and cost-benefit analysis. The VE process follows a logical sequence of steps:

1) **Data Collection**: Gather comprehensive project data, including designs, cost estimates, and client requirements, to establish a strong foundation.
2) **Function-Cost Analysis**: Identify the purpose of each electrical component, and assess its cost relative to its function.
3) **Brainstorming Alternatives**: Explore innovative and cost-effective alternatives that maintain or improve system performance.
4) **Technical Evaluation**: Assess the feasibility, reliability, and compliance of proposed alternatives.
5) **Cost Comparison**: Conduct a financial analysis comparing the initial investment, operational expenses, and potential long-term savings of different options.
6) **Client Consultation**: Present recommendations to stakeholders, ensuring alignment with project goals.
7) **Risk Management and Contingency Planning**: Identify potential risks associated with design changes, and develop strategies to mitigate them.
8) **Implementation**: Integrate approved VE modifications into the final design and construction phases.
9) **Post-Implementation Review**: Evaluate the effectiveness of the VE process by analyzing actual performance, cost savings, and lessons learned.

By following these steps, electrical design teams can systematically enhance project value, prevent unnecessary expenditures, and optimize system performance. Each phase of the VE process plays a crucial role in achieving an efficient and cost-effective electrical infrastructure.

The next sections will explore each step in detail, beginning with data collection, which serves as the foundation for the entire VE process

2.4.1 Data Collection

Objective: To establish a comprehensive and accurate data foundation that enables an effective VE process, by gathering all relevant project details upfront, teams can make informed decisions that enhance cost efficiency, system functionality, and project feasibility.

Practical Steps for Data Collection:

1) Gather Existing Designs and Schematics

What to Collect:

- Electrical layout drawings, schematics, and design documents
- Revisions, stakeholder feedback, and meeting notes
- Digital records with version tracking

Objective:

This ensures that the VE team fully understands the current design context and avoids duplicating work or overlooking improvement opportunities.

- *Best Practices:* Use Building Information Modeling (BIM) or Document Management Systems for storage and control
- Verify availability of one-line diagrams, load calculations, panel schedules, fixture schedules, and control architecture documents

2) Catalog Electrical Components and Systems

What to List:

- Key system elements (switchgear, Uninterruptible Power Supply (UPS), transformers, emergency power, lighting systems, Extra-Low Voltage (ELV) systems, etc.)
- Quantities, rated capacities, and reference drawings

Objective:

A well-structured inventory highlights areas of potential cost redundancy or over-specification.

- *Best Practices:* Utilize system inventory templates
- Cross-reference documents with actual site or design intent verification

3) Secure Cost Estimates and Budget Allocation

What to Document:

- Budget allocations per system/component
- Unit material and labor costs
- Supplier/subcontractor bids
- Current labor rates (electricians, foremen, PMs)

Objective:

Accurate costing enables credible comparison of alternatives in later VE phases.

- *Best Practices:* Benchmark using historical project data
- Include quotes and use dynamic spreadsheets for lifecycle tracking

4) Document Client Requirements and Regulatory Compliance

Key Areas to Cover:

- Load, reliability, and redundancy expectations
- Energy goals: LEED, Net Zero, WELL, Energy Star
- Codes: *National Electrical Code (NEC)*, *IEC 60364*, and any local regulations

Objective:

Ensures that all VE alternatives meet functional and regulatory thresholds and support the client's stated performance vision.

- *Best Practices:* Use structured interview templates for stakeholder input
- Include a compliance matrix to track key obligations and certification prerequisites

5) Site and Schedule Considerations

What to Include:

- Electrical room space, riser/shaft placement, and ceiling plenum constraints
- Equipment lead times, installation durations, and construction phasing
- Critical path coordination with other trades

Objective:

Physical and scheduling constraints directly influence which VE options are viable or cost-effective.

- *Best Practices:* Document sequencing and logistics limitations
- Consult installation and supplier lead time forecasts

6) Operational and Maintenance Constraints

Key Data Points:

- Service life expectations
- Maintenance burden and intervals
- Local support availability and client facility team capabilities

Objective:

Ensures that selected alternatives are operable and sustainable post-implementation.

- *Best Practices:* Interview O&M staff
- Compare long-term operating cost implications

7) Performance and Energy Metrics

What to Document:

- Design baselines, power quality needs, and redundancy strategies
- Energy usage, peak demand forecasts, and utility rate structures

Objective:

Supports alignment with sustainability goals and resilient system design.

- Best Practices: Use energy modeling or utility data
- Capture both required and aspirational performance goals

8) Data Validation and Gaps Identification

Checklist Items:

- Missing drawings, unknown specifications, and unverified assumptions
- Plan to fill gaps, responsible parties, and next steps

Best Practices:

- Use the form's data gap fields to document known unknowns
- Schedule site walks and interviews proactively if required (Figure 2.3).

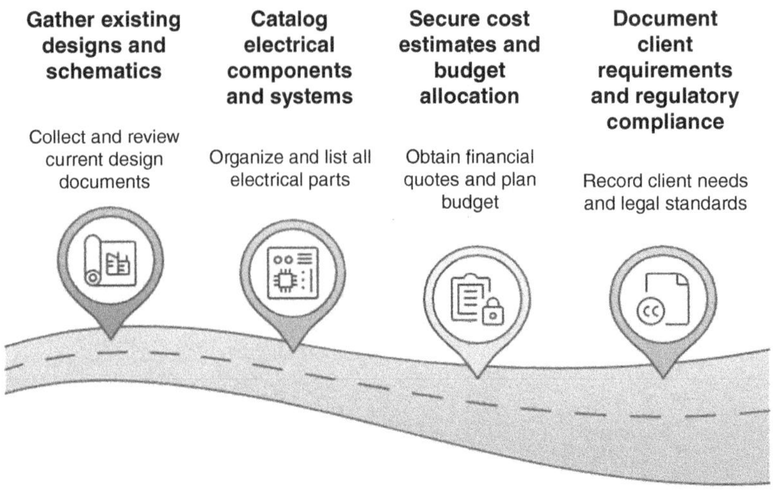

Figure 2.3 Data Collection for VE.

Implementation Note
*For structured implementation, users should complete the standardized template provided in **Section 6.2: Data Collection Sheet for VE Exercise**. This form systematically captures all categories outlined above, ensuring consistency across VE initiatives and completeness of design context.*

2.4.2 Function-Cost Analysis

Objective: Function-cost analysis (FCA) is an engineering discipline used to evaluate the purpose and financial footprint of each electrical component or system in a project. In VE, this analysis provides a structured means of determining whether the cost of a component is justified by the value of the function it performs (Kaplan and Cooper, 1998; Miles, 1972).

When applied effectively, FCA allows project teams to distinguish between *essential functionality* and *redundant or over-specified elements*, facilitating rational decisions on redesign, substitution, or elimination.

Practical Steps for Conducting Function-Cost Analysis:

1) **Function Definition in Electrical Design**

In electrical systems, components are often selected based on standard practices or vendor preferences without reexamining their *primary technical purpose*. FCA begins by clearly defining that purpose.

- **Primary Function**: the essential role (e.g., *Distribute Power* for a panel board)
- **Secondary Functions**: value-adding or supportive roles (e.g., *Enhance Redundancy*)

Functions should be stated using a **verb–noun structure** and evaluated by their contribution to project performance.

Example:

- A **transformer's** primary function is *Step Down Voltage*
- Its secondary function may be *Isolate Subsystems* or *Provide Ground Reference*

Best Practice:

- Use engineering judgment to distinguish *what the component does* (function) from *how it does it* (design method). Avoid conflating design features with functions.
- Focus on what the component achieves, not how it does it.
- Use a function classification system (e.g., Basic Required, Secondary, Supporting, Unnecessary).

2) **Linking Cost to Function**

Electrical engineers must associate **all cost categories**—capital, operational, and lifecycle—with the functions performed (Table 2.1).

Worth is the *minimum reasonable cost* to fulfill a function acceptably. Calculating a **cost-to-worth ratio** (C/W) helps identify inefficiencies (Miles, 1972):

$$\text{C/W ratio} = \frac{\text{Current cost}}{\text{Estimated worth}}$$

Example:

- **UPS System (200 kVA)**
  - Cost: $50,000
  - Worth: $35,000
  - C/W ratio = 1.43 → Acceptable
- **Busbar System (LV, 2500 A)**
  - Cost: $80,000
  - Worth: $45,000
  - C/W ratio = 1.78 → High priority for review

Best Practices:

- Use **historical data**, supplier quotes, and activity-based costing (Kaplan and Cooper, 1998)

3) **Highlight Optimization Opportunities**

FCA enables teams to identify high-cost functions that are:

- **Over-specified** (e.g., 400 A rated panel serving a 160 A load)
- **Duplicated or redundant** (e.g., multiple backup systems without critical need)
- **Served by outdated technology** (e.g., magnetic contactors vs. VFDs (Variable Frequency Drives))

Table 2.1 Type of Electrical Costs with Examples.

Cost Category	Typical Electrical Examples
Direct Costs	Procurement, installation labor, testing
Indirect Costs	Downtime from failure, access limitations
Lifecycle Costs	Maintenance frequency, obsolescence, disposal

VE Actions May Include:

- Redesign for cost-effective specification
- Replace with more efficient alternatives
- Consolidate systems to eliminate unnecessary redundancy

Electrical Example:

- Switching from copper busbars to aluminum equivalents in noncritical panels could maintain performance while reducing cost by 20–30%.

Best Practices:

- Prioritize functions with high C/W ratios (>1.5)
- Use energy modeling, total cost of ownership (TCO), or risk-impact trade-offs to validate alternatives

4) **Prioritize High-Impact Areas**

After functions have been defined and their costs evaluated, the final—and most critical—step of FCA is to determine where VE efforts should be focused. This requires balancing **technical criticality** against **cost inefficiency**, using the following two supporting tools.

A) **Function Dependency Analysis**

This tool helps engineers assess whether modifying a function would cause negative ripple effects across other systems. This is particularly important in complex electrical environments, such as facilities with integrated life-safety, automation, and mission-critical operations (Table 2.2).

B) **Function Prioritization Scoring**

This matrix ranks each function by:

- Its importance to project success (e.g., reliability, safety, compliance), and
- The opportunity for improvement (e.g., over-specification, costly method).

Table 2.2 Function Dependency Analysis Example.

Function	Dependent Functions	Impact if Modified
Distribute Power	Lighting, HVAC, fire alarm	Building-wide failure or operational downtime
Provide Emergency Power	Exit lighting, alarm systems	Safety compliance breach and hazard during outages
Monitor Energy Usage	Demand control, billing, alerts	Loss of visibility for energy management or submetering

Dependency analysis ensures that VE proposals do not inadvertently degrade system integrity, especially in safety-critical or interdependent functions.

Multiplying these scores provides a **priority index** that helps target high-value VE candidates.

Function	Importance (1–10)	Improvement Opportunity (1–10)	Priority Score
Lighting Control System	6	9	54
Busbar Trunking System	8	6	48
Emergency Generator	10	2	20

Functions with high scores are often ideal for reevaluation. Low-scoring functions may be essential and, therefore, better left unchanged.

Scoring Example

To illustrate how prioritization scores are applied in practice, consider the following three functions evaluated during a VE workshop for a hospital electrical system:

1) **Lighting Control System (Lobby and Corridors)**
   - Importance to Project Success: 6
   - Opportunity for Improvement: 9
   - Priority Score: $6 \times 9 = \mathbf{54}$

The system offers energy savings but is noncritical. High-end control methods could be replaced with more economical alternatives.
Decision: *Recommended for VE consideration*

2) **Busbar Trunking System (Podium LV Distribution)**
   - Importance to Project Success: 8
   - Opportunity for Improvement: 6
   - Priority Score: $8 \times 6 = \mathbf{48}$

The system is essential for power delivery but may be oversized relative to peak load data.
Decision: *Candidate for load-based reevaluation*

3) **Emergency Generator (Critical Power Backup)**
   - Importance to Project Success: 10
   - Opportunity for Improvement: 2
   - Priority Score: $10 \times 2 = \mathbf{20}$

While costly, the generator meets essential life-safety and code-mandated functions.
Decision: *Excluded from VE consideration*

Scores above 40 generally indicate worthwhile targets for VE exploration, assuming minimal dependency risk. Lower scores often signal functions that are either too critical or already optimized.

Best Practices:

- Apply dependency analysis first to screen out risky candidates.
- Use prioritization scores to rank the remaining options.
- Revalidate scores with input from stakeholders and operations teams (Table 2.3).

Table 2.3 Sample Completed Function Analysis Table (UPS and Distribution Example).

Component/ Subsystem	Primary Function	Secondary Functions	Class.*	Performance Requirement	Cost ($)	Worth ($)	C/W Ratio	Dependencies
UPS System (200 kVA)	Provide backup power	Improve power quality	BR	≥10 min runtime at 100% load	50,000	35,000	1.43	Controls, emergency loads
Busbar Trunking (2500 A)	Conduct power	Minimize voltage drop	BR	≤2% voltage drop at full load	80,000	45,000	1.78	LV switchboards
Lighting Controls (Digital)	Manage illumination	Save energy, enhance comfort	SR	Scene control, 10% dimming accuracy	30,000	18,000	1.67	Lighting circuits
Fire Alarm Cabling	Transmit signals	Provide redundancy	SR	Dual path communication per NFPA 72	5000	4500	1.11	Control panel, terminals
Panel Boards (400 A)	Distribute circuits	Segregate loads	S	80% loading capacity max	15,000	10,000	1.50	Feeders, load branches

Class. = function classification, BR = basic required, SR = secondary required, S = supporting, C/W ratio = cost-to-worth ratio.

Engineering Note:

This completed table illustrates how to apply the FCA methodology for key electrical components. It uses the same format as the blank worksheet provided in Section 6.4. The analysis helps prioritize VE targets—such as lighting controls or busbar trunking—based on cost justification and system criticality.

Implementation Note

The Function Analysis Worksheet provided in ***Section 6.4*** *supports this process with structured fields for function definition, cost/worth scoring, and priority setting. It should be completed after data collection (Section 2.4.1) and prior to idea generation. Use it as a reference, but ground the analysis in engineering logic, performance goals, and system interaction.*

2.4.3 Brainstorming Alternatives

Objective: This phase of the VE process is focused on generating a wide range of alternative solutions to reduce costs or improve performance of electrical systems. The goal is not to evaluate feasibility immediately, but to unlock creative potential across the design team, using structured and open-ended methods to explore new ways of fulfilling essential functions.

2.4.3.1 Best Practices for Electrical Engineering Teams

1) **Organize Focused Brainstorming Sessions**

VE brainstorming in electrical design typically focuses on high-cost systems such as:

- Power distribution networks
- Lighting control architecture
- Generator and UPS sizing
- Fire alarm and BMS(Building Management System) integration

Each session should:

- Begin with a clearly defined **target function** (e.g., *Distribute Power*)
- Include cost data and performance limits from prior FCA and data collection
- Use tools such as **mind mapping**, **SWOT**, or **SCAMPER*** to facilitate creative flow

Keep the team focused on the function, not locked into the method.

*Creativity Tools for VE Brainstorming
During brainstorming sessions described in Section 2.4.3, structured creativity methods can help generate wider solution sets. The following widely accepted techniques may be applied selectively depending on project scale and team expertise: 1) Mind Mapping A visual tool that starts with a core function (e.g., *Power Distribution*) and expands outward into branches covering design alternatives, performance factors, risks, and costs (Buzan, 2006). 2) SWOT Analysis Assesses each alternative's strengths, weaknesses, opportunities, and threats. Helps compare competing VE options by balancing technical advantages and implementation risks (Helms & Nixon, 2010). 3) SCAMPER A structured checklist encouraging innovation by systematically challenging existing designs using seven prompts: Substitute, Combine, Adapt, Modify, Put to other use, Eliminate, Reverse (Eberle, 1996).

2) **Encourage Diverse Suggestions**

Creative ideas may come from:

- Alternative materials (e.g., aluminum busbars vs. copper)
- System-level simplification (e.g., combining branch panels)
- Technology substitution (e.g., wireless controls)
- Vendor solutions with optimized modularity

Example:

A team may propose replacing a central UPS with distributed plug-in UPS units in noncritical zones—reducing cost and improving flexibility.

Use anonymous suggestion tools or breakout formats to capture input without hierarchy bias.

3) **Utilize Cross-Functional Expertise**

While electrical engineers lead technical analysis, brainstorming should include:

- **Architects** (for spatial constraints)
- **Mechanical engineers** (for integration/coordination)
- **Sustainability specialists** (for energy-related functions)
- **Cost consultants and clients** (for commercial viability)

Diverse perspectives uncover dependencies and opportunities that isolated teams often miss.

4) **Document and Categorize Ideas**

Use a shared log to capture (Table 2.4).:

- The idea
- The component or function it addresses
- Anticipated benefits (e.g., cost, reliability, maintainability)
- Concerns (e.g., code compliance, user impact)
- Preliminary feasibility (high/medium/low)

This sets the stage for formal evaluation

Practical Example—VE Idea Log (Abbreviated)

Post-Session Actions:

- Consolidate ideas
- Assign for further evaluation using function-cost and risk analysis
- Maintain the log as a repository for future VE cycles

Implementation Note

*For structured documentation, use the standard **VE Idea Generation Log** provided in **Section 6.5**. This tool mirrors the structure above and is essential for ensuring idea traceability and facilitating later evaluation.*

2.4.4 Technical Evaluation

Objective: The purpose of the technical evaluation phase is to verify the feasibility and effectiveness of proposed VE alternatives before implementation. This includes evaluating each idea's impact on system performance, reliability, maintainability, regulatory compliance, and operational viability. The analysis must ensure that any approved changes enhance value without introducing technical risks or compromising safety.

2.4.4.1 Key Evaluation Dimensions

1) **Performance Testing and Simulation**

Use modeling and electrical analysis tools to assess system behavior under proposed VE alternatives. Simulation tools such as ETAP®, SKM PowerTools®, DIGSILENT, or company-developed spreadsheets support evaluations for:

Table 2.4 Sample VE Idea Generation Log (Excerpt) (Format Aligned with Section 6.5).

Idea ID	Target System	Function Addressed	Current Solution	Alternative Proposal	Benefits	Challenges	Feasibility	Contributor
E-001	UPS system	Provide backup power	200 kVA centralized	Localized 10–20 kVA rack UPS	Reduced capex, scalable	Maintenance coordination	M	Electrical lead
E-002	Lighting control	Manage illumination	DALI system	Wireless relay-based control	Simpler install, lower cost	Limited zoning flexibility	H	Architect
E-003	Busbar trunking	Conduct power	2500 A copper busbar	2000 A aluminum with riser	Lower cost, easier routing	Thermal capacity validation	H	Engineer
E-004	Emergency generator	Supply life safety loads	Single 800 kVA unit	Separate smaller gensets	Redundancy, fuel efficiency	More switchgear needed	L	Contractor

Feasibility rating: H = high, M = medium, L = low.

- Load flow distribution
- Short circuit performance
- Harmonic distortion
- Equipment sizing and thermal limits

These analyses help verify whether a lower-cost or alternative design still meets performance, protection, and safety requirements.

Example:
Replacing a traditional feeder with a compact cable bus system should be validated by analyzing voltage drop under full load and evaluating thermal stability during peak operation.

➢ **Engineering Note**:
Full-scale simulation and technical modeling may be time-intensive. As a best practice, **initial submission of a VE alternative should include basic supporting data and a clear functional rationale**. If the concept appears technically sound and cost-justified, only then should detailed simulations and modeling be authorized.

This approach ensures efficient use of resources while preserving engineering integrity. It also avoids unnecessary delays by filtering unviable proposals early in the process.

2) **Longevity and Maintenance Analysis**

Evaluate the **service life**, **failure modes**, and **maintenance demands** of alternatives. Look for:

- Reduced preventive maintenance schedules
- Simpler accessibility for repairs
- Fewer single points of failure

Use **life-cycle cost analysis (LCCA)** tools to quantify long-term costs and reliability metrics.

Example: Switch to modular UPS with hot-swappable batteries to reduce Mean Time To Repair (MTTR).

3) **Energy Efficiency Assessment**

Quantify expected reductions in:

- Annual kWh consumption
- Peak kW demand
- Energy use intensity (EUI) per m^2

Model alternatives are using tools like **EnergyPlus™, Dialux®, or vendor energy models**. Use consistent baseline and unit cost assumptions for valid comparisons.

Example: Replacing fluorescent lighting with high-efficiency LEDs and wireless controls may yield a 25% drop in lighting energy use.

Use the structured format in ***Section 6.9: Energy Efficiency Impact Assessment*** *to document energy projections, environmental benefits (e.g., CO_2 reduction), payback calculations, utility incentives, and code compliance.*

4) **Regulatory Compliance Verification**

Ensure all alternatives conform to:

- **NEC (NFPA 70)** and local codes
- **IEC 60364 or ASHRAE 90.1** as applicable

- Project-specific design standards (e.g., LEED prerequisites)
- Fire alarm, emergency, or life-safety codes

Use compliance checklists or model cross-checks. If load or equipment changes are proposed, assess compliance using results from the Electrical Load and Capacity Form (Section 6.8).

2.4.4.2 Practical Methods for Evaluation

A) **Targeted Validation Studies (Design Stage Appropriate)**

Rather than physical prototyping, use design-stage validation methods to confirm technical viability of promising alternatives. These may include:

- **Vendor-supported lab testing results or case studies**
- **Mock-up simulations or comparative product data**
- **Internal parametric modeling or spreadsheet-based sensitivity analysis**

These methods allow engineers to test viability without the need for installed infrastructure or field deployment.

Example:
For a proposed wireless lighting control system, request manufacturer simulation data showing signal range and system load limits, or run an internal model comparing device response times and redundancy behavior.

These lightweight evaluations ensure design decisions are supported by credible data without disrupting VE timelines or exceeding scope boundaries.

B) **Expert Reviews**

Bring in neutral technical experts (internal or third-party) to independently assess risks, performance, or assumptions.

Example: External review of grounding system changes for a high-rise facility.

C) **Stakeholder Feedback**

Share findings with:

- Maintenance supervisors
- Safety officers
- Building operations teams

Feedback should address **operability**, **risk perception**, and **user acceptance**.

D) **Documentation of Findings**

Maintain a complete record of:

- Simulation reports
- Design comparisons
- Regulatory checklists
- Stakeholder comments

This forms the evidence base for final VE decisions and supports traceability (Table 2.5).

Table 2.5 Sample Evaluation Matrix (Aligned with Section 6.6).

Idea ID	Brief Description	Initial Cost Savings	Lifecycle Impact	Energy Efficiency	Code Compliance	Safety	Maintainability	Score (1–5)[a]	Decision[b]
E-001	Replace MV cable trays with bus ducts	4	5	3	5	5	4	4.5 (avg)	A
E-002	Modular switchgear replacement	3	4	3	4	4	5	4.1	M
E-003	Digital relay upgrades	5	3	2	5	5	3	4.0	A
E-004	Eliminate secondary backup pump	2	2	4	3	2	3	2.5	R

[a] Score reflects average or weighted total across all evaluation dimensions.
[b] Decision key: A = accept, M = modify, R = reject, C = combine.

Implementation Note

To ensure a structured and defensible evaluation process, use the ***VE Idea Evaluation Matrix*** *provided in* ***Section 6.6****. This tool allows for multi-criteria scoring and transparent decision-making across technical, financial, and regulatory dimensions.*

Use **Section 6.7: Electrical Load and Capacity Analysis Form** to validate right-sizing proposals, feeder reductions, or transformer reclassifications.

Use **Section 6.8: Energy Efficiency Impact Assessment** to quantify energy and environmental impacts, utility incentives, and financial outcomes.

2.4.5 Cost Comparison

Objective: This phase quantifies the financial impact of proposed VE alternatives, comparing not just initial investment but also lifecycle performance, energy use, and long-term operational costs. Cost comparison validates whether an idea provides true value or merely shifts cost from capital to operations.

2.4.5.1 Core Analytical Categories

1) **Initial Investment Analysis**

Calculate all direct implementation costs: materials, labor, installation logistics, equipment procurement, subcontractor scope, and adjustments to existing systems.

Example: A proposed switchgear revision may reduce initial cost by $35,000 through vendor substitution and footprint reduction.

2) **Operational Expense Evaluation**

Assess recurring costs tied to energy use (lighting, HVAC(Heating, Ventilation, and Air Conditioning)controls, motors), maintenance (labor, spares, scheduled inspections), and downtime risks (loss of function, repair delays). Use maintenance planning software, energy modeling platforms, or historical data to estimate annualized OPEX.

3) **Long-Term Savings Projection**

Estimate financial savings over the expected service life of the system, including reduced utility costs, extended equipment life, and fewer reactive service calls. Apply **TCO** (Fuller and Petersen, 1996) and **net present value (NPV)** analysis (Brealey et al., 2019) for meaningful comparisons.

Example: Upgrading from traditional contactors to solid-state relays may cost more upfront but reduce maintenance and energy losses over 15 years.

2.4.5.2 Practical Steps for Effective Cost Comparison

A) **Gather Accurate Market Quotes**

Request vendor quotations or bids for proposed alternatives and validate cost deltas with quantity surveyors or procurement teams.

B) **Apply TCO Analysis**

TCO is a structured financial methodology used to evaluate all costs associated with a system over its entire lifecycle (Fuller and Petersen, 1996), including:

- **Initial Cost**: Equipment, labor, commissioning
- **Operating Costs**: Energy consumption, controls, user interface loads
- **Maintenance Costs**: Routine service, spare parts, technician labor
- **Downtime Costs**: Operational interruptions, service loss risks
- **End-of-Life Costs**: Decommissioning, recycling, and replacement

$$TCO = \text{Initial Investment} + \sum(\text{Operating} + \text{Maintenance Costs}) + \textit{End}\text{-of-Life Costs}$$

Objective:
TCO helps teams make long-term value-based decisions rather than focusing only on upfront savings. It's especially useful in VE when an alternative has a higher initial cost but delivers superior financial performance over time.

Electrical Example:
A modular UPS system may cost **$60,000** vs. a traditional one at **$50,000**, but if it reduces annual energy and maintenance costs by **$2000** over 10 years, the TCO is lower and the payback period shorter—justifying the investment.

Note: TCO figures should be estimated using consistent financial assumptions (e.g., inflation rate, energy cost escalation, and system service life).

C) **Perform Sensitivity Analysis**

Use sensitivity models (Park, 2011) to test the impact of:

- Energy price inflation
- Maintenance intervals or labor rates
- Changes in system utilization or performance

This is critical when comparing similar alternatives with different operational profiles.

D) **Engage Financial Analysts**

Financial experts can:

- Validate discount rates and payback assumptions
- Integrate VE analysis into overall project financial models
- Ensure alignment with internal approval criteria (e.g., ROI (Return on Investment) thresholds)

E) **Involve Stakeholders**

Review results with:

- Project finance teams
- Facility managers
- Procurement leads

This ensures acceptance and supports budgeting alignment.

F) **Document Financial Justification**

Summarize results in a structured cost-benefit format:

- Methodology used (e.g., TCO, ROI)
- Key assumptions and modeling inputs
- Summary of comparative results and preferred option (Tables 2.6 and 2.7)

Table 2.6 VE Evaluation Criteria and Weighting.

Criteria	Description	Weight (%)
Initial Cost Savings	Reduction in capital cost	30
Lifecycle Cost Impact	Maintenance, reliability, and service life costs	20
Energy Efficiency	Impact on consumption and utility cost	15
Performance/Reliability	Effect on uptime, resilience, redundancy	10
Maintainability	Access, serviceability, parts availability	5
Code Compliance	Conformance with NEC, IEC, and local codes	5
Schedule Impact	Delays, lead times, install sequence risk	5
Constructability	Installation difficulty, trade conflicts	3
Safety	Electrical and operational risk avoidance	3
Owner Acceptance	Stakeholder confidence and buy-in	2
Total		**100%**

Use this structure to assign importance across technical and financial dimensions.

Table 2.7 Projected Savings Summary (Example).

Idea ID	Initial Cost Savings	Lifecycle Cost Savings (15 years)	Annual Energy Savings
E-001	$35,000	$60,000	$2500
E-002	$12,000	$25,000	$1200
E-003	$8000	$10,000	$750
E-004	$0	$6000	$1000
Total	**$55,000**	**$101,000**	**$5450**

These totals are used to support decision-making in the final VE recommendation package.

Implementation Note

To formally complete this stage, populate the ***VE Idea Evaluation Matrix*** *provided in* ***Section 6.6****. This allows multi-dimensional scoring and cost justification that supports objective, defendable decisions.*

Link to Financial Evaluation Strategies

For advanced financial modeling techniques—such as ROI forecasting, NPV, lifecycle cash flow modeling, and capital expenditure planning—refer to ***Chapter 13: Financial Strategies and Economic Impact of VE on Electrical Systems****. This section provides deeper guidance on aligning engineering alternatives with financial viability, long-term investment goals, and portfolio-level risk management.*

2.4.6 Client Consultation

Objective: Client consultation is the critical bridge between technical analysis and final decision-making in the VE process. It ensures that proposed alternatives not only deliver measurable value but also reflect the client's goals, risk tolerance, and operational preferences.

This stage transforms technical insights into actionable, client-approved decisions that shape the final design direction.

2.4.6.1 Key Practices for Effective Engagement

1) **Prepare Detailed Comparison Presentations**

Build presentations or summary packs that clearly outline:

- Original design intent and specifications
- VE proposal summaries, including technical performance
- Side-by-side cost comparisons (capital and lifecycle)
- Visual aids like one-line diagrams, 3D layouts, and cost graphs

Use these materials to facilitate comprehension and highlight tangible benefits and trade-offs.

2) **Schedule Consultative Meetings**

Set up structured sessions with:

- Client representatives (owners, end users, facility operators)
- Core design team and VE lead
- Financial stakeholders (if applicable)

Ensure:

- Clear agenda
- Time for Q&A
- Documented decisions and action items

3) **Highlight Value Beyond Cost**

Frame VE proposals in terms of:

- Functionality improvements
- Operational efficiency
- Maintainability
- Environmental performance

Case studies from comparable projects can help clients visualize the practical implications of proposed changes.

Practical Steps for Effective Consultation

- **Develop a Consultation Agenda**:
  Include each proposed change, rationale, impact (cost, performance, risk), and key decisions required
- **Use Clear and Accessible Language**:
  Avoid engineering jargon unless necessary. Use analogies, visuals, and simplified metrics where possible
- **Provide Comparative Analysis Documents**:
  Include side-by-side comparisons with original vs. proposed outcomes—technical, financial, and sustainability-focused
- **Engage in Active Listening**:
  Understand client concerns, operational habits, long-term plans, or political factors that may shape preferences
- **Offer Flexibility and Options**:
  Present trade-offs or alternative VE options. Some changes may be accepted with conditions (e.g., alternate suppliers, schedule phasing)
- **Document Feedback in Real-Time**:
  Capture all comments, conditions, and decisions using a structured template or meeting minutes
- **Send Follow-Up Summary**:
  Summarize all agreed-upon modifications, open questions, and next steps in a post-meeting communication

Implementation Note

Client feedback plays a decisive role in shaping final VE recommendations. Ensure that all consultation outcomes are formally documented in the ***VE Idea Evaluation Matrix*** *provided in* ***Section 6.6****. Use the "Owner Acceptance" score and decision fields to reflect stakeholder input, and capture feedback in the evaluation notes section. This ensures traceability, accountability, and alignment with project objectives.*

2.4.7 Risk Management and Contingency Planning

Objective: To proactively identify and mitigate risks associated with VE alternatives, ensuring that approved changes do not jeopardize system performance, schedule, budget, or compliance. Structured risk analysis enables teams to make informed decisions about which proposals are worth pursuing and what safeguards are required.

2.4.7.1 Key Risk Management Activities

1) **Conduct Comprehensive Risk Analysis**

Identify technical and execution risks related to:

- System performance degradation
- Code noncompliance
- Supply chain or delivery delays
- Cost overruns and stakeholder resistance

Use structured frameworks like a **risk probability-impact matrix**, supported by expert input and lessons from similar projects.

2) **Develop Contingency Plans**

For each moderate to high-risk VE proposal, define:

- Backup solutions or components
- Budgetary reserves or material alternatives
- Schedule buffers or accelerated review protocols

Example: For a new UPS vendor, include a contingency for delayed testing certifications or substitute equipment availability.

3) **Establish Risk Mitigation Strategies**

Mitigation can include:

- Sourcing from prequalified vendors
- Early stakeholder validation
- Adding QA/QC hold points during implementation
- Conducting more extensive simulation or factory testing

Practical Risk Planning Tools:

- **Risk Assessment Workshops**: Facilitate open forums with engineering, Project Management, and client teams

- **Risk Registers**: Use structured logs to track risks, mitigation owners, and live status
- **Quantitative Risk Analysis**: Run scenarios using tools like Monte Carlo *(Metropolis and Ulam, 1949) or three-point estimating
- **Integration into** Project Management **Systems**: Embed mitigation into schedules, submittal trackers, and commissioning plans
- **Training and Culture**: Promote open identification and documentation of risks across stakeholders
- **Communication Protocol**: Define how and when risks are reported, escalated, and resolved

***Quantitative Risk Analysis**

For high-impact or high-uncertainty VE proposals, conduct a quantitative risk analysis to understand potential cost and schedule implications. One effective method is **Monte Carlo Simulation**.

Monte Carlo Simulation is a statistical modeling technique that analyzes the behavior of a system under uncertainty by running thousands of randomized iterations based on defined probability distributions. In a VE context, it allows teams to simulate a wide range of outcomes for schedule delays, cost fluctuations, or performance risks.

Example

A proposed switchgear redesign might introduce variability in both procurement time and installation effort. A Monte Carlo model can simulate thousands of scenarios using realistic input ranges, producing a probability distribution of total project delay. This helps quantify the likelihood of exceeding deadlines or cost thresholds.

Monte Carlo results can inform **contingency reserves**, **risk-adjusted schedules**, and **proposal acceptance decisions**, making it a powerful tool in the VE evaluation process.

Note: While Monte Carlo simulation is rarely used in smaller VE efforts, it becomes valuable in complex or high-risk projects—particularly where multiple uncertain variables may interact to affect schedule or cost. Its use should be guided by project scale, risk exposure, and availability of reliable input data.

2.4.7.2 Three-Point Estimating

Another commonly used method within quantitative risk analysis is **three-point estimating**, which models uncertainty by defining a range of possible outcomes rather than a single deterministic value.

For each cost or schedule element, three estimates are identified:

- **Optimistic (O)**: The best-case scenario, assuming minimal issues.
- **Most Likely (M)**: The expected outcome based on historical data and expert judgment.
- **Pessimistic (P)**: The worst-case scenario, considering potential complications or delays.

A weighted average expected value (E) is often calculated using the following formula:

$$E = (\mathrm{O} + 4\mathrm{M} + \mathrm{P})/6$$

This technique provides a more realistic basis for assessing potential variation in cost or schedule forecasts. Three-point estimates can also serve as input distributions when conducting Monte Carlo simulations for more complex risk analysis.

Implementation Note
*All VE-related risks should be logged and monitored using the structured **Risk Register** embedded within the **VE Implementation Tracking Log** provided in **Section 6.10**. This tool ensures that mitigation plans, contingency actions, and status indicators are consistently tracked from approval through implementation.*

2.4.8 Implementation

Objective: To execute approved VE changes in a controlled and traceable manner. This includes updating design documentation, coordinating with implementation teams, maintaining compliance, and verifying that the intended value outcomes are realized.

Implementation Workflow:

1) Update Design Documentation
   - Revise schematics, schedules, and equipment specs
   - Ensure updated documents are fully coordinated
   - Record document references in submittal systems
2) Coordinate Project Team
   - Conduct briefings with all affected parties: engineers, contractors, and installers
   - Highlight scope impacts, new requirements, and sequence adjustments
3) Provide Targeted Training
   - For new technologies or procedures, train O&M or construction teams
   - For design-stage VE, provide rationale so construction understands design intent
4) Quality Control and Compliance Checks
   - Insert QA/QC checkpoints at key milestones
   - Conduct site inspections and testing aligned with VE-specific changes
5) Adjust Schedules and Interfaces
   - Modify timelines as required for procurement lead times or sequencing
   - Minimize disruption through look-ahead planning and trade coordination
6) Maintain Stakeholder Alignment
   - Provide regular progress updates to client and internal stakeholders
   - Capture feedback and adjust course if needed
7) Documentation and Recordkeeping
   - Maintain a central VE implementation log with design references, issues encountered, and resolutions
   - Archive all correspondence, approvals, and cost verification records
8) Final Review and Handover
   - Validate installed systems via commissioning and performance tests
   - Complete final documentation: as-builts, O&M manuals, and warranty records
   - Formally hand over to the client with clear records of VE impacts

Special Notes for Existing Projects:

- Not all VE proposals are retrofit-suitable
- Evaluate changeability, risk, and operational impact
- Use **incremental deployment** to validate impact and ease transition

For procedural guidance on implementing VE in live environments, refer to ***Chapter 11: Value Engineering for Existing Electrical Systems (Renovation, Brownfield Integration, and Retrofit Optimization)***

Implementation Note
*Track VE execution activities using the structured **VE Implementation Tracking Log** in **Section 6.10**. This log captures approval references, submittal and installation status, savings verification, and issues encountered. It also includes a **risk register**, progress report templates, and fields for lessons learned—ensuring full traceability from decision through final turnover.*

2.4.9 Post-Implementation Review

Objective: To evaluate the real-world impact of the implemented VE proposals, verify whether they met technical and financial expectations, and capture key lessons for future application. This step ensures accountability, improves VE methodology, and contributes to ongoing knowledge development within the organization.

2.4.9.1 Post-Implementation Review Activities

1) **Performance Assessment**

Review the as-installed electrical systems to assess whether:

- Targeted functions were achieved
- Operational performance matches simulation or design intent
- Systems deliver reliability, safety, and control quality as expected

Compare **baseline projections vs. actual data** (e.g., energy savings, fault performance, uptime metrics).

2) **Cost Analysis**

Compare final financial outcomes with:

- VE-approved cost savings
- Final installation and commissioning invoices
- Any variances from the projected lifecycle model

Document causes of variances such as change orders, vendor delays, or field conditions.

3) **Stakeholder Feedback**

Gather input from:

- Clients and end users
- Maintenance teams and building operators
- Contractors and design consultants

Focus on usability, satisfaction with design outcomes, and any issues during operation or maintenance.

4) **Documentation Review**

Ensure all records are complete, including:

- Updated as-builts and O&M manuals
- Commissioning reports
- VE proposal outcomes and change logs
- Product submittals and warranty data

5) **Lessons Learned Session**

Host a dedicated session with:

- Design, construction, and commissioning teams
- VE study participants
- Project managers and client representatives

Topics include:

- What worked as planned
- What required unexpected intervention
- Process inefficiencies and improvements
- Internal communication gaps and resolution quality

6) **Training and Development**

Based on review outcomes:

- Identify training needs (e.g., new system technologies)
- Improve internal procedures and design practices
- Recommend role-based skill development for future VE efforts

7) **Regulatory Compliance Verification**

Confirm:

- Final system configuration complies with NEC, IEC, and local codes
- All deviations have documented approval
- Compliance forms or third-party signoffs are archived

8) **Sustainability and Energy Performance Review**

Compare projected energy and carbon savings with:

- Actual consumption and demand data
- Environmental performance metrics
- LEED or green building verification records (if applicable)

9) **Final Report Compilation**

Prepare a comprehensive **VE Post-Implementation Report**, including:

- Performance summary
- Financial reconciliation
- Stakeholder commentary

- Lessons learned
- Recommendations for future studies

10) **Dissemination and Knowledge Sharing**

Distribute review findings to:

- Project stakeholders and leadership
- Internal VE and design teams
- The broader organization or professional network (as appropriate) (Figure 2.4)

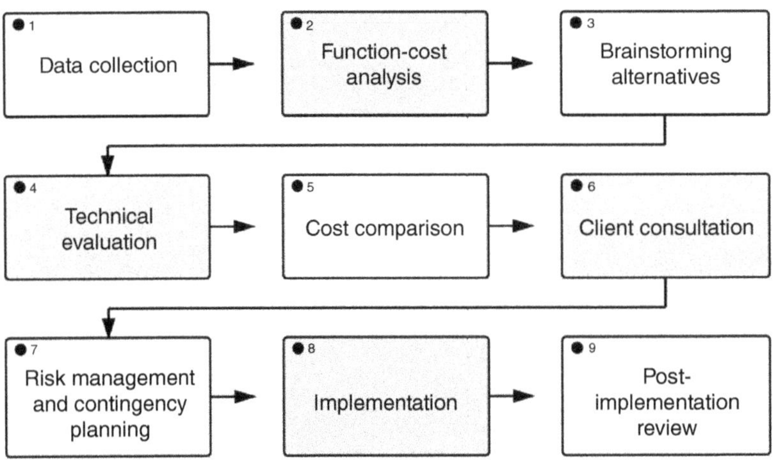

Figure 2.4 VE Process for Electrical Services.

Implementation Note

To support long-term knowledge retention and VE program development, complete both the ***VE Summary Report Template (Section 6.9)*** *and the* ***VE Lessons Learned Register (Section 6.11)*** *after the implementation phase. These tools ensure structured reflection, provide a reference for future projects, and help institutionalize the most effective VE strategies and avoid recurring issues.*

2.5 Summary and Key Takeaways

Chapter 2 has outlined the structured methodology for conducting VE during the electrical design stage. It demonstrated how VE principles are systematically applied—from data gathering to technical evaluation and client consultation—to ensure optimized performance, reduced costs, and regulatory compliance.

Key Takeaways:

- **VE is most effective when initiated early** in the design process, allowing for broad solution space and flexibility without compromising project momentum.

- A **structured VE process** includes critical phases: data collection, FCA, brainstorming alternatives, evaluation, client engagement, implementation, and post-implementation review.
- **FCA** provides the analytical foundation to identify underperforming components and prioritize cost-saving opportunities.
- **Stakeholder involvement and cross-functional expertise** are vital throughout the VE process—from brainstorming through approval.
- **Technical evaluation and risk management** ensure that all proposals are vetted not just for cost-effectiveness but also for operational, safety, and compliance performance.
- **Effective implementation and tracking** are essential to ensure proposed benefits are realized and documented.
- **Post-implementation review and lessons learned** complete the feedback loop, driving continuous improvement and institutional knowledge building.

With the framework established in Chapter 2, the following part of this guide will focus on applying this VE process across specific project stages and system types, demonstrating how it can be adapted to real-world challenges in electrical design and construction.

Closing Reflections on the VE Process

While the structured VE process presented in this chapter may appear complex in its entirety, it is important to recognize that in real-world practice, VE is often a **targeted, time-bound exercise** adapted to the specific needs and scale of each project. A typical VE workshop may span **a few days to a few weeks**, depending on scope and timing within the project lifecycle.

Not every step must be applied in full detail. Depending on **project complexity, budget constraints, or client priorities**, steps may be **merged, simplified, or omitted**, provided the core VE principles of function-driven analysis and value optimization are preserved.

Critically, the **benefits of VE almost always outweigh the resources invested**. When applied thoughtfully, VE contributes to cost certainty, improved system functionality, and better stakeholder alignment—delivering measurable value not only during construction, but over the system's full lifecycle.

References

ASHRAE/IES. (2022). *ANSI/ASHRAE/IES Standard 90.1-2022: Energy Standard for Buildings Except Low-Rise Residential Buildings*. Atlanta, GA: ASHRAE.

Brealey, R. A., Myers, S. C., & Allen, F. (2019). *Principles of corporate finance* (13th ed.). McGraw-Hill Education.

Buzan, T. (2006). *The mind map book*. BBC Active.

DIALux. (2024). *DIALux evo lighting design software*. DIAL GmbH. https://www.dialux.com

SKM Systems Analysis Inc. (2025) SKM PowerTools® for Windows. Available at: https://www.skm.com

DIgSILENT GmbH. (2025) PowerFactory. Available at: https://www.digsilent.de

Eberle, B. (1996). *SCAMPER: Games for imagination development*. Prufrock Press.
ETAP. (2024). *ETAP electrical power system analysis software*. Operation Technology, Inc. https://etap.com
Fuller, S. K., & Petersen, S. R. (1996). *Life-cycle costing manual for the federal energy management program (NIST handbook 135)*. National Institute of Standards and Technology.
Helms, M. M., & Nixon, J. (2010). Exploring SWOT analysis. *Journal of Strategy and Management*, **3**(3), 215–251.
IEC. (2022). *IEC 60364: Low-voltage electrical installations*. International Electrotechnical Commission. https://www.iec.ch
Kaplan, R. S., & Cooper, R. (1998). *Cost & effect: Using integrated cost systems to drive profitability and performance*. Harvard Business School Press.
Metropolis, N., & Ulam, S. (1949). The Monte Carlo method. *Journal of the American Statistical Association*, **44**(247), 335–341. https://doi.org/10.2307/2280232
Miles, L. D. (1972). *Techniques of value analysis and engineering* (2nd ed.). McGraw-Hill.
NFPA. (2023). *NFPA 70: National electrical code (NEC)*. National Fire Protection Association.
Park, C. S. (2011). *Contemporary engineering economics* (5th ed.). Pearson Educatio.
U.S. Department of Energy. (2023). EnergyPlus™ whole building energy simulation program. https://energyplus.net
USGBC. (2023). *LEED v4 for building design and construction*. U.S. Green Building Council. https://www.usgbc.org/leed

3

Value Engineering Strategies for Electrical Systems Design: *Applied Methods for Cost, Performance, and Lifecycle Improvement*

3.1 Introduction

Incorporating value engineering (VE) into electrical design is a detailed process that requires a deep understanding of both the theoretical and practical aspects of electrical systems. This part of the guidebook goes into the specific components and systems within electrical design that can be examined and optimized through VE. By focusing on these practical steps, professionals can pinpoint areas for cost savings, efficiency improvements, and innovative solutions without compromising on quality.

As the **longest and most technically detailed chapter** in the book, this section forms the **practical backbone** of electrical VE implementation. It is intended to be used both as a reference and as a decision-making guide throughout the project lifecycle.

3.2 Scope of VE Implementation in Electrical Design

VE in electrical design is most effective when applied with precision to the subsystems, components, and design decisions that exert significant influence over project cost, performance, and constructability. This section outlines the specific scope of VE implementation in the electrical domain, providing clarity on the evaluation targets, design levers, and system areas where VE measures can yield tangible value enhancements.

3.2.1 Overview of Evaluated Elements

In electrical engineering, VE does not seek to reduce quality or compromise safety; instead, it aims to refine the design process by challenging assumptions, optimizing configurations, and eliminating inefficiencies. The elements typically evaluated through VE include:

- System configurations and topologies
- Equipment selection and sizing
- Installation methods and construction logistics
- Redundancy and scalability strategies
- Compliance interpretations and specification alignment

Value Engineering for Electrical Systems in Construction, First Edition. Hazem Hussein.

Companion website: www.wiley.com/go/hazem

Each of these elements presents opportunities to reduce capital expenditure, optimize operational performance, and streamline future maintenance, without undermining functional requirements or regulatory standards.

3.2.2 Key Targets for VE in Electrical Design

To operationalize VE within an electrical project, designers and reviewers must identify design decisions that are cost-driving or layout-intensive. The areas that are commonly targeted during VE assessments are listed in Table 3.1.

These targets are not system-specific; rather, they span multiple electrical systems and are evaluated based on their contribution to cost, safety, and functionality.

Table 3.1 Key Targets for VE in Electrical Design

Target Area	VE Focus
Load Estimation and Demand Profiles	Preventing oversizing through refined diversity factors and load assessments
System Architectures	Simplifying topologies (e.g., radial vs. looped, centralized vs. distributed)
Equipment Sizing	Right-sizing transformers, switchboards, breakers, and conductors
Containment Layouts	Route optimization, shared service trays (where permissible)
Installation Methods	Modular assemblies, prefabricated cableways, skid-mounted systems
Controls and Automation	Rationalized sensor and control placement, cost-effective integration
Compliance Interpretation	Reassessing how code intent can be met through alternative technical means

3.2.3 Systems Commonly Subject to VE

While all parts of an electrical design may benefit from VE, certain systems consistently offer higher value returns due to their cost intensity, spatial complexity, or potential for modularization. The following systems are typically subject to VE during design development:

- **Power Distribution Systems (MV/LV):** Including transformers, main switchboards, subdistribution, and feeder layouts
- **Earthing and Lightning Protection Systems (LPS):** With focus on grounding grid optimization and surge protection coordination
- **Lighting and Lighting Control Systems:** Including general, task, and emergency lighting with integrated controls
- **Emergency Power Systems (Uninterruptible Power Supplies [UPS], Generators):** Evaluation of topologies, autonomy times, and redundancy levels
- **Information and Communications Technology/Extra-Low Voltage (ICT/ELV) Systems:** Structured cabling, data racks, security, building management systems (BMS), and communication networks
- **Fire Alarm and Life Safety Systems:** Including detection loops, panel locations, integration with safety strategy
- **Cable Management and Containment Systems:** Trunking, tray sizing, vertical risers, and underfloor/overhead routing

Each of these systems is analyzed in detail in Section 3.4, where system-specific VE applications are presented.

3.2.4 Key VE Strategies

The effective application of VE in electrical system design requires the identification of specific, quantifiable opportunities to improve value. This is achieved by implementing design strategies that reduce capital and operational costs, enhance constructability, streamline integration, and preserve system performance. The following strategies are commonly applied to balance performance requirements against cost and complexity:

- **Reviewing Performance Specifications:** Critically assess project specifications to eliminate brand-specific or overly conservative requirements. Where feasible, propose performance-based or functionally equivalent alternatives that satisfy the original design intent while reducing cost, procurement lead times, or vendor lock-in.
- **Optimizing Load Estimation Methodologies:** Refine load calculations by applying realistic demand profiles, diversity factors, and simultaneity assumptions. Avoid equipment oversizing by calibrating transformer, generator, and switchgear capacities in accordance with operational requirements and regulatory margins.
- **Calibrating System Redundancy:** Analyze the reliability needs of each system segment and apply redundancy levels (e.g., N + 1, 2N, dual-feed) only where operationally and financially justified. Excessive redundancy adds cost, complexity, and footprint without corresponding value in noncritical applications.
- **Standardizing Components and Systems:** Adopt uniform hardware platforms—including distribution panels, circuit protection devices, metering solutions, and communication interfaces—to reduce engineering time, simplify maintenance protocols, and facilitate spare part management across the facility.
- **Implementing Modular and Prefabricated Assemblies:** Utilize factory-built components such as modular switchboards, power distribution units (PDUs), prefabricated skids, and preterminated cabling systems to minimize on-site labor, reduce quality risks, and accelerate project timelines. Modularization also enhances change flexibility during late-stage coordination.
- **Optimizing Installation and Routing Strategies:** Coordinate containment routing, cable tray layouts, and equipment access strategies with other disciplines to prevent spatial conflicts. Use structured installation sequencing and prefabricated mounting arrangements to reduce construction complexity, especially in congested service corridors.

These strategies form the foundation for the detailed VE evaluation areas presented in Section 3.3, where each design aspect is analyzed through cost–function–constructability lenses.

3.3 VE Evaluation Areas in Electrical Design

This section outlines the key evaluation areas used in the implementation of VE within electrical design. Each area represents a strategic focus where engineers can apply targeted reviews and technical adjustments to achieve cost savings, improved performance, and simplified implementation—without compromising functionality or compliance. These evaluation areas are applicable across various electrical systems and form the foundation for system-specific VE measures presented in Section 3.4.

3.3.1 System-Level Design Considerations

System-level design decisions form the foundation of an electrical installation and carry the most significant cost and performance implications. These decisions are typically made during the early stages of the project (e.g., concept or preliminary design) and often remain unchallenged through to construction. Applying VE at this level involves a critical review of the overall system architecture, initial load assumptions, and project specifications to ensure the design aligns with actual functional requirements.

3.3.1.1 Load Estimation and Demand Profiles

Load estimation forms the foundation of all downstream electrical design decisions. It directly influences the sizing of transformers, generators, switchgear, cables, containment, and protective devices. Therefore, inaccuracies or overly conservative assumptions in early load profiles can lead to significant overdesign, cost escalation, and inefficient system operation.

VE in this area aims to establish **accurate, data-driven, and functionally justified** demand estimates, tailored to the specific use case of the building or facility. This ensures that electrical infrastructure is **right-sized**, not overbuilt, while still maintaining safety, redundancy, and regulatory compliance.

■ ***Common Overdesign Drivers***

Several factors often contribute to inflated load estimates during early design stages:

- **Generic rules of thumb** applied without consideration for building typology or occupancy profiles
- **Neglect of diversity factors** and simultaneity in system usage
- **Overly conservative safety margins** added without quantitative justification
- **Failure to account for energy-efficient equipment** and smart control systems
- **Stacking of peak demands** across systems that are unlikely to operate concurrently

■ ***VE Strategies for Load Estimation***

The recommended strategies to optimize demand profiles are listed in Table 3.2.

Table 3.2 Recommended Strategies for Optimizing Load Profile

Strategy	Implementation
Use Empirical Data	Refer to actual load curves from similar operational buildings or industry benchmarks
Apply Diversity and Simultaneity Factors	Adjust peak loads using building-specific usage patterns, zoning, or phased operation schemes
Incorporate Load-Shedding Strategies	Account for controlled shutdown of noncritical loads during generator or UPS sizing
Update as Design Evolves	Revise initial assumptions at each design milestone (e.g., 30%, 60%, Issued For Construction (IFC)) with real user input
Consult with MEP/ICT Teams	Coordinate loads across disciplines to avoid duplication (e.g., HVAC motors counted in both packages)

■ ***Load Profiling by Building Type***

Load density assumptions should be context-specific. Table 3.3 provides indicative figures (which must be refined per project) and the potential VE impact.

Note:

- Values are indicative and vary by climate zone, operational hours, building efficiency targets, and equipment schedules.
- Refined estimates should consider actual space utilization, equipment schedules, lighting design, and plug load diversity.
- Always cross-reference with local design codes (e.g., NEN 1010 in Netherlands, National Electrical Code(NEC) in the United States [NFPA], BS 7671 in the United Kingdom [BSI]) and client-specific briefs.

Table 3.3 Sample Load Densities Based on Building Type

Building Type	Typical Initial Estimate	Refined Estimate via VE	Key VE Considerations
Office (Open Plan)	100–120 W/m^2	60–80 W/m^2	LED lighting, efficient HVAC, limited socket use
Hospital (General Areas)	150–180 W/m^2	100–130 W/m^2	Load shedding for noncritical zones
Data Center	800–1000 W/m^2 (IT only)	Site-specific (tier level)/client brief	Tier classification, UPS topology, actual Power Usage Effectiveness(PUE) performance
Residential	6–8 kVA per dwelling	3–5 kVA per dwelling	Modern appliances, diversity across floors
Educational	80–100 W/m^2	60–75 W/m^2	Daylight control, efficient equipment usage

Additional VE Considerations

- Revisit power factor assumptions: Avoid oversizing upstream systems based on uncorrected loads.
- Differentiate essential and nonessential loads early to optimize backup power system sizing.
- Account for usage patterns (e.g., business hours vs. 24/7 zones) when evaluating simultaneity.

3.3.1.2 Topology and System Architecture

The topology and architectural configuration of an electrical distribution system directly affect system reliability, spatial requirements, capital cost, and operational flexibility. These high-level decisions—such as choosing between radial and looped distribution, centralized vs. decentralized switchgear, or implementing selective redundancy—are typically made early in design, often without rigorous evaluation.

VE at this level focuses on challenging default assumptions and tailoring the system configuration to the actual functional, operational, and redundancy needs of the project, ensuring that investment is focused where it provides measurable value.

Common Overdesign Scenarios

Without VE, system topologies may default to:

- **Fully redundant architectures** (e.g., 2N or dual-feed) even where criticality doesn't warrant it
- **Decentralized distribution boards** in every zone, regardless of space constraints or load density
- **Overlapping supply paths** for all loads, including nonessential or intermittent ones
- **Overuse of looped MV or LV feeders** even where a radial system with sectionalizing is sufficient

These decisions can lead to:

- Increased conductor lengths and material costs
- Larger switchrooms and electrical risers
- Higher fault levels and complex protection coordination
- More maintenance points and increased system complexity

VE Strategies for Topology Optimization (Table 3.4)

Table 3.4 VE Optimization Approaches for Power Distribution Topology Elements

Design Element	VE Optimization Approach
Main Distribution Topology	Evaluate radial vs. looped vs. meshed distribution based on load criticality and layout
Switchboard Architecture	Use centralized LV distribution where spatially and electrically efficient
Feeder Redundancy	Provide dual feeds selectively to essential zones only
Power Zoning	Match distribution architecture to building zones or usage profiles for load separation and flexibility
Segmented Power Paths	Design for isolation and sectionalization without full duplication

Example Applications

Example 1—Loop vs. Radial Distribution

A campus-style university project proposed MV ring main distribution for all academic blocks. VE analysis showed that only the data center and main lecture theatres required dual-feed reliability. Remaining buildings were shifted to radial feeders with local protection.

Outcome: 1.4 km of MV cable and 3 ring main units (RMUs) eliminated, saving €120,000 and reducing fault complexity.

Example 2—Switchboard Centralization

An industrial facility originally designed with one LV board per process area (eight total). VE analysis revealed that centralized switchboards with distributed panelboards via prefabricated busbars reduced material cost and allowed shared protection and monitoring.

Outcome: Switchroom count reduced from five to two, enabling process area expansion and simpler cable routing.

■ ***Redundancy Right-Sizing***
Redundancy is often overspecified without quantifiable risk assessments. VE focuses on **selective redundancy**, ensuring that:

- Essential life safety systems (e.g., emergency lighting, smoke extraction) are fully supported
- Noncritical or easily restorable systems (e.g., small offices, stairwell sockets) are served by single-feed or delayed backup
- Generator or UPS autonomy is sized only for **true essential** loads, not the entire building

VE Tip: Integrate N + 1 redundancy at system level (e.g., UPS), not component level (e.g., double breakers per board), unless operational risk demands it.

■ ***Coordination with Other Disciplines***
System topology should be developed in coordination with:

- **Architects** to optimize room layouts and avoid oversized service zones
- **Mechanical engineers** for shared routing and shaft planning
- **Fire engineers** to confirm separation of redundant paths where required (e.g., via protected routes)

Key Takeaways:

- VE in system topology challenges traditional "one-size-fits-all" approaches.
- Selective application of redundancy and distribution complexity results in significant savings.
- Early coordination is key to integrating efficient topologies into architectural and structural layouts.

3.3.1.3 Project Specification Review

Project specifications play a critical role in defining the performance, safety, and quality expectations of electrical systems. However, they often contain outdated references, conservative assumptions, or prescriptive requirements that drive unnecessary cost and design complexity. VE at this level seeks to align specifications with project needs, current standards, and practical implementation realities, without compromising safety or compliance.

Key Focus Areas and VE Actions:

1) **Review of Current Standards:** Ensure alignment with the latest updates from regulatory and industry bodies (e.g., IEC, IEEE, NEC, NEN). Maintain a register of applicable standards and systematically update specifications during each project phase.
2) **Component Specifications:** Evaluate specified ratings, materials, and brand-specific clauses for necessity and cost-efficiency. Replace proprietary requirements with performance-based descriptions where feasible.
3) **Energy Efficiency Requirements:** Verify that specified equipment supports project energy goals. Consider alternative systems with higher energy efficiency ratings and favorable lifecycle costs.
4) **System Integration and Compatibility:** Confirm interoperability of electrical systems with mechanical, fire, ICT, and BMS subsystems. This reduces coordination issues, eliminates redundant infrastructure, and prevents retrofitting costs.

5) **Futureproofing:** Incorporate modularity and scalability into specifications to support phased expansions or technology upgrades without significant redesign.
6) **Safety and Compliance Assurance:** Ensure that all components comply with applicable safety and local authority regulations and are certified by recognized bodies (e.g., CE, UL, KEMA).
7) **Feedback from Past Projects:** Integrate lessons learned from previous works and industry feedback. Use this insight to revise generic clauses and adopt tested cost-saving practices.

■ *Factory Acceptance Test Optimization*

Factory acceptance test (FAT) procedures are essential for quality control, particularly for high-value or critical systems. However, not all components warrant the cost of extensive FAT protocols. A value-driven approach defines FAT scope based on equipment criticality, balancing quality assurance with cost control.

Recommended Equipment for FAT and Witness Testing (Table 3.5)

Table 3.5 Recommended FAT Scope and Justification for Major Electrical Equipment Types

Equipment Type	Reason for FAT (Recommended When)	VE-Based Consideration Against FAT
1. High-Voltage Switchgear and Panels	Critical for primary distribution; FAT verifies internal wiring, protection schemes, and arc containment	Standardized switchgear (IEC 62271 compliant) may not need FAT unless modified or custom-built
2. Transformers (Large Capacity)	High capital cost and essential for voltage transformation; ensures thermal performance and losses	FAT may be skipped for dry-type transformers <1 MVA if manufacturer provides type test data and certificates
3. Backup Generators	Essential for critical power continuity; FAT confirms performance under load and control system operation	FAT may be replaced with on-site load testing for standard generator sets with known reliability and certified compliance
4. Ring Main Units (RMUs)	Validates switching and protection functions for MV reliability and fault management	FAT adds limited value for factory-standard RMUs (IEC 62271-200); can rely on routine test reports if unmodified
5. Power Factor Correction (PFC) Units	Verifies correction capacity and controller operation to maintain target power factor	Modular PFCs often pass factory routine tests; on-site verification may be sufficient unless the unit is large/customized
6. Motor Control Centers (MCCs)	Ensures motor protection, control logic, and interlocking operate correctly under load	FAT may be omitted for standard MCCs unless complex PLC integration or hazardous area classification is involved
7. Automatic Transfer Switches (ATS)	Ensures automatic switchover between utility and generator sources under simulated failure	Pre-certified ATS (IEC 60947-6-1) with simple logic can be validated via on-site tests; FAT adds marginal value
8. Uninterruptible Power Supplies (UPS)	Verifies runtime, switchover time, and bypass operation—especially important for data centers, hospitals, etc.	FAT may not be necessary for single-module or rack-mounted UPS units with internal QA protocols from Tier 1 vendors

Notes:

- FAT should be prioritized for equipment with **high complexity, long lead times, or direct impact on operational continuity**.
- For standard, precertified equipment, **routine factory testing with documentation may suffice**.
- Decisions should be based on **project-specific criticality**, **client risk tolerance**, and **contractual requirements**.

Implementation Guidelines:

- Establish FAT scope based on risk and capital cost.
- Consolidate testing procedures where practical.
- Prepare detailed FAT checklists and protocols.
- Witness testing selectively to reduce travel and logistics costs.
- Ensure all noncompliances are resolved prior to shipment.

Example: In a regional data hub project, the original FAT scope included all LV panels. Following VE review, FAT was limited to only the main MV switchgear, UPS system, and generator automatic transfer switches (ATS). This resulted in testing cost reductions exceeding €25,000, while preserving critical quality oversight.

■ ***Virtual FAT (vFAT) as a VE-Driven Alternative***

Traditional FAT procedures often require on-site witness testing by consultants, clients, or third parties, incurring travel, scheduling, and administrative costs. With advancements in digital infrastructure, **vFAT** has become a viable and cost-effective alternative for many systems.

What Is vFAT?

vFAT involves the **remote witnessing and documentation** of factory testing procedures using live-streamed video, prerecorded test sessions, synchronized documents, and real-time communication tools.

VE Benefits of vFAT (Table 3.6)

Table 3.6 VE Benefits of vFAT in Electrical Projects

Value Engineering Benefit	How vFAT Contributes
Cost Reduction	Eliminates travel, accommodation, and daily fees for witnessing teams
Schedule Flexibility	Easier to coordinate across time zones and teams; supports fast-track projects
Documentation Clarity	Video recordings and shared test logs improve traceability and reduce disputes
Reduced Environmental Impact	Lowers carbon footprint associated with intercontinental travel
Enhanced Stakeholder Access	Enables broader participation (e.g., client PMs, QA teams) from remote locations

vFAT Implementation Guidelines:

- Define test scope and expectations clearly in advance (same as physical FAT).
- Use stable video conferencing platforms (e.g., MS Teams, Zoom, WebEx) with multiple camera angles.
- Share real-time data from test instruments and Supervisory Control and Data Acquisition (SCADA)/Human Machine interface (HMI)displays.
- Provide annotated test protocols in advance for reference.
- Record sessions and store them in the project Quality Assurance(QA) archive.
- Include provisions for corrective action verification (if required).

Suitable Use Cases for vFAT (Table 3.7)

Table 3.7 Use Case Suitability Matrix for vFAT vs. Traditional FAT Methods

Recommended for vFAT	Not Ideal for vFAT
Standardized LV Switchgear and Panels	HV equipment with live arc testing or large power simulation
UPS Systems, ATS Panels, Distribution Boards	Systems requiring high-voltage primary injection testing
Prewired MCCs with Factory-Integrated I/O Checks	Equipment needing hands-on inspection or interactive troubleshooting
BMS or SCADA Configuration and Logic Testing	Vendor workshops with low connectivity or undocumented test procedures

Example: In a healthcare expansion project during travel restrictions, the consultant approved vFAT for modular UPS systems and LV panels. All procedures were live-streamed with prereviewed checklists, and recordings were stored in the project archive.
Outcome: FAT duration was reduced by 2 days, €7000 was saved in logistics costs, and client QA staff from three locations were able to participate.

■ ***Spatial Coordination and Integration***

Electrical infrastructure has a substantial impact on building space planning, particularly in dense service environments such as risers, plantrooms, corridors, and ceiling voids. Without early coordination, electrical designs often result in **oversized rooms, inefficient routing**, and **conflicts with other disciplines**, which later manifest as rework, clashes, or underutilized space.

VE in spatial integration aims to **align the electrical layout with architectural, structural, and mechanical systems** from the early stages. This not only improves constructability and service accessibility but also reduces the spatial footprint and associated construction costs.

Common Coordination Issues Identified During VE

- **Oversized electrical rooms** based on default layout templates, not project-specific needs
- **Redundant risers** created due to discipline-specific separation instead of integrated planning
- **Lack of vertical alignment** between stacked floors, causing complex cable routing
- **Duplication of service zones** (e.g., separate containment for power, lighting, and ICT where shared trays are possible)
- **Inadequate allowance for maintenance clearance**, requiring last-minute redesigns or work-around solutions

VE Strategies for Optimizing Spatial Use (Table 3.8)

Table 3.8 Targeted VE Strategies for Optimizing Spatial Use of Electrical Infrastructure

Target Area	VE Strategy
Switchroom Design	Right-size based on actual panel dimensions, clearances, and manufacturer shop drawings
Riser Planning	Combine electrical, ICT, and low-voltage systems into shared vertical risers with compliant segregation
Containment Routing	Use multiservice trays or integrated trunking to reduce ceiling congestion
Modularization	Use factory-built switchboards, MCCs, or UPS systems with known dimensions to define room layout early
3D Coordination	Incorporate BIM for spatial clash detection and maintain coordination with MEP and structural models

Example Applications

Example 1—Switchroom Footprint Reduction

A 15 m^2 LV switchroom in a commercial building was allocated based on typical block layout standards. VE coordination with the switchgear vendor showed that wall-mounted boards with top-entry cabling and rear maintenance access reduced the required area to 9 m^2.

Outcome: 6 m^2 freed for rentable space, with reduced cooling and lighting demand in the switchroom.

Example 2—Integrated Riser Design

An office tower was designed with three separate risers for power, ICT, and lighting systems. Through VE workshops, containment routes were consolidated using multicompartment cable trays with compliant segregation barriers.

Outcome: Two risers eliminated per floor, increasing net usable area and reducing vertical containment by 35%.

■ ***Maintenance and Operational Considerations***

VE must ensure that spatial optimization does not compromise system accessibility or maintainability. This includes:

- Adhering to statutory and manufacturer-specified working clearances (e.g., IEC 61439, NEC 110.26)
- Allowing sufficient space for cable pulling, breaker operation, and future modifications
- Ensuring adequate ventilation and heat dissipation in confined spaces (especially for UPS, VFDs, or switchgear)

VE Tip: Always validate proposed spatial optimizations using equipment general arrangement drawings and manufacturer-provided layout recommendations.

■ ***Cross-Disciplinary Collaboration***

Spatial coordination is most effective when executed as a collaborative effort. VE reviews should include representatives from:

- **Architecture**—to understand wall placements, shaft locations, and space standards
- **Structural engineering**—to align heavy equipment loads with slab reinforcements
- **Mechanical systems**—to coordinate ceiling heights, duct crossings, and access zones
- **ICT/fire protection**—to validate shared containment and clearances for other critical services

Key Takeaways:

- VE in spatial integration improves both cost efficiency and functional building design.
- Early use of manufacturer data and digital modeling ensures accurate room and riser planning.
- Coordination across disciplines avoids spatial conflicts and unlocks potential for space savings.

3.3.2 Equipment and Component Optimization

Electrical systems are built from a wide range of components—switchboards, breakers, enclosures, cabling accessories, protection devices, and system controllers. These components directly influence capital cost, installation labor, equipment footprint, and system reliability. Yet, in many projects, equipment is selected or sized based on legacy templates, conservative margins, or unnecessary brand prescriptions.

VE in this area focuses on **right-sizing and optimizing the selection** of key electrical components to align with **actual system demands**, **installation constraints**, and **operational priorities**—while maintaining full compliance with standards and performance requirements.

Common Overdesign or Misselection Issues

VE studies frequently uncover the following:

- **Oversized breakers and panels** due to exaggerated fault current or load assumptions
- **Unused extra poles or space** in switchboards leading to oversized enclosures
- **Redundant metering, protection, or automation features** not required for system operation
- **Overspecified enclosure ratings** (e.g., IP65 where IP44 suffices)
- **Unjustified brand or component exclusivity**, reducing competition and increasing lead time

VE Strategies for Component Optimization (Table 3.9)

Table 3.9 Component-Level VE Optimization in Electrical Equipment Selection

Focus Area	VE Optimization Approach
Circuit Breakers	Use molded case circuit breakers (MCCBs) with integrated metering/protection when feasible
Switchgear Panels	Right-size boards using actual load data and verified outgoing circuit counts
Protection Devices	Optimize selectivity and cascading design; avoid duplication across devices
Enclosures and IP Ratings	Match IP ratings to environment classification per IEC 60529 or EN 62208
Busbar and Cable Systems	Evaluate the use of prefabricated busbars for risers or tap-off systems in high-density areas
Functional Integration	Choose multifunctional components (e.g., combined controller and protection relay units)

Example Applications

Example 1—Circuit Breaker Rationalization

An airport terminal design included main breakers rated for 65 kA at 400 V. VE analysis revealed that the actual fault level was 38 kA, allowing for downsizing to a lower interrupting capacity without loss of protection.

Outcome: Cost reduction of ~€7500 across main and feeder breakers, and reduced panelboard width by 30%.

Example 2—Enclosure Rating Optimization

All distribution boards in a large warehouse were initially specified as IP66. After VE review, boards in internal, dry, and secure locations were reclassified to IP43 per IEC 60529.

Outcome: Saved ~€900 per board across 40 units, simplified vendor options, and reduced delivery time.

Example 3—Panel Consolidation

A multizone retail development had 12 small panelboards serving adjacent tenants. VE studies proposed combining six zones into shared boards with tap-off circuits and lockable MCBs.

Outcome: Five boards eliminated, 80 m of cable reduced, and switchroom space cut by 10 m^2.

■ ***Procurement and Vendor Flexibility***

Excessively specific component selections can limit procurement flexibility and reduce price competitiveness. VE aims to:

- Specify based on **performance criteria** rather than brands (e.g., "Type B MCCB, 25 kA, IEC 60947 compliant")
- Accept equivalent offers if they meet the required certification and functional needs
- Encourage factory-integrated assemblies to reduce on-site labor and coordination risk

VE Tip: When using design-and-build or EPC * delivery models, allow for **preapproved alternates** to reduce design delays and unlock better commercial options during bidding.

* ***Engineering, Procurement, and Construction.***

It's a common project delivery model where a single contractor is responsible for the ***design (engineering)***, ***procurement of materials and equipment***, *and* ***construction*** *of the project. This model is widely used in large infrastructure and industrial projects because it provides a single point of responsibility and often a fixed price and schedule.*

■ ***Lifecycle and Maintainability Considerations***

VE should not only target capital savings but also consider operational efficiency. Optimized component selection must address:

- **Ease of replacement** or modular upgrades
- **Local availability** of spares and technical support
- **Compatibility with existing or legacy systems** (especially for phased refurbishments)
- **Monitoring or integration capability**, if required for energy management or remote diagnostics

Example: Selecting breakers from a globally supported brand enabled a hospital operator to reuse existing service contracts and parts, reducing long-term operational expenditure (OPEX).

Key Takeaways:

- VE in component selection improves cost, space, procurement lead times, and system maintainability.
- Standardization, functional integration, and rating optimization are key tools.
- Design should avoid unnecessary duplication, overprotection, or enclosure overrating.
- Procurement flexibility must be preserved through performance-based specification writing.

3.3.3 Power Distribution Path Optimization

The power distribution path—comprising cable routes, containment systems, and busbar configurations—plays a pivotal role in both capital cost and spatial efficiency. VE in this area focuses on optimizing **routing**, **containment selection**, **conductor sizing**, and **infrastructure integration** to reduce material usage, simplify installation, and enhance coordination with other disciplines.

Distribution path decisions often go unchallenged due to default assumptions or late-stage layout development, leading to overcomplicated, costly, or inflexible installations. This section outlines how VE can be applied to streamline and optimize the physical distribution of power.

■ *Common Inefficiencies in Distribution Paths*

VE workshops often uncover the following avoidable issues:

- **Excessive cable run lengths** due to indirect routing or lack of early coordination
- **Overuse of large containment systems** sized for conservative cable scenarios
- **Redundant parallel routes** for systems that could be consolidated
- **Missed opportunities to use busbar systems** for risers or dense vertical zones
- **Inadequate coordination with structure or HVAC**, resulting in rerouting and spatial conflict

VE Strategies for Distribution Optimization (Table 3.10)

Table 3.10 Optimization Strategies for Power Distribution Path Elements

Focus Area	Optimization Strategy
Cable Routing	Use direct, shortest-path routing where feasible; coordinate early with structure to minimize bends and supports
Containment Sizing	Base tray/trunking size on refined cable schedules after diversity and load balancing
System Integration	Combine LV, ICT, and control cabling into shared containment with compliant segregation
Busbar Trunking Systems	Use prefabricated busbar risers in vertical cores to reduce installation time and containment congestion
Prefabricated Assemblies	Apply modular skids or tap-off busbars for repetitive distribution nodes (e.g., hotel rooms, data halls)

Example Applications

Example 1—Cable Route Simplification

In a high-rise office tower, initial cable layouts followed structural grids, resulting in excessive horizontal detours. VE coordination enabled direct core-to-core riser connections through preformed slab openings.

Outcome: Reduced cable length by 18%, containment runs by 25%, and installation labor by ~€30,000.

Example 2—Busbar Use in Vertical Distribution

A hospital used traditional multicore cables for vertical power distribution. VE analysis recommended replacing these with busbar trunking.

Outcome: Faster installation, reduced riser width, and improved maintainability. Initial busbar cost was offset by a 40% reduction in labor and containment materials.

Example 3—Shared Trunking for ELV and Power

In a commercial building, initial designs had separate trunking for power and ICT. VE introduced multicompartment trays with fire-rated dividers.

Outcome: Reduced horizontal trunking by 150 m per floor, saving on materials, ceiling space, and coordination time.

■ ***Integration with BIM and Coordination Models***

Early coordination using 3D modeling tools (e.g., Building Information Modeling (BIM), Navisworks) allows:

- Accurate routing based on real structural geometry
- Clash detection with HVAC, sprinklers, and architecture
- Proper alignment of risers and horizontal pathways
- Realistic containment sizing verified against actual cable schedule

VE should ensure that the containment model is based on updated load and equipment layouts—**not generic placeholder content**—to avoid overprovisioning.

■ ***Installation Methodology and Labor Savings***

VE can also improve installation efficiency by:

- Standardizing cable tray support systems
- Using prefabricated brackets or trapeze hangers
- Implementing color-coded tray systems for easier installation and inspection
- Sequencing installation per floor or zone to avoid rework

VE Tip: Consider using BIM-linked quantities and modular routing kits to align procurement with actual layout, reducing on-site waste.

■ ***Strategic Equipment and Panel Placement***

In addition to route optimization and containment design, the physical **placement of electrical equipment**—such as transformers, panelboards, and distribution boards—has a major impact on routing efficiency, voltage drop, and total material usage. Strategic placement decisions during design development can yield measurable value in cost, performance, and constructability.

a) **Position Transformers and Main Distribution Boards Near Load Centers:** Reducing the distance between large electrical consumers and their supply sources (e.g., transformers or switchboards) minimizes voltage drop and energy loss due to cable resistance. This is especially beneficial in high-load zones such as industrial processes, data centers, or central HVAC plants.
b) **Place Panelboards Close to Majority of Loads:** Locating final distribution boards within or near tenant zones, lighting zones, or equipment rooms reduces branch circuit lengths, simplifies containment layouts, and improves electrical performance.
c) **Choose Shortest Conduit and Cable Routes:** During layout coordination, designers should seek the most direct feasible paths—horizontally and vertically—to reduce conduit and cable quantities. Short routes reduce installation time and cost and ease future inspection.
d) **Minimize Bends and Intersections:** Straight conduit runs and gentle sweep bends facilitate easier cable pulling, reduce mechanical stress on cables, and allow faster installation. Fewer intersections also reduce the number of accessories and points of failure.

Example: In a healthcare facility, VE review identified that relocating four LV panelboards 6 m closer to patient care areas reduced total cable length by over 700 m and eliminated 16 bends across three floors—resulting in labor and material savings of ~€12,000.

3.3.4 System Reliability, Modularity, and Expandability

Electrical systems must be designed not only for present-day performance but also to accommodate operational resilience and future growth. VE in this domain seeks to balance **system reliability** with **cost-effectiveness**, while enabling **modular deployment** and **scalable expansion** where appropriate.

Rather than defaulting to fully redundant or overbuilt systems, VE promotes **selective reliability strategies**, **modular configurations**, and **flexible infrastructure** that can adapt to change—without excessive initial capital investment or unnecessary spatial commitment.

3.3.4.1 Reliability and Redundancy Optimization

Designers often implement high-reliability features such as 2N redundancy, dual feeds, or generator-backed systems across entire buildings without evaluating the actual **risk tolerance** or **criticality of loads**.

VE Strategies:

- Apply N + 1 or 2N redundancy **only to critical systems** (e.g., data centers, life safety loads, ICU suites).
- Use **selective redundancy** at the subsystem level, rather than replicating entire distribution architectures.
- Incorporate **ATS** and **manual bypass options** for operational flexibility.

3.3.4.2 Modular and Scalable Design Applications

Modular and scalable design is a foundational strategy within VE for optimizing long-term performance, reducing initial capital expenditure, and enabling future adaptability. By breaking down systems into standardized, prefabricated units, electrical infrastructure can be deployed, expanded, or reconfigured with minimal disruption and reduced cost over the building lifecycle.

This approach supports phased construction, minimizes on-site labor, enables just-in-time procurement, and improves coordination through standardized interfaces. The following subsections present practical applications of modular and scalable design principles in modern electrical systems.

■ ***Modular Switchgear***

Description: Modular switchgear consists of standardized, pre-engineered units that can be added or removed based on evolving power distribution needs. Each unit includes disconnects, metering, or protection and can be reconfigured as needed.

Benefits:

- Simplified expansion or reconfiguration without full panel replacement.
- Reduced downtime during maintenance or upgrades.
- Standardization of components simplifies spare part management.

Implementation:

- Select switchgear models that offer true modularity and top/bottom entry flexibility.
- Design switchrooms with space allowances for future outgoing modules.
- Coordinate busbar and termination zones to support future additions.

■ ***Scalable UPS Systems***

Description: Scalable UPS systems are modular backup power platforms designed to grow with the facility's needs by adding power or battery modules over time. This allows right-sizing from the outset while preserving future upgrade capacity.

Benefits:

- Delays capital investment until capacity is needed.
- Reduces risk of oversizing and unused autonomy.
- Simplifies maintenance through module-level servicing.

Implementation:

- Choose UPS systems with modular topology (e.g., rack-mounted, block-configurable).
- Reserve floor space, airflow clearance, and electrical capacity for future modules.
- Use VE assessments to match initial autonomy to current critical load only.

■ ***Plug-and-Play Lighting Systems***

Description: These systems feature standardized connectors and modular luminaires that allow for fast installation, easy replacement, and simplified future upgrades without rewiring.

Benefits:

- Reduced installation time and labor.
- Supports efficient upgrade to new technologies (e.g., LED, Power over Ethernet [PoE]).
- Enhances maintainability with tool-free fixture replacement.

Implementation:

- Specify systems with universal plug connectors (e.g., Wieland, digital addressable lighting interface (DALI]-compatible).
- Plan for load grouping, dimming control, and smart integration from the start.
- Reserve capacity in lighting circuits for fixture additions where applicable.

■ *Modular Data Centers*

Description: Modular data centers use prefabricated, self-contained units with integrated power, cooling, and network infrastructure. These units can be deployed incrementally, enabling agile response to changing IT demands.

Benefits:

- Scalable growth without facility-wide redesign.
- Consistent quality from factory-assembled, tested modules.
- Reduced deployment time and site disruption.

Implementation:

- Align modular unit sizing with IT growth projections.
- Ensure compatibility with base building utilities (power, HVAC, comms).
- Allocate space for future units and ductwork/cabling integration paths.

■ *Scalable BMS*

Description: Scalable BMS platforms support gradual integration of additional building systems (e.g., HVAC zones, lighting groups, access control) as facilities expand or evolve.

Benefits:

- Allows phased investment aligned with occupancy growth.
- Supports integration of emerging technologies (Internet of Things [IoT], energy analytics).
- Improves long-term system resilience through distributed architecture.

Implementation:

- Choose open-protocol, modular BMS platforms (e.g., BACnet/IP, KNX).
- Establish a roadmap for staged integration of systems.
- Provide staff training and documentation for modular expansion workflows.

■ *Modular Electrical Outlets and Sockets*

Description: These are reconfigurable systems that allow electrical outlets and specialized connectors (e.g., USB, HDMI, smart ports) to be added or relocated without extensive rewiring.

Benefits:

- Simplifies tenant fit-out or functional changes in use.
- Reduces cost and time for future upgrades.
- Supports customized functionality for different zones.

Implementation:

- Use trunking or underfloor boxes with interchangeable outlet modules.
- Plan layout for future socket additions based on room function flexibility.
- Ensure compatibility with modular wiring infrastructure.

■ ***Modular Wiring Systems***

Description: Modular wiring systems use factory-assembled cables and connectors for power and lighting circuits. These plug-and-play systems minimize site wiring errors and support fast deployment.

Benefits:

- Reduces site labor and installation time.
- Improves QA/QC through pretested assemblies.
- Simplifies future maintenance and circuit modifications.

Implementation:

- Select modular wiring systems compatible with planned containment routes.
- Coordinate with lighting and power layout early in design.
- Document all circuits and connector types for future reference.

■ ***DfMA Integration in Modular Electrical Systems***

Many of the modular strategies described above benefit directly from **Design for Manufacture and Assembly (DfMA)** principles. These include prefabricated switchgear, UPS skids, modular cabling kits, and containerized data center pods.

Applying DfMA during the design phase offers significant VE benefits:

- Reduced on-site labor and installation errors
- Improved QA/QC through factory testing
- Faster project timelines via parallel fabrication
- Simplified logistics and standardization

DfMA Design Methodology for Electrical Systems

DfMA in electrical engineering is a design-centric methodology that emphasizes modularization, repeatability, and off-site integration. This approach aims to shift as much work as possible into controlled environments, thereby reducing on-site time, cost, and risk (Kamrani & Abouel Nasr, 2010).

Key Principles:

- **Standardize components and interfaces** to reduce part count and simplify assembly, which enhances efficiency and lowers production costs (Boothroyd & Dewhurst, 1990).
- **Avoid site-specific customization** where feasible, enabling scalability and reducing design complexity (Kamrani & Abouel Nasr, 2010).
- **Coordinate early with suppliers** to ensure physical integration and alignment with prefabrication strategies, which is essential for concurrent engineering success (Boothroyd & Dewhurst, 1990).
- **Use BIM and 3D tools** for accurate spatial validation and clash detection, improving constructability and reducing rework (Kamrani & Abouel Nasr, 2010).
- **Plan for installation sequencing and maintenance access** to streamline on-site operations and ensure long-term serviceability (Kamrani & Abouel Nasr, 2010).

DfMA Implementation Checklist for Engineers (Table 3.11)

Table 3.11 DfMA Implementation Checklist for Electrical Systems

Action	Objective
Identify DfMA Opportunities in System Layout	Target repeatable or high-density systems: switchboards, UPS, cabling, and controls
Select Modular or Prefabricated Solutions	Engage with manufacturers who offer tested, plug-and-play options
Coordinate Dimensions and Access Early	Validate delivery path, installation clearance, and maintenance space
Define QA/QC and FAT Protocols	Ensure pretested assemblies meet project specifications
Plan Logistics and Sequencing	Align factory delivery with site readiness milestones
Document Connection Interfaces and Expansion	Ensure future add-ons can be integrated without site redesign

When embedded into VE workflows, DfMA transforms how electrical systems are planned and delivered—unlocking faster, safer, and more cost-efficient installations while preserving flexibility and quality.

Summary Benefits of Modular and Scalable Systems (Table 3.12)

Table 3.12 Modular and Scalable Design Contributions to Key Electrical Design Objectives

Design Objective	How Modular/Scalable Design Supports It
Reduce First Cost	Defer investment through phaseable systems
Accelerate Construction	Use prefabricated modules to minimize on-site time
Simplify Maintenance	Use replaceable modules with plug-in architecture
Support Expansion	Enable seamless addition of circuits, devices, or systems without redesign
Improve Quality	Leverage factory-tested components with consistent assembly standards
Enable Flexibility	Accommodate changing occupancy needs, technologies, or space usage with minimal disruption

3.3.4.3 Infrastructure Provisioning for Future Expansion

An essential aspect of VE in electrical design is enabling future expansion and system upgrades **without requiring invasive rework or significant redesign**. Infrastructure provisioning involves anticipating future needs and integrating the necessary physical and electrical capacity during the initial design and construction phase.

By identifying opportunities to **reserve space, capacity, and pathways**, engineers can reduce the cost and complexity of future modifications, support evolving functional requirements, and extend the system's useful life.

VE Objectives in Infrastructure Provisioning:

- Avoid oversizing today, but allow for growth tomorrow.
- Reduce disruption during future phases or tenant fit-outs.
- Prevent premature obsolescence of panels, risers, or backbone systems.
- Align infrastructure with realistic growth projections and industry trends (e.g., Electric Vehicle (EV)charging, smart systems).

Provisioning Techniques (Table 3.13)

Table 3.13 Infrastructure Provisioning Techniques for Future Electrical System Expansion

Provisioning Area	VE Implementation Strategy
Electrical Room Sizing	Design switchrooms, UPS rooms, and comms spaces with clear future equipment zones
Spare Panel Capacity	Include blanked-off outgoing feeders or space for additional modules in panels and UPS systems
Busbar and Cable Routing	Install larger containment or leave spare trays for future circuits
Conduit and Raceway Allowance	Provide spare conduits (capped or labeled) between floors or key areas
Utility Coordination	Align with utility providers for future capacity reservation (e.g., transformer plinths, LV entry ducts)
ICT and BMS Infrastructure	Install spare fiber cores, empty conduit paths, and extra backbone ports for future connections

Example Applications

Example 1—Reserved LV Panel Capacity

A research facility was expected to expand its lab spaces within 5 years. VE proposed specifying a main LV panel with 30% blanked outgoing slots and documented cable trunking with spare space.

Outcome: Future expansion was completed without new switchgear procurement, minimizing shutdown risk.

Example 2—Transformer Expansion Planning

A commercial campus designed with a single 1000 kVA transformer initially included space and ductwork for a second unit.

Outcome: When demand increased after 5 years, the second transformer was installed without civil modifications or service disruptions.

Example 3—EV Charger Readiness

A business park installed oversized cable containment and spare breakers in anticipation of future electric vehicle charger installations.

Outcome: The client was able to quickly deploy 12 EV chargers within 2 years at minimal additional cost.

■ *Documentation and Communication*

Provisioning elements must be clearly documented in:

- As-built drawings
- O&M manuals
- Labeling systems (e.g., "Spare for future UPS #2")
- BIM models with asset metadata

VE requires this information to be handed over to facility managers and embedded in asset management systems, ensuring future teams can easily identify available provisions.

■ *Strategic Planning with Stakeholders*

VE teams should engage early with:

- **Clients**—to identify realistic growth or change scenarios
- **IT/Facilities teams**—to plan backbone infrastructure capacity
- **Utilities**—to preapprove staged connection upgrades
- **Architects**—to allocate and protect required physical space

Key Takeaways:

- Infrastructure provisioning is a low-cost, high-value VE strategy to futureproof electrical systems.
- It avoids overspending today while enabling smart, efficient growth tomorrow.
- Reserved space, spare capacity, and clear documentation ensure flexibility without disruption.
- Collaboration with stakeholders and thoughtful phasing are critical to successful implementation.

3.3.5 Control Strategy Optimization

Control systems—including lighting controls, automation interfaces, and building integration protocols—are essential for managing electrical energy usage, enhancing occupant comfort, and supporting facility operations. However, these systems are frequently overdesigned or unnecessarily complex, leading to excessive capital cost, commissioning delays, and underutilized features (Roy & Das, 2025).

VE in control strategies focuses on delivering **fit-for-purpose solutions** that balance functionality, user needs, energy savings, and lifecycle cost. VE encourages simplification, standardization, and selective integration—particularly where advanced control adds little operational value (Guan et al., 2025).

Common VE Issues in Control Systems

- Overspecification of centralized control for systems that could operate autonomously.
- Integration of lighting, HVAC, and access systems into BMS without clear functional need.
- Redundant sensors or controllers across adjacent zones.
- Use of proprietary or nonscalable control protocols.
- Complex user interfaces that require ongoing support.

VE Strategies for Optimizing Control Systems (Table 3.14)

Table 3.14 Control Strategy Optimization for Lighting and HVAC Systems by Target Area

Target Area	VE Optimization Approach
Lighting Controls	Apply zoning, daylight harvesting, and occupancy-based logic only where justified
HVAC and BMS Integration	Limit full BMS control to major plant or public zones; use stand-alone logic for secondary areas
Sensor Distribution	Combine occupancy, temperature, and daylight sensors where possible to reduce hardware
Protocol Selection	Use open, scalable communication protocols (e.g., KNX, BACnet, Modbus)
User Interfaces	Provide simple, intuitive interfaces for users; avoid over-automation that complicates operations

Example Applications

Example 1—Simplified Lighting Control

An educational facility design included a fully addressable DALI system for all classrooms, corridors, and staircases. VE assessment reduced scope to local Passive Infrared(PIR) sensors in classrooms and time-based logic for circulation areas.

Outcome: Reduced lighting control hardware cost by 45% and simplified commissioning.

Example 2—Selective BMS Integration

A high-rise office tower initially required BMS integration for every distribution board. VE review limited BMS control to HVAC plant, common area lighting, and utility metering—excluding tenant panels.

Outcome: Shortened commissioning timeline and reduced software integration costs.

■ ***Control Strategy Alignment with Functional Requirements***

VE requires clear alignment of control logic with actual space use, operational schedules, and user expectations.

- **Office spaces:** Occupancy-based lighting and HVAC zones
- **Retail zones:** Time-based control with user override
- **Data centers or labs:** High automation, environmental control, and alarm integration
- **Parking and back-of-house:** Low-cost timers or standalone sensors without BMS interface

■ ***Maintainability and Lifecycle Considerations***

Complex controls often require specialized support. VE encourages:

- Use of vendor-independent platforms with local support
- Modular components with easy reprogramming or replacement
- Clear documentation and simple user manuals
- Avoiding control systems that require proprietary commissioning tools

VE Tip: Where possible, specify systems that can be maintained by in-house staff or general contractors, not requiring vendor intervention for every change.

■ ***Scalability and Future Integration***

A good control system should support future needs without major reconfiguration. VE supports:

- Installing controllers with spare I/O for future zones
- Using cabling and backbone that supports additional devices or features
- Selecting control protocols compatible with future BMS, EMS, or IoT platforms

Key Takeaways:

- VE in control strategies focuses on simplicity, relevance, and long-term usability.
- Full-scale automation is not always appropriate—fit-for-purpose logic provides better value.
- Avoiding proprietary or over-engineered systems reduces cost and improves maintainability.
- Lighting and HVAC controls should be aligned with space function and actual operational needs.

3.3.6 Vendor and Procurement Strategy

Procurement practices have a direct impact on the cost, quality, delivery schedule, and maintainability of electrical systems. While specifications define technical requirements, vendor engagement and purchasing strategies determine whether these requirements are delivered in the most efficient and cost-effective way.

VE in procurement strategy ensures that **performance-based specifications**, **market competitiveness**, and **lifecycle considerations** are embedded into vendor selection and purchasing decisions. This enables the project to realize technical requirements without being constrained by unnecessary brand exclusivity or rigid specifications that limit value-adding alternatives.

Common VE Issues in Procurement

- Brand-specific or vendor-exclusive specifications without technical justification.
- Use of high-tier equipment in noncritical applications.
- Absence of approved equals or alternative submittal clauses.
- Lack of early engagement with suppliers, leading to coordination issues.
- Late-stage changes due to misalignment between design and market availability.

VE-Based Procurement Strategies (Table 3.15)

Table 3.15 VE Approaches to Procurement Focus Areas in Electrical Projects

Procurement Focus	VE Approach
Specification Flexibility	Write performance-based specifications that allow for equivalent product submissions
Vendor Prequalification	Select vendors based on capability, local support, past performance, and financial stability
Tiered Equipment Selection	Apply differentiated equipment grades based on system criticality (e.g., hospital vs. parking garage)
Framework or Bulk Procurement	Leverage bulk orders or project bundling for cost savings on high-quantity items
DfMA-Compatible Vendors	Choose suppliers offering factory-assembled, modular, or integrated solutions

■ ***Vendor Evaluation and Qualification***

Objective: To ensure that selected vendors can reliably meet the project's performance, quality, and schedule requirements while offering the best long-term value. Vendor evaluation minimizes risks such as delivery delays, cost overruns, and quality issues, while also fostering long-term supplier relationships.

VE Action Steps (Table 3.16)

Table 3.16 Vendor Evaluation Criteria and Recommended VE Practices for Electrical Procurement

Evaluation Area	Recommended Practice
Review the Approved Vendor List (AVL)	Validate past performance, current capability, and compliance with project requirements
Regional Sourcing	Prioritize compliant vendors closer to site to reduce lead time and logistics cost
Vendor Specialization	Match vendors with niche systems (e.g., BMS, UPS, modular wiring) to improve reliability
Samples or Demonstrations	Use for new technologies or untested vendors to assess product compatibility
Pricing and Terms	Negotiate commercial terms, batch discounts, and after-sales support
Certifications	Verify product and process certifications (e.g., ISO 9001, IEC, UL)
Warranty and Support	Evaluate warranty coverage, local spare availability, and service response structure
Factory Audits *(if feasible)*	Assess manufacturing quality and capacity, especially for critical components

VE Tip: Vendor evaluation should be integrated into both the **design development phase** and the **tender review phase**, especially when long-lead equipment or complex systems are involved.

■ ***Approved Alternatives and Competitive Submittals***

A cornerstone of VE is to allow **technical equivalents** that meet the performance intent of the design.

- Use "**or approved equivalent**" language in tender documents.
- Define **clear evaluation criteria** (compliance with standards, references, and certifications).
- Establish a **review process** that balances technical integrity with market flexibility.
- Allow alternate submissions during VE workshops or pre-award phases.

Example: A data center project originally specified a single UPS brand with long lead times. VE workshops introduced an equivalent modular UPS system from a prequalified alternate vendor—meeting all technical specs, reducing delivery by 6 weeks, and saving €40,000.

■ ***Lifecycle and Maintenance Considerations***

Selecting equipment based solely on the lowest price often overlooks long-term operational costs. VE incorporates:

- Availability of **local spares and service contracts**
- Compatibility with existing infrastructure and facility maintenance protocols

- **Training and support services** included in vendor packages
- Modular components with replaceable parts and clear documentation

VE Tip: Involve facility management during procurement reviews to validate compatibility with their maintenance strategies.

■ *Early Vendor Engagement and Technical Input*

VE encourages **proactive engagement** with key vendors during design development, especially for complex or integrated systems.

Benefits include:

- Design validation based on real product dimensions and configurations
- Optimized installation methods (e.g., pre-cabled, skidded, or busbar-based solutions)
- Early coordination of logistics, testing, and commissioning protocols
- Reduced redesign risk due to supplier-specific limitations

■ *Owner-Furnished, Contractor-Installed Equipment*

Overview: Owner-furnished, contractor-installed (OFCI) is a procurement strategy in which the client (owner) purchases specific equipment directly, and the contractor is responsible for installation and final connections. This approach is commonly used for client-preferred vendors, long-lead items, or cost-saving opportunities. While OFCI can deliver VE benefits, it requires precise coordination to avoid project disruption.

VE Considerations (Table 3.17)

Table 3.17 VE-Focused Responsibilities and Coordination Aspects for OFCI

Aspect	VE-Focused Guidance
Responsibility Clarity	Clearly define who is responsible for shipping, receiving, storage, and inspection
Program Coordination	Align OFCI delivery dates with contractor program to avoid idle labor or rework
Technical Specifications	Ensure OFCI items are fully coordinated in design (footprint, interfaces, voltage, etc.)
Warranty Management	Determine whether installation voids vendor warranty; require pre- and post-install inspections
Storage and Handling	Confirm on-site storage capacity and conditions if delivery precedes installation readiness
Insurance and Liability	Define risk ownership for damage during transport, storage, or installation
Testing and Commissioning	Coordinate FAT, SAT, or pre-commissioning roles between vendor, owner, and contractor

Example: A client-provided UPS was delivered without prior coordination on footprint and weight. The contractor had to modify cable trench layouts and install a secondary base frame—delaying commissioning by 2 weeks. VE planning in future phases mandated detailed OFCI checklists and cut sheets during IFC issuance.

VE Tip: Use a dedicated OFCI responsibility matrix in the specification to avoid ambiguity between parties during procurement, delivery, installation, and startup.

■ ***Digital Procurement and Specification Tools***

Modern VE practices also leverage digital procurement platforms and BIM-integrated product libraries to:

- Ensure specification accuracy and consistency
- Compare cost-performance ratios of products in real time
- Track lead times and product availability
- Integrate manufacturer-specific models into coordination workflows

Key Takeaways:

- Procurement strategy is a key VE lever—early planning and flexible specifications unlock cost and time savings.
- Avoiding brand exclusivity and overspecification increases market competition and improves value.
- Lifecycle performance, maintainability, and support are essential selection criteria.
- Early vendor engagement ensures alignment between design and supply-chain realities.

3.4 System-Specific VE Applications

While earlier sections focused on cross-cutting principles and evaluation areas, **system-specific VE** addresses optimization opportunities **within individual electrical systems**. Each system type—whether power distribution, lighting, ICT, or life safety—presents unique challenges and opportunities for cost savings, space efficiency, and performance enhancement without compromising compliance or reliability.

This section outlines key VE actions, risk considerations, and system-specific recommendations for major electrical subsystems commonly found in building and infrastructure projects.

3.4.1 Power Distribution (MV/LV)

The power distribution system forms the backbone of any electrical infrastructure, enabling the reliable delivery of power from incoming utility sources through medium voltage (MV) and low voltage (LV) systems to end-use equipment. It includes MV switchgear, transformers, LV main distribution boards (MDBs), sub-distribution boards (SDBs), risers, cable systems, and protection schemes.

This system is often designed conservatively due to perceived criticality and limited load certainty, making it a prime candidate for VE interventions—particularly around load rationalization, topology simplification, and equipment right-sizing.

VE Objectives:

- Optimize topology and equipment sizing to match realistic load profiles.
- Eliminate redundant feeders, oversized components, and unnecessary system complexity.
- Reduce material and installation costs through route optimization and standardization.
- Improve maintainability and space efficiency while preserving future capacity and reliability.

Typical Overdesign or Inefficiencies:

- Oversized transformers or panels due to unrealistic demand forecasts.
- Multiple parallel cables where busbar systems would suffice.
- Overspecified switchgear ratings (e.g., 65 kA where 36 kA is sufficient).
- Redundant switchboards serving lightly loaded zones.
- Rigid radial topology where selective ring or distributed schemes are more efficient.

Targeted VE Strategies (Table 3.18)

Table 3.18 VE Strategies for Main MV/LV Power Distribution Topology

Focus Area	VE Approach
Load-Based Sizing	Use actual or benchmarked demand profiles; apply diversity and simultaneity
Topology Simplification	Minimize interconnections; eliminate redundant panels in noncritical zones
Panel and Feeder Right-Sizing	Select switchgear ratings based on fault analysis and protection coordination
Riser Optimization	Use busbar risers for high-rise or dense circuits; eliminate oversizing
Modular Distribution Design	Use modular boards with provisions for future expansion where applicable
Smart Metering	Specify meters only where needed for performance or billing—not every SDB

Implementation Considerations:

- Validate transformer capacity using VE-reviewed demand estimation (ref. Section 3.3.4.1).
- Use protection studies to justify interrupting capacity of switchgear (e.g., IEC 60947, IEC 61439, IEC 60947-2).
- Consider use of **preassembled switchgear** or containerized MV substations for time and space savings.
- Align design topology with redundancy strategy (ref. Section 3.3.4.1): N + 1, dual feed, or selective backup.
- Coordinate busbar/breaker integration with spatial layout and riser design (see Section 3.3.3).

Example Application

In a commercial development, VE analysis revealed that the original transformer sizing of 2 × 1250 kVA exceeded projected demand by 60%. One transformer was deferred, and the switchgear was designed with provisions for later expansion.
Outcome: €90,000 savings on transformer and switchgear; reduced switchroom size by 20%.

Key Takeaways:

- MV/LV distribution is often conservatively overdesigned—VE enables right-sizing based on data and standards.
- Optimizing topology, feeder sizing, and panel design reduces cost, footprint, and complexity.
- Prefabricated, modular, and scalable switchgear options support futureproofing without overspending.
- Close coordination with protection studies, riser planning, and room layouts is critical for success.

3.4.1.1 MV Switchgear and Protection

System Overview

MV switchgear serves as the **primary interface between utility supply and internal electrical infrastructure**, typically operating in the 11–33 kV range It ensures safe switching, isolation, and protection of the power distribution network. MV switchgear configurations may include RMUs, air-insulated switchgear (AIS), gas-insulated switchgear (GIS), or metal-clad switchgear, depending on space, fault levels, and reliability requirements.

Protection is managed through relays, current transformers (CTs), voltage transformers (VTs), and circuit breakers, ensuring fast fault clearance and system selectivity.

VE Objectives:

- Select the most appropriate switchgear configuration (AIS, GIS, RMU) based on functional and spatial needs.
- Avoid overrating switchgear in terms of fault level and insulation class.
- Eliminate unnecessary duplication in ring or radial systems.
- Optimize panel count and busbar sections.
- Integrate protection schemes without overspecifying relays or requiring custom programming unless justified.

Typical Overdesign or Inefficiencies:

- Selection of Gas-Insulated Switchgear(**GIS) over** Air-Insulated Switchgear **(AIS**)in noncritical indoor settings without spatial constraints.
- Unused feeder bays specified "just in case" without justified growth plan.
- Overrated short-circuit withstand ratings beyond utility fault level at the point of common coupling.
- Overuse of programmable protection relays where simpler overcurrent + earth fault protection suffices.
- Redundant or underutilized ring main arrangements in small or medium projects.

Targeted VE Strategies (Table 3.19)

Table 3.19 VE Strategies for MV Switchgear Configuration and Protection Selection

Focus Area	VE Strategy
Switchgear Type Selection	Use AIS where space allows; RMUs for compact setups; GIS only in space-constrained or harsh environments
Feeder Optimization	Remove spare feeder bays not supported by documented future expansion needs
Fault Level Rating	Base ratings on short-circuit study results using realistic utility fault levels
Protection Simplification	Use standard Inverse Definite Minimum Time (IDMT) and overcurrent relays unless selectivity or load profile complexity requires advanced logic
Configuration Rationalization	Avoid looped systems unless justified by redundancy or maintenance criteria

Implementation Considerations:

- Ensure compatibility with local utility interface requirements (e.g., incoming breaker type, protection schemes, arc suppression).
- Base short-circuit ratings on IEC 62271 series (IEC 62271-100) and verified short-circuit calculations.
- Consider future transformer additions or capacity upgrades only if supported by a phased masterplan.
- Modular switchgear assemblies can be factory-tested (FAT) and expanded later via plug-in bays.
- Document ring operation procedures if implemented; otherwise, consider radial as a cost-saving default.

Example Application

In an airport substation project, the initial design specified GIS panels for all MV rooms due to perceived reliability needs. VE analysis showed that AIS would suffice for 80% of the application, with GIS retained only for the main intake room with footprint constraints.
Outcome: System reliability maintained, total MV equipment cost reduced by €120,000, and lead time shortened by 6 weeks.

Key Takeaways:

- MV switchgear is a major cost driver; VE encourages fit-for-purpose configuration based on space, load, and reliability.
- Standard AIS or RMU solutions are often more economical than GIS in typical commercial or institutional projects.
- Right-sizing fault ratings and avoiding unnecessary feeders avoids overinvestment.
- Simplifying protection strategies reduces cost and commissioning complexity while maintaining system safety.

■ ***Sustainability Consideration—SF_6-Free Switchgear***
Traditional GIS relies on **SF_6 (sulfur hexafluoride)**—a potent greenhouse gas with a global warming potential ~23,500× that of CO_2 (Siemens Energy, 2023). While effective for insulation and arc quenching, its environmental impact and growing regulatory restrictions are driving a shift toward SF_6-free alternatives.

SF_6-free MV switchgear solutions now use vacuum interruption and eco-friendly insulating mediums such as:

- Clean air or synthetic air (Siemens Energy, 2023)
- Fluoronitrile-based gas mixtures (e.g., g^3) (GE Grid Solutions, 2024)
- Solid insulation systems

VE Recommendation: For projects with strong environmental goals, regulatory sensitivity, or public-sector ownership, consider SF_6-free switchgear early in the design. This supports corporate ESG commitments, lowers lifecycle environmental cost, and aligns with future legislation on F-gases and greenhouse emissions.

Note: SF_6-free solutions are now competitively priced and available up to 36 kV, making them viable for most building-scale and infrastructure applications (European Commission., 2024).

3.4.1.2 MV Cabling

System Overview

MV cabling connects the MV switchgear to downstream components, such as power transformers or large mechanical loads. These cables typically operate at voltages between 6 and 33 kV, depending on regional standards and utility interfaces. MV cables are critical for transmitting power across distances and through complex site topologies while maintaining safety, voltage stability, and thermal performance.

MV cable types may include single-core or three-core configurations with XLPE (cross-linked polyethylene) insulation and metallic sheathing or screening for earthing and fault containment.

VE Objectives:

- Match cable sizing to actual load and installation conditions rather than applying conservative assumptions.
- Minimize route length and cable quantity through optimized equipment placement.
- Evaluate cost/performance trade-offs between cable types, materials, and installation methods.
- Reduce long-term maintenance and losses through proper jointing and termination strategies.
- Avoid unnecessary parallel runs and excessive derating.

Typical Overdesign or Inefficiencies:

- Oversized cables due to overly conservative derating or ambient temperature assumptions.
- Use of multiple parallel cable sets when a single higher-capacity cable would suffice.
- Unnecessarily long routing between switchgear and transformer due to poor layout planning.
- Overuse of armored cables indoors where mechanical protection is not required.
- Lack of route planning resulting in excessive bends, thermal hotspots, or crossing utilities.

Targeted VE Strategies (Table 3.20)

Table 3.20 VE Approaches for Optimizing MV Cable Design

Focus Area	VE Approach
Cable Sizing Optimization	Use detailed load calculations and installation method-based derating (per IEC 60502, IEEE Std 835)
Route Shortening	Position transformers closer to switchgear to minimize run lengths
Cable Type Selection	Use nonarmored or single-core cables in protected ducts/trays where permitted
Parallel Runs Evaluation	Evaluate whether single, higher-rated cable sets are more cost-effective
Thermal Coordination	Avoid grouping dissimilar load cables; verify grouping factors per site conditions
Joint and Termination Strategy	Limit joints where possible; use pretested terminations for reliability

Implementation Considerations:

- Confirm voltage drop and thermal performance using cable modeling tools (e.g., CYMCAP, Amtech).

- Plan routes early in coordination with civil and mechanical disciplines to avoid duct congestion.
- Validate bending radius and pulling tension limits, especially for long duct banks or tunnels.
- Use cable ladders or cleats for secure spacing and fault restraint in high-current systems.
- Ensure cable selection complies with fire resistance or low-smoke zero-halogen (LSZH) requirements where applicable.

Example Application

In a data center utility yard, VE analysis reconfigured the switchgear-transformer layout to reduce average MV cable length by 28 m per run. Armored cable was replaced with ducted non-armored XLPE types, compliant with site protection strategy.
Outcome: €18,000 saved on materials and terminations; reduced voltage drop by 1.1%; improved installation schedule by 5 working days.

Key Takeaways:

- MV cabling presents significant VE potential through route optimization and cable type selection.
- Overdesign in cable sizing, armoring, or redundancy often goes unchecked without system-level review.
- Smart planning of layout, bends, and joint locations reduces losses and improves system reliability.
- Use recognized sizing standards (e.g., IEC 60287, IEEE 835) tailored to actual site conditions.

3.4.1.3 Transformers

System Overview

Transformers are essential components in the MV/LV distribution system, stepping down MV (typically 11–33 kV) to LV (400/230 V) for downstream utilization. Common configurations include oil-immersed or dry-type transformers, supplied as stand-alone units or integrated into packaged substations. Transformer capacity selection, quantity, and placement have significant implications on both capital cost and operational efficiency.

VE Objectives:

- Right-size transformer ratings based on realistic, diversity-adjusted demand.
- Optimize transformer count and placement to balance redundancy, load management, and footprint.
- Select appropriate cooling type (oil vs. dry) based on site environment and maintainability.
- Minimize losses through efficient loading and layout.
- Avoid overcommitment to transformer provisions without substantiated future demand.
- Optimize transformer impedance and vector group to enhance system coordination and reduce downstream equipment cost.

Typical Overdesign or Inefficiencies:

- Oversized transformers based on conservative or unvalidated demand forecasts.
- Redundant transformers deployed without business-criticality or phase-based justification.
- Excessive copper loss due to low loading factor (<30%) in multitransformer setups.

- Inappropriate selection of dry-type transformers in high-capacity, cost-sensitive outdoor applications.
- Lack of physical clearance or ventilation considerations causing forced derating.
- Low-impedance transformers increasing fault levels and oversizing protection equipment.
- Improper vector group selections leading to harmonics, grounding issues, or paralleling incompatibility.

Targeted VE Strategies (Table 3.21)

Table 3.21 Transformer Selection and Sizing Optimization Based on VE Principles

Focus Area	VE Approach
Transformer Sizing	Use realistic demand profiling and diversity factors to avoid oversizing
Transformer Count	Reduce number of units where redundancy is not justified (e.g., from 2 × 1000 kVA to 1 × 1600 kVA)
Technology Selection	Use dry-type for indoor, public, or high-safety applications; oil-immersed for high-efficiency or outdoor use
Placement Optimization	Locate transformers close to major loads to minimize LV cable runs and losses
Loss Evaluation	Use total ownership cost analysis (TOC) to select higher-efficiency models only where payback is justified
Impedance Optimization	Select impedance to reduce downstream fault levels and avoid overrating LV switchgear
Vector Group Selection	Use Dyn11 for general use; avoid nonstandard groups unless functionally necessary
Ventilation Planning	Ensure transformer rooms allow sufficient airflow to avoid derating (especially for dry types)
Protection Coordination	Use resistance earthing in large systems to limit fault levels and reduce downstream breaker sizing
Cable Sizing and Damage Risk	Lower fault currents mean smaller earth conductors and less thermal stress on cables
Touch Voltage Control	Resistance earthing limits voltage rise during faults, improving personnel safety
Cost of Fault Containment	Avoids massive mechanical stress and arc flash energy in solidly earthed systems

Implementation Considerations

- Base design on load estimates from demand calculations (ref. Section 3.3.4.1).
- Refer to IEC 60076-1 for transformer design, testing, and installation.
- Align transformer impedance with fault level management and breaker rating optimization.
- Maintain consistency in **vector group** (e.g., Dyn11) across systems for reliable phase coordination.
- Provide forced or natural ventilation per manufacturer requirements to prevent overheating or thermal derating.
- Verify utility and civil compatibility for oil containment, access, and fire separation.
- For **TN-S** systems: Ensure proper separation of neutral and earth throughout.

- For **resistance earthed systems**: Use neutral earthing resistor (NER) rated per **IEC 60076-5**, with insulation coordination and thermal trip.
- For **IT systems**: Specify insulation monitoring devices and alarm logic for first fault detection.
- Always verify **utility interface** if MV side neutral is grounded (e.g., Yyn0 transformers with earthed star point).
- **Evaluate the need for transformer enclosures** based on location and risk profile:
  - Omit for dry-type units in indoor electrical rooms with access control.
  - Required for outdoor installations, public access areas, or zones with environmental exposure.
  - For packaged substations, avoid duplicating enclosure where the housing provides adequate protection.

Example Application

In a commercial business park, the original design specified 3 × 1000 kVA dry-type transformers. A VE workshop consolidated loads into 2 × 1250 kVA oil-filled units with future provision for a third unit, reallocating indoor switchroom space to revenue-generating tenant space.

An 8% impedance was selected (vs. default 6%) to reduce LV switchgear interrupting rating by one tier. Ventilation was confirmed with passive louvres and overhead ducting.

Outcome: €45,000 saved in equipment cost, 15 m^2 of space recovered, and downstream protection savings of €12,000.

Key Takeaways:

- Transformers are frequently oversized "just in case"—VE focuses on demand-aligned right-sizing.
- Oil vs. dry selection should be based on site, maintainability, and safety, not preference alone.
- Proper impedance selection can reduce fault level and avoid overspecifying LV gear.
- Vector group standardization prevents phasing errors, grounding problems, and harmonics.
- Transformer rooms must be designed for thermal performance, access, and servicing needs.

3.4.1.4 Transformer LV Panels/Main Incomers

System Overview

The transformer LV panel, often referred to as the **main incomer or transformer secondary panel**, serves as the first point of protection and control on the low-voltage side of the transformer. It houses the main circuit breaker (or fused switch), CTs, metering devices, and connections to the MDB or main risers.

This panel plays a critical role in **fault isolation**, **protection discrimination**, and **metering**, and it often sets the tone for downstream equipment sizing and layout.

VE Objectives:

- Right-size main breaker capacity and interrupting rating based on transformer output and protection studies.
- Avoid overspecification of panels beyond their actual fault level or loading scenario.
- Standardize panel layouts across multiple transformers to simplify design, installation, and maintenance.

- Use plug-in or withdrawable circuit breakers only when truly necessary.
- Minimize complexity and cost of CTs, metering, and busbar arrangements while ensuring compliance.

Typical Overdesign or Inefficiencies:

- Main breakers specified with excessive interrupting capacity (e.g., 100 kA vs. required 50 kA).
- Oversized panels with unnecessary spare ways or oversized busbars.
- Full withdrawable breakers used where fixed breakers would suffice.
- Overuse of complex metering schemes without integration into the building's actual energy management strategy.
- Separate panels provided instead of integrating incomers directly into the MDB when layout permits.

Targeted VE Strategies (Table 3.22)

Table 3.22 Optimization of LV Transformer Panel (Main Incomer) Design Parameters Through VE

Focus Area	VE Approach
Breaker Sizing	Base on transformer full load current + margin; confirm short-circuit rating via protection study
Breaker Type Optimization	Use fixed or plug-in breakers unless withdrawal is needed for maintenance or critical continuity
CT and Metering Rationalization	Eliminate duplicate metering when downstream BMS-integrated energy meters are already used
Panel Standardization	Apply common layout for transformer panels across multiple substations or levels
Panel Integration	Where possible, integrate incomer function directly into the MDB to reduce footprint and cost

Implementation Considerations:

- Refer to IEC 61439-1 for panel construction, busbar sizing, and type testing.
- Coordinate protection settings with upstream MV protection to ensure selectivity.
- Validate panel dimensions and breaker clearances based on room layout and maintenance space.
- Ensure panel internal separation (Form 3b/4b) aligns with project's operational and safety requirements.
- Confirm short-circuit withstand of panel and breakers per calculated fault current (typically ≤50 kA for 6% impedance transformers).
- **Optimize ingress protection (IP) rating** based on actual environmental conditions:
  - Use **IP31/IP41** for indoor electrical rooms with controlled dust and moisture levels.
  - Reserve **IP54/IP65** for dusty, semi-exposed, or corrosive environments.
  - Avoid overspecifying high IP ratings in clean plantrooms to improve panel heat dissipation, accessibility, and reduce costs.

Example Application

In a hospital project with six 1600 kVA transformers, the original design called for withdrawable Air Circuit Breaker(ACB)s in individual transformer panels. VE review found fixed-type ACBs with Form 3b segregation sufficient, as each transformer had individual isolation and maintenance could be sequenced.
Outcome: Saved €9500 per panel (~€57,000 total); reduced switchroom footprint and simplified maintenance training.

Key Takeaways:

- LV transformer panels should be right-sized and configured based on transformer load and system coordination.
- Avoid overspecified breaker types and unnecessary metering duplications.
- Panel standardization and integration with MDBs improve efficiency and reduce cost.
- Breaker selection must consider accessibility, safety, and maintenance—not just equipment rating.

3.4.1.5 Main LV Distribution Boards

System Overview

MDBs serve as the central hub of LV power distribution in a facility. They typically receive power from one or more transformers, backup generators, or UPS systems, and distribute it to downstream SDBs, risers, mechanical panels, and essential services. MDBs may include multiple incoming sections, bus couplers, outgoing feeders, power monitoring systems, ATS, and communication interfaces.

MDB configuration directly impacts **system selectivity**, **space usage**, **operational continuity**, and **future expandability**—making it a critical target for VE.

VE Objectives:

- Optimize the configuration, form type, and sizing of MDBs based on real load profiles and distribution topology.
- Eliminate unnecessary form-of-separation levels or oversized copper busbars.
- Rationalize the number of outgoing feeders and eliminate spare ways without a documented expansion plan.
- Select the right level of monitoring and control for the operational model of the facility.
- Improve maintainability and access without compromising safety.

Typical Overdesign or Inefficiencies:

- Oversized MDBs with high form separation (e.g., Form 4b) in low-risk, controlled-access environments.
- Excessive spare feeders or oversized busbars not justified by load diversity or growth forecast.
- High interrupting capacity ACBs specified without reference to calculated fault levels.
- Multiple incomers, tie-breakers, or ATS systems in buildings that do not require true dual supply logic.
- Duplication of monitoring and metering systems without integration into a central platform.

Targeted VE Strategies (Table 3.23)

Table 3.23 VE Considerations for MDB Configuration

Focus Area	VE Approach
Form Type Optimization	Use Form 3b for most commercial applications; reserve Form 4 for critical or industrial zones
Feeder Rationalization	Align number and size of outgoing feeders with validated load schedules
Busbar Sizing	Size based on calculated peak demand + diversity factor; avoid default oversizing
Integrated Functionality	Combine ATS, metering, and communication modules to reduce panel footprint
Maintenance Access	Prioritize front access only where rear access is not feasible or needed

Implementation Considerations:

- Design per IEC 61439-1/2 for LV switchgear assemblies, temperature rise, clearances, and type testing.
- Confirm maximum prospective short-circuit current using detailed protection study.
- Apply modular panel design with vertical sections for scalability.
- Validate operational and fault clearance zones in electrical rooms (typically ≥1200 mm front clearance).
- Confirm coordination with emergency power sources, tie-breaker logic, and auto-transfer sequences.
- Avoid form-of-separation escalation unless operational maintenance, continuity of service, or safety standards require it.
- **Specify cable entry type (top or bottom)** based on infrastructure strategy and flexibility needs:
  - **Bottom entry** offers neat routing and shorter runs when using underfloor trenches or raised floor systems, but increases civil works cost and reduces future reconfigurability.
  - **Top entry** provides better adaptability and is more cost-effective where overhead trays or risers exist, but it may increase cable length and room height requirements.
  - VE review should align entry type with containment infrastructure, future expandability, and room serviceability to avoid costly redesigns.

Example Application

In a logistics facility, the original MDB design included Form 4b segregation and multiple ACBs for each outgoing feeder. VE review simplified the board to Form 3b with MCCBs on noncritical circuits, integrated power meters into feeder sections, and eliminated spare ways without fore-casted justification.
Outcome: €24,000 saved on panel cost, improved delivery time by 4 weeks, and reduced panel width by 1.2 m, freeing up space in the switchroom.

Key Takeaways:

- MDBs often carry unnecessary complexity and cost that can be reduced through proper load analysis and layout planning.

- Use the correct **form of separation**, **busbar sizing**, and **functional configuration** based on risk, space, and operations—not default templates.
- Integrate metering and ATS functions to reduce panel size and avoid duplication.
- Design MDBs for maintainability and future growth without overcommitting to unused capacity.

■ ***Are Transformer Panels and MDBs the Same?***

Sometimes They Are... But Not Always (Table 3.24).

Table 3.24 Physical Arrangement Scenarios for Transformer Panels and MDBs

Case	Description
Separate Transformer Panel + MDB	Transformer feeds a dedicated LV incomer panel (with ACB + metering), which then connects to an MDB or main bus section
Integrated Design (Combined)	The transformer feeds directly into an MDB section with integrated incomer and busbars—no separate panel
Multiple Transformers to One MDB	In multitransformer setups, each has a **separate transformer panel**, and they **feed a common MDB**

Functional Distinction (Table 3.25)

Table 3.25 Functional Comparison of Transformer LV Panels and MDBs

Transformer LV Panel (a.k.a. Incomer Panel)	Main LV Distribution Board (MDB)
Directly connected to transformer secondary	Receives power from one or more incomers
Includes main ACB, CTs, metering, isolation	Includes outgoing feeders to SDBs, risers, generators, etc.
May be isolated for transformer maintenance	May have tie breakers, generator interlocks, load shedding
Often located in transformer room	Often located in electrical switchroom

VE Implication:

- In **small or mid-sized buildings**, the transformer panel and MDB are often **physically combined** for space and cost efficiency.
- In **critical or multitransformer systems**, it is typical to have **dedicated incomer panels**, sometimes per transformer, feeding a **central MDB.**

3.4.1.6 LV Cables and Busducts

System Overview

LV cabling and busduct systems form the physical infrastructure that distributes power from MDBs to SDBs, mechanical loads, risers, and critical systems. These systems must be designed to handle thermal loads, fault currents, and installation constraints, while maintaining safety, accessibility, and serviceability.

LV power distribution typically uses **XLPE-insulated copper or aluminum cables** in conduits, trunking, or trays. **Busducts (busways)** offer an alternative, particularly for high-current risers or distribution between switchboards, where space or modularity is a priority.

VE Objectives:

- Optimize cable sizes and number of runs based on accurate loading and derating conditions.
- Select busducts where cost-effective or layout-efficient, particularly in high-rise or plant environments.
- Reduce total cable length through strategic panel placement and vertical alignment.
- Avoid overuse of armored cables, overspecification of containment, or redundant routing.
- Minimize voltage drop and energy losses through efficient routing and grouping.

Typical Overdesign or Inefficiencies:

- Oversized cable cross-sections due to conservative assumptions or generic derating.
- Multiple parallel runs specified instead of busduct or larger single-core sets.
- Redundant containment (e.g., armored cable in steel trunking) with no added benefit.
- Excessive cable tray/basket layers due to poor coordination with mechanical systems.
- Ignored grouping and ambient correction factors during sizing, leading to unnecessary cost or thermal derating.

Targeted VE Strategies (Table 3.26)

Table 3.26 VE Strategies for LV Cabling and Busduct Selection and Routing

Focus Area	VE Approach
Cable Sizing Optimization	Use detailed current, voltage drop, and derating calculations (IEC 60287/IEEE 835)
Containment Simplification	Eliminate redundant mechanical protection; avoid overuse of trunking where trays suffice
Busduct Application	Use for vertical risers, transformer-to-MDB links, or high-current plant distribution
Material and Type Selection	Choose armored vs. unarmored, fire-rated, or LSZH cables based on environmental risk and code compliance
Visual and Spatial Impact	Consider compact busducts in visible areas or congested riser spaces to minimize clutter and improve aesthetics
Installation and Accessibility	Evaluate ease of routing, connection, and long-term service access for each method
System Flexibility	Opt for systems that allow expansion or reconfiguration with minimal downtime or civil work

Implementation Considerations:

- Confirm conductor size and number of runs using approved software tools (e.g., Eplan, Amtech, Ecodial).
- Coordinate with architectural and HVAC layouts to reduce spatial conflict in risers and corridors.

- Ensure grouping, ambient temperature, and installation method corrections are applied properly.
- Validate containment design per load type, room function, and fire zoning (e.g., 2-hour fire-rated cable in escape routes).
- Evaluate cost–benefit of busbar trunking systems where:
  - Vertical risers serve multiple floors
  - Current exceeds 1000 A
  - Reconfigurability or prefabrication is prioritized
- Consider arc containment or segregation barriers for high-capacity trunking systems.
- Evaluate **fire resistance classification** and **smoke emission** (e.g., LSZH) where cables pass through occupied zones, escape routes, or plenum spaces.
- Where **aesthetic or architectural integration** is required (e.g., exposed ceilings in commercial interiors), favor busducts or discrete trunking over large cable bundles.
- In mission-critical or frequently modified facilities, favor cable ladders or plug-in busways that offer **easy future expansion or reconfiguration.**
- Consider long-term **accessibility for maintenance**, especially for cable terminations in raised floors or sealed trench systems.

Example Application

In a high-rise commercial tower, the original design specified four runs of $4 \times 240\ mm^2$ cables for the vertical riser between the transformer and main LV distribution panel. During a VE workshop, busduct was proposed as an alternative. While the **material and labor cost of the busduct system was higher**, the design team recognized benefits in **space optimization**, **modular installation**, and **improved layout coordination** within the constrained riser shaft.
Outcome: Although the busduct option added approximately €14,000 in direct cost, it reduced riser shaft width by 200 mm and improved access for future maintenance and upgrades—supporting long-term operational efficiency and system expandability.

Key Takeaways:

- LV cable and containment systems present clear opportunities for cost reduction and spatial efficiency.
- Busduct systems provide a compact and scalable alternative to cable bundles in appropriate settings.
- Accurate sizing and layout coordination reduce material waste, voltage drop, and future upgrade constraints.
- VE strategies in cabling should be integrated with architectural, HVAC, and fire safety design early in the project.

Note: Busduct systems should not be assumed to be cost-saving in all cases. While they can reduce installation time and improve layout, they often come at a **premium in material and procurement complexity**. A thorough VE assessment should compare total installed cost, future flexibility, and spatial constraints before substitution.

3.4.1.7 SDBs/Local Panels

System Overview

SDBs, also referred to as local panels, serve as the intermediary point between MDBs and final circuit outlets. They distribute power to lighting systems, small power outlets, mechanical loads, and other local systems on individual floors or zones.

Their design impacts not only electrical safety and maintainability but also space planning, containment layout, and end user functionality. SDBs are frequently replicated across floors or zones, making them a prime target for standardization and optimization.

VE Objectives:

- Standardize panel design across multiple floors or similar zones to reduce design complexity and procurement time.
- Optimize circuit grouping, feeder sizing, and protection types based on actual demand and load type.
- Eliminate excessive spare ways and oversegmented panels that drive up cost and wall space requirements.
- Align SDB configuration with final circuit layout to reduce cable length and containment routing.
- Simplify metering and monitoring strategies unless zone-level billing is required.

Typical Overdesign or Inefficiencies:

- Panels with 30–50% spare circuits despite no realistic future expansion needs.
- Use of 100 A main incomers for panels supplying only 4–5 kW of load.
- Independent SDBs specified per system (e.g., lighting vs. power), instead of integrated multiservice boards.
- Full Form 4 segregation or high ingress protection (IP54+) used unnecessarily in indoor, dry zones.
- Location choices that increase cable lengths to lighting or power clusters.

Targeted VE Strategies (Table 3.27)

Table 3.27 VE Guidelines for SDB Design and Placement

Focus Area	VE Approach
Panel Standardization	Replicate panel schedules and configurations across similar areas or floors
Spare Way Rationalization	Limit to 10–15% spare circuits unless future growth is documented [ENGINEERING PRACTICE—varies by jurisdiction][a]
Incomer Sizing	Size main MCCBs based on grouped circuit loads, not default 100–125 A ratings
Panel Consolidation	Combine lighting and power panels where circuits are few or serve the same area
Ingress Protection	Use IP31/IP41 unless the panel is located in dusty, humid, or exposed areas (IEC 61439-1: Clause 10.2.2 Table 1)
Form Segregation	Use Form 2b or Form 3a, unless operational continuity requires isolation (IEC 61439-1: Section 7.1)

[a] Spare circuit requirements vary significantly by jurisdiction and application. Some codes require 20% spare breakers (i.e., Florida Building Code for educational facilities), while others have no mandatory spare requirements (NEC 90.8 is advisory only). Local engineering practice typically ranges from 10% to 20% based on building type and future expansion plans.

Implementation Considerations:

- Refer to **IEC 61439-3** for final circuit distribution boards in commercial and residential settings.
- Ensure clear front access for maintenance (≥600 mm) and cable space at top/bottom entry zones.
- Locate SDBs close to their load centers (e.g., lighting clusters, office banks) to reduce branch circuit length.
- Coordinate board dimensions and fixing method (surface vs. recessed) with architectural and ID layouts.
- Apply consistent circuit labeling and panel schedules to simplify O&M.
- If smart metering or energy management is required, centralize to MDB level unless tenant-level billing is needed.

Example Application

In a multilevel educational facility, 48 sub-distribution panels were originally specified—one for each wing and system type (lighting/power/HVAC). VE review consolidated multiple panels on each floor using standardized 36-way boards with modular breakers and integrated surge protection.
Outcome: Fourteen fewer panels procured, saving €19,000 in hardware, and 24 m^2 of wall space; cable routing simplified by 180 linear meters, reducing labor time and complexity.

Key Takeaways:

- SDBs are frequently overspecified due to template-driven design; VE helps tailor layout to actual load and space needs.
- Panel standardization across zones or floors improves efficiency, installation time, and O&M.
- Circuit consolidation and logical placement reduce branch circuit length and improve safety.
- Eliminate unnecessary form segregation and high IP ratings unless truly justified by environmental conditions.

3.4.1.8 Final Circuits and Floor Distribution

System Overview

Final circuits represent the last leg of the power distribution system, delivering electrical power from SDBs to end user devices such as lighting fixtures, general power outlets, HVAC control points, and specialized equipment. These circuits include wiring, protective devices, junction boxes, and termination accessories. Their design directly impacts user functionality, safety, energy efficiency, and maintenance accessibility.

Final distribution also encompasses **horizontal containment systems** (e.g., trunking, conduits, cable trays) and the configuration of floor boxes, wall outlets, and lighting grids.

VE Objectives:

- Optimize circuit layout to reduce cable lengths, voltage drops, and conduit complexity.
- Eliminate redundant or oversized final circuits for low-power devices.
- Simplify routing through efficient alignment with architectural zones and ceiling systems.
- Avoid excessive circuit separation where not required by regulation.
- Promote modularity and ease of future reconfiguration for flexible spaces (e.g., offices, classrooms, labs).

Typical Overdesign or Inefficiencies:

- Use of oversized conductors (e.g., 4 mm^2 instead of 2.5 mm^2) for general sockets without voltage drop justification.
- Individual circuits for each small appliance or lighting fixture in noncritical areas.
- Over-segregation of lighting, power, and data conduits, increasing installation time and material.
- Complex ceiling or floor routing with unnecessary bends, junctions, and fittings.
- Floor boxes or sockets specified in fixed locations in adaptable-use spaces.

Targeted VE Strategies (Table 3.28)

Table 3.28 Final Circuit Design Optimization Strategies for Power and Lighting Layouts

Focus Area	VE Approach
Circuit Grouping	Combine low-load devices onto shared circuits where permissible
Conductor Right-Sizing	Apply correct size based on voltage drop, ambient temperature, and circuit length
Conduit and Cable Route Optimization	Minimize bends and intersections; use shortest safe route path
Flexible Power Distribution	Use raised floor or plug-and-play systems in adaptable-use environments
Standard Outlet Layouts	Standardize spacing and mounting height to reduce variation and speed up installation

Implementation Considerations:

- Design per **IEC 60364** or local wiring code, applying correct cable sizing, protection, and earth continuity.
- Confirm all final circuits are protected by appropriate Miniature Circuit Breaker(MCBs)or Residual Current Breaker with Overcurrent protection(RCBO)s with defined disconnection times.
- Coordinate closely with architectural ceiling and furniture layout to align routing and outlet placement.
- Use modular trunking systems or raised floor boxes to allow easy adaptation to layout changes.
- Ensure labeling and circuit charts are updated for every final circuit to support operation and maintenance.
- Avoid complex conduit intersections in ceiling voids where ceiling tiles and MEP services already compete for space.

Example Application

In an administrative office layout, VE review identified oversized conductors (4 mm^2) specified for general-purpose outlets. By correcting sizing to 2.5 mm^2 and regrouping loads based on desk clusters, cable quantity was reduced by 20%, and conduit diameters were downsized by one level. ***Outcome:*** €6500 saved on final circuit materials; reduced ceiling congestion allowed faster MEP installation; and improved ease of future changes.

Key Takeaways:

- Final circuit layout and routing are often overlooked in VE reviews, despite significant impact on cost, flexibility, and ease of install.
- Right-sizing conductors, combining circuits, and simplifying routes yield material, labor, and spatial savings.
- Modular and reconfigurable distribution systems support evolving user needs in modern facilities.
- Coordination with ceiling, wall, and floor services is essential to ensure integration and avoid rework.

3.4.1.9 Discrimination, Cascading, and Protection Studies

System Overview

Electrical system design is not complete without detailed coordination studies that validate safety and operational performance. Among these, the most critical are:

- **Discrimination (selectivity) studies**: Ensure faults are cleared by the closest protective device without affecting upstream systems (IEC (2020c)).
- **Cascading (back-up protection)**: Use of upstream breakers to protect downstream components under fault, enabling use of lower-rated, cost-effective devices ((IEC, 2021d)).
- **Short-circuit and thermal analysis**: Verify that equipment withstand ratings are adequate for the prospective fault levels (IEC, 2016)
- **Arc flash studies** (where applicable): Identify zones of high energy exposure to guide Personal Protective Equipment (PPE), system design, and mitigation methods (NFPA 70E; IEEE 1584).

These studies are essential for performance and safety—but also offer meaningful VE leverage when used as **design tools**, not just compliance checks.

VE Objectives:

- Use coordination studies to avoid blanket oversizing of protection devices or panels.
- Apply cascading selectively to reduce cost of downstream devices without compromising protection.
- Optimize breaker types and trip curves for selectivity without redundancy.
- Confirm short-circuit withstand requirements using actual fault level analysis, not conservative defaults.
- Reduce system-wide arc energy levels by adjusting protection response or segmentation.

Typical Overdesign or Inefficiencies:

- Upstream and downstream breakers of equal rating, leading to lack of selectivity and full outage on minor faults.
- All breakers specified at high interrupting capacity (e.g., 50–65 kA) regardless of position or requirement.
- Arc flash exposure not evaluated, leading to unnecessary switchgear segregation or PPE requirements.
- Use of expensive electronic trip units on low-impact panels without the need for programmable protection.

Targeted VE Strategies (Table 3.29)

Table 3.29 Coordination Study Types and Their Associated VE Opportunities

Study Type	VE Lever
Discrimination Study	Adjust breaker trip settings to ensure cascading faults do not trip upstream feeders
Cascading Study	Use validated manufacturer tables to reduce downstream MCCB interrupting capacity
Short-Circuit Analysis	Confirm actual fault level at each bus—avoid overspecifying switchgear
Arc Flash Study	Optimize system segmentation and breaker timing to reduce incident energy levels
Thermal Rating Check	Avoid grouping or derating assumptions that drive up cable or breaker size

Implementation Considerations:

- Use software tools (e.g., ETAP, Ecodial, SKM PowerTools, Amtech) for accurate simulations.
- Validate that cascading combinations are certified by the manufacturer (per IEC 60947-2 Annex A).
- Integrate studies into early design stages (post-single line, pre-procurement) for highest VE impact.
- Apply study outputs to inform panel ratings, cable sizing, equipment clearances, and breaker selection.
- Rerun studies after final equipment selection to confirm installed system compliance.

Example Application

In a pharmaceutical facility, VE workshops highlighted that all sub-distribution panels had MCCBs rated at 25 kA, despite fault levels under 6 kA. A cascading study showed that 10 kA-rated MCBs with backup protection from upstream ACBs were fully compliant.

Outcome: €18,500 saved in device and panel cost, with improved selectivity and simplified panel dimensions.

➢ VE Insight—Why Short-Circuit Studies Should Drive Breaker Selection

Many electrical designs default to specifying circuit breakers with high interrupting capacities (e.g., 25–40 kA), often without validating the actual fault levels. This practice leads to inflated equipment costs, oversized panels, and unnecessary design complexity.

A properly conducted short-circuit and protection study can reveal that fault levels at SDBs or final circuits are far lower than assumed. This opens the door for using more cost-effective 6–10 kA-rated breakers, or applying cascading (backup protection) strategies with certified upstream devices.

A single change from a 40 kA to a 10 kA breaker can result in:

- **Up to 4–6× reduction in device cost**
- **Smaller panel footprint and lower heat output**
- **Faster procurement and simplified maintenance**

Key Takeaways:

- Coordination and protection studies are not just compliance tasks—they are active VE tools.
- Discrimination and cascading enable safe cost reduction in breaker ratings and panel build.
- Arc flash and thermal studies support safer, leaner system layouts without excess segregation.
- Integrating these studies early prevents costly late-stage redesigns or overspecified procurement.

3.4.2 Earthing and Lightning Protection

System Overview

The **earthing system** is critical for protecting people, equipment, and systems from electrical faults by ensuring a safe, low-impedance path to dissipate fault current into the ground. It also stabilizes system voltage and enables correct operation of protective devices. Components include earth electrodes, main and supplementary earth bars, equipotential bonding, and connections to structural elements.

The **LPS** prevents structural damage and internal system disruptions by intercepting lightning strikes and safely diverting their energy through down conductors and grounding terminals. It comprises air termination devices, down conductors, equipotential bonding points, and earth terminations.

Both systems must be fully coordinated to meet safety regulations, minimize electromagnetic interference (EMI), and support ongoing reliability and maintainability.

VE Objectives:

- Ensure regulatory compliance and personal safety while optimizing material use, layout, and installation effort.
- Avoid overdesign in conductor sizing, redundant electrodes, or excessive lightning terminal density.
- Integrate bonding and earthing systems for power, ICT, ELV, and mechanical systems to minimize duplication.
- Align down conductor routing with structural lines to reduce cable length, visibility, and interference with other trades.
- Prioritize measurement-based grounding designs (resistivity and fault current studies) over standard templates.

Typical Overdesign or Inefficiencies:

- Duplicate earthing systems (e.g., separate for power and ELV) where common bonding is permissible.
- Oversized copper conductors and excessive electrode depth or spacing without site-specific soil data.
- LPS air terminals placed without risk assessment methodology, leading to unnecessary cost and façade clutter.
- Stainless steel or tinned copper used unnecessarily in noncorrosive environments.
- Fault loop impedance margins overly conservative, resulting in higher cable costs.

Targeted VE Strategies (Table 3.30)

Table 3.30 Earthing and LPS Optimization Strategies Under VE

Focus Area	VE Approach
Conductor Sizing	Right-size earthing conductors per IEC 60364, IEEE 80, or local code based on actual fault current and duration
Electrode Optimization	Perform site-specific resistivity tests (e.g., Wenner method) to determine electrode count and layout, IEEE 80
System Integration	Implement a **common bonding network (CBN)** for ELV, BMS, ICT, and electrical systems where code permits,
LPS Risk-Based Design	Use rolling sphere, mesh, or protection angle method per **IEC 62305-2** to justify number and placement of terminals
Material Rationalization	Use copper-clad steel or bare copper in low-corrosion environments instead of tinned or stainless steel
Architectural Coordination	Align LPS routing with vertical elements (columns, façade mullions) to conceal and minimize run length
Use of Natural LPS Elements	Where compliant, use structural steel columns or reinforced concrete rebars as natural down conductors or earth electrodes (IEC 62305-3, Cl. 5.3.6)
Eliminate Redundant Conductors	Omit separate copper or aluminum down conductors where natural paths meet continuity, cross-section, and exposure criteria
Rebar Integration	Use foundation steel reinforcement as a foundation earthing network where bonding and continuity are validated
Visual Integration	Route unavoidable down conductors along metallic cladding, columns, or gutters for concealed protection pathways

Implementation Considerations:

- Design grounding systems per **IEC 60364-5-54**, **IEC/EN 62561**, or **IEEE Std 80**, depending on region and system type.
- Confirm site conditions via early **soil resistivity testing** to avoid overly conservative default designs.
- Coordinate with structural and civil teams to incorporate **foundation earthing** where practical.
- Ensure LPS elements maintain correct separation distances from metal services and electrical systems to prevent side flashing.
- Test completed earthing systems to confirm target resistance (e.g., $<1\ \Omega$ for critical systems), and document continuity across bonded systems.
- For critical infrastructure, consider redundancy in earth paths and lightning strike counters for tracking performance.
- Confirm bonding of rebar cages at vertical and horizontal joints.
- Conduct continuity checks during early construction phases.
- Coordinate with structural engineers to identify steel members suitable for discharge paths.
- Apply this strategy primarily to **Class II–IV risk category** structures or where lightning exposure is high.

Example Application

In a high-tech logistics warehouse, the initial design included 36 lightning air terminals and 14 down conductors. A VE review used rolling sphere and mesh method simulations based on IEC 62305-3, which reduced the required air terminals to 22 and optimized conductor routes along existing steel columns.

Simultaneously, a **common bonding network** was implemented between the power system, ELV racks, and CCTV grounding via a shared main earth bar.

Outcome: €18,000 saved in copper material and labor; improved system performance; and LPS components visually integrated with the façade.

Key Takeaways:

- Earthing and LPS systems are often overdesigned due to reliance on default assumptions or lack of site data—VE identifies realistic, code-compliant alternatives.
- Integrated bonding of electrical and ELV/ICT systems improves performance and avoids duplication.
- Site-specific resistivity testing and lightning risk assessment are essential VE tools for optimized design.
- Coordination with architecture and structure enables discreet, efficient, and maintainable LPS routing.

VE Insight—Using Natural Structural Components as Lightning Protection

IEC 62305-3 permits the use of **natural conductors** such as structural steel, metal roofs, or rebar cages as part of the LPS, provided they meet continuity, cross-sectional area, and corrosion resistance requirements.

Examples of Natural Lightning Protection Components:

- Steel columns or beams used as down conductors.
- Reinforced concrete with electrically continuous rebars used as foundation earth electrodes.
- Metallic roof decks or sheeting used as air-termination elements.
- Gutter systems or parapet cappings integrated as part of the discharge path (with proper bonding).

Project Example: In a logistics facility with a steel portal frame and corrugated metal roof, a VE review replaced 14 standalone air terminals and external down conductors with bonded roof trusses and columns, validated through continuity checks.

Outcome:

- Saved €12,000 in copper tape, saddles, and fixings.
- Reduced visual impact on the building envelope.
- Enhanced long-term durability by eliminating exposed LPS components.

3.4.2.1 Surge Protection Devices

System Overview

Surge Protection Devices (SPDs) are essential components of electrical safety systems, designed to protect installations from transient overvoltages caused by lightning strikes, grid switching, and internal switching surges. They work by diverting excess voltage to earth before it reaches sensitive equipment.

SPDs are typically classified into (IEC 61643-11—Section 4) IEC (2011c):

- **Type 1**: For direct lightning currents, installed at the origin of the electrical installation (main incomer).
- **Type 2**: For indirect surges and switching transients, installed at distribution boards.
- **Type 3**: For fine protection of sensitive loads, installed close to terminal equipment.

SPDs must be coordinated with the **earthing system**, **LPS**, and **equipment insulation levels** to function effectively.

VE Objectives:

- Ensure proper SPD selection and placement for effective protection and cost-efficiency.
- Avoid over-application of SPDs where protection overlaps or minimal surge risk is present.
- Reduce unnecessary material and panel space by applying a risk-based SPD hierarchy.
- Integrate monitoring and maintenance-friendly features where required by operational profile.
- Maintain SPD-earth bonding impedance within manufacturer-recommended limits.

Typical Overdesign or Inefficiencies:

- SPDs installed at every subpanel regardless of distance or load sensitivity.
- Use of oversized Type 1 SPDs in buildings not exposed to direct lightning currents.
- Lack of coordination between SPD levels, leading to overlapping or ineffective protection.
- SPD earth connections routed via long or shared paths, compromising discharge performance.
- Installing SPDs without clear equipment risk or exposure justifications.

Targeted VE Strategies (Table 3.31)

Table 3.31 VE Approaches for Integrating SPDs into Electrical Design

Design Focus	VE Approach
Correct Typing	Apply Type 1 at service entrance (if LPS present); Type 2 at key SDBs; Type 3 only for sensitive loads (IEC 61643-11)
Selective Application	Omit SPDs in boards feeding only mechanical loads or lighting unless required by code
Grounding Optimization	Keep SPD bonding leads short and direct to earth bar (ideally ≤0.5 m) (IEC 60364-4-44—Section 443.4)
Smart Monitoring	Use SPDs with visual indicators or dry contacts at critical nodes instead of redundant installations
Insulation Coordination	Ensure SPD up (voltage protection level) is ≤ equipment withstand voltage per IEC 60364-4-44

Implementation Considerations:

- Specify SPDs according to **IEC 61643-11** (low-voltage AC) and **IEC 60364-4-44** (protection against overvoltages).
- Conduct risk assessments as per **IEC 62305-2** to determine SPD need based on structure, exposure, and service type.

- In TN systems, install SPDs between each live conductor and neutral, and between neutral and earth.
- For IT or TT systems, verify correct configuration to prevent improper clamping or floating neutral issues.
- Coordinate SPD selection with equipment surge withstand category (Category I–IV, per IEC) (IEC 60364-4-44—Table 44.3).

Example Application

In a corporate office building, VE assessment removed noncritical SPDs from mechanical panels and stairwell lighting boards. Type 1 + 2 SPDs were installed only at the main switchboard, with Type 2 SPDs reserved for AV, data, and automation panels.
Outcome: €9000 saved in SPD devices and installation time, with no reduction in equipment protection. A centralized maintenance plan was introduced using SPDs with built-in status contacts for remote monitoring.

Key Takeaways:

- SPDs are essential for overvoltage protection but must be **selectively and correctly applied** to deliver value.
- Risk-based application, correct coordination, and direct earthing are critical to SPD performance.
- Avoid redundancy by evaluating actual surge exposure, load criticality, and insulation requirements.
- Integrate SPD planning with **earthing design**, **panel layout**, and **system segmentation** early in design.

3.4.3 Lighting and Controls

System Overview

Lighting systems are a fundamental component of every building and infrastructure project, contributing to safety, comfort, productivity, and energy efficiency. The system comprises general lighting (indoor and outdoor), emergency lighting, façade illumination, and specialized task lighting.

Lighting controls include manual switches, presence detectors, daylight sensors, timers, DALI/KNX-based automation, and centralized lighting control panels. These systems aim to balance user needs, building operations, and energy performance targets.

As lighting systems are repeated across large areas and often involve high fixture counts, they are a prime target for VE optimization in both design and controls architecture.

3.4.3.1 Normal Lighting Systems

System Overview

Normal (general-purpose) lighting systems provide the baseline illumination for all functional, habitable, and circulation spaces in a facility. These systems are governed by standards that dictate required illuminance levels, uniformity, glare control, and color rendering, with a growing emphasis on energy performance and visual comfort.

Components typically include ceiling-mounted or recessed luminaires, wall-mounted fixtures, linear LED systems, and specialty task lighting. The lighting layout, spacing, and circuiting impact electrical loads, user comfort, and long-term operation.

Due to their repetitive nature and high fixture counts, normal lighting systems represent a major opportunity for VE in design standardization, specification strategy, and performance-based layout optimization.

VE Objectives:

- Right-size lighting designs to meet illuminance requirements without over-lighting.
- Standardize fixture types across zones to reduce procurement and maintenance complexity.
- Select efficient and cost-effective LED luminaires suitable for space function and usage patterns.
- Integrate daylight availability into layout strategy to reduce fixture counts.
- Apply appropriate lighting quality metrics (UGR, CRI, CCT) for each space type without overspecification.

Typical Overdesign or Inefficiencies:

- Use of excessive fixture densities due to lack of simulation-based lux calculations.
- Multiple fixture types specified across similar spaces, increasing spares and replacement costs.
- Oversized wattage or lumen output not matched to actual room size or reflectivity.
- Premium architectural fixtures used in utility areas without visual or functional benefit.
- Ignoring daylight availability, resulting in unnecessary luminaires near windows or skylights.

Targeted VE Strategies (Table 3.32)

Table 3.32 Lighting System Design Elements and Associated VE Opportunities

Design Focus	VE Approach
Fixture Standardization	Limit luminaire families to 3–5 types (where feasible) across a project to simplify procurement, installation, and spare-part management
Illuminance Calculation	Use lighting design software (i.e., DIALux/Relux) to simulate required lux levels based on EN 12464-1, avoiding arbitrary fixture spacing
Lumen Right-Sizing	Match fixture output to room surface reflectance and geometry—avoid overspecification "for margin"
Daylight Integration	Implement daylight-responsive layout in perimeter zones to reduce energy use and fixture count
Off-the-Shelf Selection	Favor commercially available, noncustom luminaires unless architectural impact justifies the upgrade
Visual Comfort Metrics	Apply CRI and UGR values aligned with space function (EN 12464-1—Table 5.1) (e.g., CRI ≥80, UGR ≤19 for office tasks) and appropriate CCT values (3000–4000 K)
Fixture Efficiency and Layout	Use higher-lumen, high-efficacy luminaires with better optics to reduce total fixture count while maintaining uniform coverage

Implementation Considerations:

- Design to **EN 12464-1** standards for interior spaces, with lux levels tailored to space usage (e.g., 300 lx for circulation, 500 lx for general office).

- Validate layout using lighting simulation software; include reflectance values of ceilings, walls, and floors for accurate modeling.
- Ensure luminaire mounting height, orientation, and cut-off angles control glare (UGR) and maintain even coverage.
- Specify drivers with appropriate dimming capabilities (e.g., 1–10 V, DALI) for adaptable output control.
- Coordinate layout with HVAC diffusers, fire devices, ceiling features, and access panels to avoid conflicts.
- Consider lifecycle costs, not just initial fixture cost—include energy consumption and maintenance intervals.
- Consider using fewer, higher-output luminaires with high luminous efficacy (≥120 lm/W) and good optical distribution to reduce the number of fixtures needed per zone, provided uniformity and UGR remain within target thresholds. For example, replacing six 2400 lm fixtures with four 3600 lm fixtures may reduce material and labor without affecting coverage.

Example Application

In an institutional office block, initial lighting layouts used one fixture type per zone and defaulted to 600 × 600 LED panels at 3 m spacing throughout. VE review rationalized to four luminaire types for the entire project and adjusted fixture output per zone via simulation. ***Outcome:*** 17% reduction in fixture count, €19,000 saved on lighting supply and install, improved uniformity in work zones, and lighting power density brought below regulatory maximum.

Key Takeaways:

- Lighting systems are often overdesigned by habit—simulation and standardization are key VE tools.
- Right-sizing fixture count and wattage based on space function improves both efficiency and comfort.
- Overuse of architectural or premium fixtures in back-of-house areas adds cost with minimal return.
- Consider maintenance access, visual uniformity, and daylight contribution early in design.

3.4.3.2 Emergency Lighting Systems

System Overview

Emergency lighting systems are critical life safety elements designed to provide illumination during power outages or emergency conditions, ensuring safe evacuation and operational continuity in critical areas. These systems are governed by stringent codes and must meet specific performance criteria related to illuminance levels, autonomy (typically 1–3 hours), and reliability.

Emergency lighting includes:

- **Escape route lighting** for corridors, stairwells, and exits
- **Open area (anti-panic) lighting** in large spaces to prevent confusion
- **High-risk task area lighting** in zones where visibility is critical during failure (e.g., labs, control rooms)
- **Exit signage** to guide evacuation

System configurations vary between:

- **Self-contained (individual battery-powered)** luminaires
- **Central battery systems** distributing DC power to connected fittings
- **UPS-backed emergency lighting** in specific sectors (e.g., mission-critical control rooms)

VE Objectives:

- Ensure full compliance with safety standards while minimizing cost, complexity, and maintenance effort.
- Select system type (self-contained vs. central) based on building layout, lifecycle cost, and accessibility.
- Reduce luminaire count by optimizing layout through lighting simulations.
- Avoid overspecification of autonomy, photometric output, or backup infrastructure.
- Minimize manual testing burdens through use of auto-test or addressable monitoring systems.

Typical Overdesign or Inefficiencies:

- Uniform use of central battery systems across wide building layouts, resulting in long cabling runs and voltage drop issues.
- Overuse of high-output emergency luminaires where lower lumen versions suffice.
- Redundant circuiting or standalone wiring when integration with general lighting is code-permitted.
- Inefficient or nonmonitored systems that require costly manual inspection.
- Autonomy specified at 3 hours by default in low-risk areas where 1 hour meets legal requirements.

Targeted VE Strategies (Table 3.33)

Table 3.33 VE Strategies for Emergency Lighting System Optimization

Design Focus	VE Approach
System Type Selection	Use self-contained systems in spread-out or retrofitted buildings; central systems in compact or easily accessed layouts
Layout Optimization	Use photometric simulations (per EN 1838) to confirm lux levels and spacing—avoid default rules of thumb
Dual-Purpose Wiring	Where allowed, integrate emergency and general lighting on maintained luminaires with dual supply inputs
Test Automation	Use luminaires with self-test or group-monitoring features to reduce maintenance labor
Signage Consolidation	Combine emergency lighting and exit signage where permitted to reduce device count and wiring complexity
Right-Sized Autonomy	Apply 1-hour or 3-hour autonomy based on building type and fire strategy (ref. local fire codes or IEC 60598-2-22)

Implementation Considerations:

- Design to **EN 1838** for photometric performance and **IEC 60598-2-22** for luminaire and system characteristics.
- Confirm required illuminance: typically ≥1 lx on escape route centerline, ≥0.5 lx for anti-panic areas.
- Validate battery life, recharge times, and failure indication method in submittals.
- Evaluate maintenance implications:
  - **Self-contained** systems: lower initial cost, higher distributed maintenance.
  - **Central battery**: higher capex, but easier centralized monitoring and testing.
- Plan risers and emergency containment layout early for central battery or UPS-fed systems.
- Document testing strategy, schedules, and maintenance responsibilities in the O&M manual.
- Consider using combined mains/emergency luminaires with integrated battery packs where appropriate. These fixtures can:
  - Reduce fixture count and ceiling clutter.
  - Simplify wiring and containment.
  - Lower overall cost if testing and emergency output are compliant.
  - Be most effective in standard-height open spaces using DALI-EM or self-test drivers.
  - Require careful thermal and photometric validation to ensure battery reliability and output meets code.

Example Application

In a four-story commercial office building, the base design specified separate emergency lighting fixtures throughout the open-plan office floors, with a central battery system located in the basement. During the VE review, the team proposed using combined mains-emergency LED panels with integrated batteries in open areas, retaining a central battery only for stairwells and corridors.

A lighting simulation verified that the emergency mode lumen output from the dual-purpose fittings met EN 1838 escape route requirements. The revised design significantly reduced circuit complexity, cabling routes, and battery maintenance scope.

Outcome:

- Eliminated 46 standalone emergency luminaires.
- Reduced containment length by over 200 m.
- Saved €21,500 in equipment and installation cost.
- Simplified maintenance via self-test-enabled luminaires in all open office zones.

Key Takeaways:

- Emergency lighting design must balance **code compliance**, **installation efficiency**, and **long-term maintainability.**
- Selecting the correct system type is critical—one size does not fit all.
- Simulation-based layout design prevents overuse of luminaires while ensuring legal compliance.
- Auto-testing and monitoring reduce operational cost and support robust safety management.

3.4.3.3 Lighting Control Systems

System Overview

Lighting control systems govern the operation, scheduling, dimming, and automation of lighting circuits. They range from simple manual switching to advanced, addressable digital systems such as DALI, KNX, or BACnet-integrated BMS controls. Effective control strategies support user comfort, energy conservation, and code compliance, while offering flexible operation.

Control methods include:

- Manual switches or push-buttons
- Occupancy/vacancy sensors
- Daylight harvesting with lux-level sensors
- Time-based scheduling or astronomical clocks
- Centralized or networked control systems (wired or wireless)

Control infrastructure can significantly impact project cost, especially in large-scale buildings. Therefore, **VE** is critical in ensuring control scope matches functional need.

VE Objectives:

- Match control system complexity to occupancy type, usage pattern, and performance goals.
- Avoid over-engineered control systems in static or low-use areas.
- Optimize sensor placement to reduce quantity and avoid overlap.
- Standardize control interfaces for end user clarity and commissioning efficiency.
- Enable integration with BMS only where needed (e.g., public zones, energy benchmarking, critical load shedding).

Typical Overdesign or Inefficiencies:

- Use of fully addressable systems (e.g., DALI/KNX) in back-of-house or noncritical zones.
- Overspecification of sensors (e.g., both PIR and ultrasonic) in small or enclosed areas.
- Lack of zone consolidation, leading to fragmented circuits and unnecessary panels.
- Redundant local and central controls without justification.
- Overly complex control logic requiring specialist programming for simple lighting needs.

Targeted VE Strategies (Table 3.34)

Table 3.34 Lighting Control System Design Optimization Based on Usage Patterns and Control Needs

Design Focus	VE Approach
Control Type Selection	Use manual or PIR-based control in single-use or low-traffic areas; reserve addressable systems for complex or open zones
Zoning Optimization	Group lighting loads by function and usage patterns to minimize control hardware and simplify logic
Sensor Rationalization	Apply one sensor per zone type (e.g., PIR or microwave), with proper coverage range and placement
Dimming Strategy	Use 1–10 V or DALI dimmable drivers only where daylight or task variability warrants it
Interface Simplification	Standardize on-off or dimming switches across similar zones to ease user operation and reduce training time
Integration Logic	Integrate only key zones (e.g., public lobbies, meeting rooms) with BMS—avoid system-wide integration without added value

Implementation Considerations:

- Comply with local energy codes (e.g., **ASHRAE 90.1**, **BREEAM**, **EPBD**) requiring occupancy control, daylighting, or automatic shut-off.
- Use DALI-2 for addressable control zones requiring scene setting or remote monitoring.
- Include override functions in conference rooms, auditoria, and multipurpose spaces for user control.
- Confirm sensor placement with ceiling layout and HVAC system to avoid detection of blind spots or interference.
- Label control panels clearly; ensure wiring paths support future reprogramming or re-zoning where flexibility is expected.
- Define commissioning protocol—include timeouts, dimming curves, occupancy thresholds, and test procedures in the design brief.

Example Application

In a business center with open-plan offices and enclosed service rooms, the base design proposed KNX control throughout. VE analysis segmented the control logic:

- **Open office areas** retained DALI-based dimming with daylight and occupancy sensors.
- **Toilets, storerooms, and corridors** were shifted to PIR sensors and time-delay relays.
- **Meeting rooms** included manual override push-buttons with presence detection.
- Resulting system was easier to commission, used fewer control modules, and reduced wiring complexity.

Outcome: €14,200 saved on control modules, cabling, and programming; faster handover and lower maintenance complexity.

Key Takeaways:

- Lighting controls are often overspecified—VE aligns functionality with space usage and actual performance needs.
- Sensor selection, zoning logic, and interface simplicity are key cost drivers.
- Networked systems (e.g., DALI/KNX) should be applied strategically, not uniformly.
- Clear commissioning procedures and wiring flexibility improve long-term adaptability.

3.4.4 Emergency Power and UPS Systems

System Overview

Emergency power systems provide electrical continuity and operational resilience during utility power interruptions, grid disturbances, or planned shutdowns. They are essential in mission-critical facilities, life-safety applications, and control-dependent infrastructure where uptime, safety, or regulatory compliance cannot be compromised.

These systems can be composed of multiple backup elements working in coordination, including:

- UPS systems for instantaneous power without transfer delay
- Standby and prime-rated generators for extended backup duration
- DC power systems for protection, communication, and control loads
- Battery banks and energy storage for bridging gaps and enabling autonomous operation
- ATS or static transfer switches (STS) for source switching
- Load prioritization, shedding logic, and zone segregation

Depending on the facility type, load profile, and regulatory requirements, emergency power architecture may vary significantly in scope, topology, and redundancy.

VE Focus

Due to their **high capital cost** and infrastructure demands (space, fuel, HVAC, testing), emergency power systems offer substantial VE potential, including:

- **Accurate classification of critical vs. noncritical loads** to avoid system oversizing
- **Right-sizing** UPS and generator capacities based on realistic autonomy targets
- **Modular and scalable topologies** that support future growth without up-front overinvestment
- **Battery and runtime optimization**, selecting the right technology per application
- **Avoiding redundant systems** where a unified backup strategy is viable
- **Lifecycle cost evaluation**, including maintenance, cooling, replacement intervals, and test protocols

Properly value-engineered systems not only **maintain uptime performance and safety compliance** but also reduce capex, OPEX, and lifecycle environmental footprint.

3.4.4.1 Load Classification and Backup Strategy

System Overview

Correctly classifying electrical loads and defining their backup strategy is the foundational step in designing an efficient and compliant emergency power system. Not all loads require the same level of continuity or redundancy, and misunderstanding these differences often leads to oversized, overly complex, and cost-intensive backup systems.

Backup strategies must distinguish between:

- **Life Safety Loads:** Legally mandated loads that must remain operational to protect life during emergencies. This includes emergency lighting, fire alarm systems, exit signage, smoke extraction fans, and fire-fighting lifts. These loads are required by local building codes, fire safety regulations, and standards such as IEC 60364-5-56, NFPA 70, or BS 8519.
- **Uninterruptible Loads:** Loads that require seamless power with zero interruption, such as data servers, communication hubs, or critical medical equipment.
- **Essential Loads:** Loads that must remain operational during outages but can tolerate short interruptions. This includes HVAC supporting critical spaces, BMS, elevators in high-occupancy buildings, or key mechanical services.
- **Nonessential Loads:** General or comfort-based loads that do not impact safety or core functions. These are typically excluded from backup (e.g., workstation outlets, decorative lighting)

This classification informs decisions regarding:

- UPS sizing and autonomy
- Generator capacity and fuel system design
- Load shedding logic
- Power zoning and switchgear segmentation

VE Objectives:

- Prevent over-application of UPS or generator coverage to low-importance loads.
- Optimize system architecture by functionally grouping loads.
- Define clear, tiered response levels for power outages.

- Reduce capex and operating costs by right-sizing backup systems to realistic demand profiles.
- Avoid overdesign driven by default "backup everything" mentalities.

Typical Overdesign or Inefficiencies:

- UPS applied to entire floors or panels without identifying which loads truly require uninterrupted power.
- All systems treated as equally critical, leading to unnecessary generator sizing or fuel autonomy.
- Absence of zoning logic, with all loads on the same transfer switch or distribution board.
- Redundant DC systems installed for the same loads already protected by UPS.
- Applying 2N redundancy universally, even in facilities with low business continuity requirements.

Targeted VE Strategies (Table 3.35)

Table 3.35 Load Classification and Backup Strategy Recommendations for Emergency Power Systems

Design Focus	VE Strategy
Load Tiers Definition	Classify loads into Tier 1 (uninterruptible), Tier 2 (essential), and Tier 3 (nonessential) based on life-safety and operational impact
Functional Grouping	Separate UPS and generator-backed loads physically and electrically via dedicated DBs and risers
Zonal Backup Strategy	Apply N + 1 redundancy only to high-value zones (e.g., data center, ICU); use generator-only backup for general areas
Load Shedding Logic	Define load-shedding sequences for long outages to preserve generator fuel and UPS runtime
Risk-Based UPS Allocation	Provide UPS only where system failure has unacceptable consequences (e.g., network core, alarm systems, critical control)

Implementation Considerations:

- Conduct early workshops with stakeholders to define **criticality levels per space/function.**
- Apply standards such as **IEC 60364-5-56**, **NFPA 70**, or **BS 8519** for emergency system design thresholds.
- Document backup requirements in a **Load Matrix**, mapping all circuits to their power source (normal/UPS/generator).
- Use **color-coded single line diagrams** to visualize emergency vs. normal power paths.
- Ensure compatibility between load type and backup source (e.g., HVAC controls may not tolerate UPS waveform distortion).
- Incorporate **black-start considerations** where generator reinitiation is required after full outage.
- Clearly distinguish **life safety loads** from operationally important but nonmandatory loads; ensure they are:
  - Fed from dedicated emergency distribution boards.
  - Powered through code-approved sources (e.g., generator, battery-backed UPS).

  - Routed via **fire-rated cables** with protected pathways.
  - Confirm alignment with **authority having jurisdiction (AHJ)**, local fire code, and applicable standards for emergency system reliability and response time (e.g., ≤15 seconds for generator start-up).

Key Takeaways:

- Backup power design must start with accurate load classification to avoid unnecessary investment and complexity.
- Functionally grouping loads enable right-sizing of UPS and generator systems.
- Zoning and segmentation help enforce backup policies and improve maintainability.
- Load shedding and redundancy should be applied based on operational risk, not uniformity.

3.4.4.2 UPS System Selection and Architecture

System Overview

A **UPS** system provides instantaneous backup power to critical loads during utility failure or voltage disturbance. It bridges the power gap between loss of normal supply and generator startup, or maintains continuous operation in environments where **zero interruption** is permissible.

UPS systems are commonly used to support:

- Data centers and ICT infrastructure
- BMS and automation
- Life-safety systems requiring no-break power (e.g., fire alarms, evacuation interfaces)
- Industrial control systems and lab equipment
- Healthcare diagnostic and support systems

UPS systems are categorized by **topology**, **redundancy**, and **deployment architecture**—each with distinct performance, cost, and scalability implications.

VE Objectives:

- Select a UPS topology that aligns with actual risk profile, not simply defaulting to double-conversion.
- Avoid unnecessary redundancy (e.g., 2N) unless required by client Service Level Agreement (SLA)s or regulatory frameworks.
- Use **modular or scalable UPS designs** that match current load and allow future growth.
- Determine optimal placement strategy: **centralized**, **zonal**, or **rack-level** UPS based on space, cooling, and access.
- Balance CAPEX and OPEX by selecting UPS types with suitable **efficiency curves**, **maintenance needs**, and **battery compatibility.**

Typical Overdesign or Inefficiencies:

- Applying 2N (fully redundant) topology for general-purpose or low-impact loads.
- Centralized UPS systems used across multiple floors without considering voltage drop, containment cost, or failure impact.
- Oversizing UPS capacity with excessive margin beyond actual runtime requirement.
- Using double-conversion UPS for loads that tolerate short power interruptions.
- Lack of modularity, requiring complete UPS replacement during upgrades.

Targeted VE Strategies (Table 3.36)

Table 3.36 VE Guidelines for UPS System Topology, Redundancy, and Deployment Strategy

Design Focus	VE Strategy
Topology Selection	Use **line-interactive** or **offline UPS** for low-tier systems; apply **double-conversion** only where true continuity is critical
Redundancy Matching	Apply N + 1 or 2N only for loads with defined SLAs (e.g., data centers, hospitals); use **single-path** UPS with generator backup for less critical systems
Modular Architecture	Use modular UPS units with hot-swappable power modules to allow staged investment and simplified maintenance
Zonal vs. Central UPS	Use zonal UPS near load centers (e.g., ICT closets) to reduce cabling length and heat concentration in central rooms
Efficiency Optimization	Select UPS systems with >94% part-load efficiency and **ECO mode** features when suitable for the application

UPS Topologies at a Glance (Table 3.37)

Table 3.37 Comparison of UPS Topologies: Functionality, Application Suitability, and VE Considerations

Type	Functionality	Applications	VE Notes
Offline (Standby)	Switches to battery during outage	Low-cost PC, minor AV	Not suitable for life-safety or industrial loads
Line Interactive	Voltage regulation with battery switchover	Retail, small office, network closets	Good efficiency; short switching delay (~5–10 ms) (IEEE)
Online (Double Conversion)	Continuous double conversion via inverter	Critical IT, medical, automation	Highest protection, always on, higher heat, and losses (IEEE)
Modular UPS	Scalable architecture with plug-in modules	Data centers, hospitals, Tier 2/3	Ideal for VE—right-size now, expand later
Decentralized (Rack-Level)	Individual UPS per cabinet or small zone	Small facilities, distributed sites	Avoids single point of failure, but harder to maintain

Implementation Considerations:

- Confirm output waveform, transfer time, and harmonic distortion levels are compatible with connected loads.
- Account for UPS internal losses (~5–8%) in load calculations and HVAC planning (IEEE).
- Use **static bypass paths** to allow for maintenance without interrupting load.
- Ensure clear access for **battery replacement**, cooling airflow, and service access.
- Coordinate with ICT engineers and facility managers regarding monitoring and Simple Network Management Protocol (SNMP) integration.
- Include UPS **monitoring interface** with alarm reporting, battery health tracking, and power event logging.
- Provide environmental controls for battery performance and UPS room heat rejection (especially for large Valve-Regulated Lead-Acid (VRLA) or Li-ion banks).

Example Application

In a high-specification office building, the original design included a centralized 250 kVA double-conversion UPS with 2N redundancy to feed IT closets across five floors. A VE workshop reclassified floors based on IT risk level and replaced the solution with:

- One 80 kVA modular N + 1 UPS for floors with critical server rooms
- Decentralized 3 kVA line-interactive UPS units in low-risk floors
- Generator-backed panels for lighting and HVAC controls

Outcome: €95,000 saved on UPS capacity and cabling, reduced UPS room footprint by 50%, and simplified long-term maintenance without sacrificing availability.

Key Takeaways:

- UPS selection must be based on **load criticality**, not assumed importance or commercial default.
- Modular and decentralized UPS systems enable flexibility, growth, and reduced initial investment.
- UPS system efficiency and integration must be coordinated with HVAC, monitoring, and load behavior.
- VE success lies in applying **the right protection to the right load**, not treating all loads equally.

3.4.4.3 Battery Technology and Runtime Planning

System Overview

Battery banks are the **core energy storage component** of UPS and DC backup solutions. Their sizing, type, and configuration directly influence the **reliability, runtime autonomy, cost**, and **space planning** of emergency power systems.

Battery systems may serve:

- UPS systems (5–15 minute autonomy typical)
- Central emergency lighting inverters
- DC systems for 110/220 V switchgear and control panels
- Telecommunication and SCADA systems

Battery selection must consider not only **initial cost** but also **efficiency**, **lifecycle**, **replacement intervals**, **space constraints**, and **thermal behavior**—all of which carry implications for long-term value.

VE Objectives:

- Match battery autonomy to actual system needs (not default values).
- Select battery chemistry based on lifecycle cost, space, environment, and maintenance profile.
- Avoid redundant battery systems by consolidating backup topologies where feasible.
- Consider modular battery architecture to simplify expansion, replacement, and integration.
- Reduce air conditioning and ventilation loads by selecting low-heat emission chemistries.

Typical Overdesign or Inefficiencies:

- Defaulting to 30-minute autonomy where 5–10 minutes suffices (especially with generator support).
- Oversized battery strings with unused runtime, adding weight, heat, and cost.
- Multiple overlapping battery systems (UPS + 110 VDC) serving similar control or BMS functions.

- VRLA batteries used in high-heat or high-cycle environments, reducing lifespan (Clark & Harris, 2024).
- Battery rooms lacking proper monitoring and thermal isolation, increasing failure risk.

Targeted VE Strategies (Table 3.38)

Table 3.38 Battery System Design and Runtime Optimization Strategies for Emergency Power

Design Focus	VE Strategy
Autonomy Optimization	Match runtime to functional need: e.g., 5–7 minutes for generator ride-through, 10–15 minutes for orderly shutdown
Chemistry Selection	Use **VRLA** for short autonomy in conditioned spaces; use **Li-ion** for high-performance, compact, low-maintenance setups (Clark & Harris, 2024)
Thermal Management	Reduce HVAC and ventilation load by using batteries with wider operating temperature range (e.g., Li-ion, NiCd) (Clark & Harris, 2024)
Lifecycle Costing	Consider full **TCO** (total cost of ownership), including replacement interval, charging efficiency, and maintenance effort
Redundancy Consolidation	Avoid duplicating battery banks for loads already covered under UPS or DC schemes unless code mandates separation

Battery Chemistry Comparison (Table 3.39)

Table 3.39 Comparison of Battery Technologies for UPS and DC Systems with VE Considerations

Type	Typical Use	Advantages	VE Considerations
VRLA (AGM/Gel)	UPS, emergency lighting	Low cost, proven tech, sealed design	Shorter life (~3–5 years), heat-sensitive (Clark & Harris, 2024)
Li-ion	Modern UPS, space-constrained	High energy density, long life (~8–12 years), fast charging	Higher capex, requires BMS, often no ventilation needed (Clark & Harris, 2024)
NiCd	Harsh environments, DC loads	High lifecycle, temperature-tolerant	Expensive, bulky, cadmium disposal requirements
Flooded Lead-Acid	Utility and industrial	Longest life (~15 years), economical per Ah	Requires large ventilated space, regular maintenance (Petrovic, 2020)

Battery Chemistry–Space and Weight Considerations (Table 3.40)

VE Insight: In retrofit or high-density environments, switching from VRLA to Li-ion can reduce battery room size by up to 60%, freeing space for additional UPS capacity or eliminating the need for a dedicated enclosure (Clark & Harris, 2024). Similarly, structural load considerations become critical for elevated floors—particularly for NiCd or flooded cells in multistory buildings (Petrovic, 2020).

Table 3.40 Battery Chemistry Performance Characteristics and Spatial Impacts

Battery Type	Energy Density	Relative Footprint	Weight per kWh	VE Impact
VRLA (AGM/Gel)	Moderate (50–70 Wh/L)	Moderate–large	~25–35 kg/kWh	Standard sizing, but high footprint in long-autonomy applications
Li-ion	High (150–250 Wh/L)	Very compact	~10–15 kg/kWh	Excellent for space-constrained areas (e.g., telecom rooms, small UPS closets)
NiCd	Low–moderate	Bulky	~35–45 kg/kWh	Tolerates harsh conditions but requires substantial floor space
Flooded Lead-Acid	Low (30–40 Wh/L)	Very large	~40–50 kg/kWh	Long life, but very high spatial and structural load demand

Implementation Considerations:

- Use manufacturer battery sizing tools or IEEE 485/1115 to calculate autonomy based on kW load and discharge profile.
- Apply derating for **temperature, aging, and real-world inefficiencies** (typically 1.1–1.3 safety factor).
- Provide battery room layout with:
  - Adequate **clearances.**
  - **Fire-rated enclosures.**
  - **Temperature control** (typically 20–25 °C for VRLA).
- Include **battery monitoring systems** for status, aging, fault detection, and proactive replacement.
- For large UPS systems, consider **rack-based modular battery solutions** to simplify deployment and expandability.
- Coordinate **ventilation and spill containment** if using open or flooded cells.

Example Application

In a hospital expansion project, the initial design called for 20-minute autonomy using VRLA batteries in three UPS rooms. VE review determined that generator startup was under 10 seconds, and equipment ride-through required only 7 minutes. The design was revised to:

- Ten-minute autonomy using **Li-ion** battery strings (rack-integrated)
- Central UPS room HVAC downsized by 30% due to lower thermal load
- Reduced footprint by 40%, allowing elimination of one UPS room

Outcome: €62,000 in capex savings, improved battery lifespan (10+ years), and simplified long-term maintenance.

Key Takeaways:

- Batteries are major cost, space, and lifecycle drivers—optimize autonomy and technology per application.

- VE gains come from **chemistry selection**, **thermal efficiency**, and **consolidated design.**
- Avoid over-designing autonomy "just in case"—base sizing on actual outage and recovery conditions.
- Battery systems must be integrated early with ventilation, electrical coordination, and asset tracking strategies.

3.4.4.4 Standby Generator Systems and Fuel Infrastructure

System Overview

Diesel generators serve as the primary backup power source during extended outages that exceed the autonomy of UPS or battery systems. They ensure continuity for essential and life-safety systems and support uninterruptible loads through integration with UPS systems.

Generators are classified per **ISO 8528-1** based on their operational use:

- **Emergency Standby Power (ESP):** For utility outages only (≤200 hours/year)
- **Prime Power (PRP):** For extended or variable load operation with limited annual hours
- **Continuous Power (COP):** For uninterrupted base-load applications, often in off-grid or utility-replacement scenarios

In most commercial buildings, **ESP-rated generators** are used and are activated by **ATS** upon grid failure. Generator systems must be supported by:

- Fuel storage and refilling infrastructure
- Exhaust and noise attenuation systems
- Control panels and synchronization logic
- Ventilation and fire safety systems

VE Objectives:

- Size generators based on actual emergency load demand, not total installed capacity.
- Select generator class (ESP, PRP, COP) based on site risk and utility reliability.
- Apply redundancy (e.g., N + 1) only where justified by continuity requirements.
- Optimize fuel tank size, refueling interval, and containment design.
- Consolidate generator use across functions (e.g., life-safety, HVAC, ICT) where permitted by code.

Typical Overdesign or Inefficiencies:

- Generator sized for full building load regardless of emergency load scope.
- Prime-rated generator applied where standby is sufficient, increasing cost and emissions compliance burden.
- Separate generator sets for UPS and essential loads instead of unified backup bus.
- Oversized fuel tanks with infrequent use, leading to fuel degradation and unnecessary volume.
- Acoustic enclosure or room designs with excessive reserve margin or custom builds.

Targeted VE Strategies (Table 3.41)

Table 3.41 Generator Sizing, Classification, and System Design Optimization via VE

Design Focus	VE Strategy
Load-Based Sizing	Size generator for actual emergency and legally required loads (not entire electrical load); apply diversity only to non-life-safety loads, where permitted
Class Selection	Use **ESP-rated** gensets for standard commercial backup; reserve **PRP** only where extended runtime is common
Redundancy Optimization	Use **N + 1** or **2N** only in Tier 3/4 facilities or mission-critical systems—avoid defaulting to 2N in commercial settings
Fuel System Right-Sizing	Size tanks for required autonomy (e.g., 6, 8, 24 hours) based on refueling logistics and code; use bunded or double-wall tanks
Multisystem Consolidation	Feed multiple backup loads (e.g., life safety and HVAC) from shared emergency switchboard unless separation is required by code
Future-Proofing	Select models that support paralleling or capacity expansion
Cooling Type	Consider water-cooled gensets for compact urban installations with ventilation constraints
Dual Fuel Systems	Use only where permitted by code and AHJ; may be applicable where fuel security or emission compliance is an issue
Installation Constraints	Roof-mounted units require vibration isolation and structural coordination

Implementation Considerations:

- Apply **ISO 8528-1**, **NFPA 110**, and **IEC/IEEE 63332-387** for classification, startup time (≤10 seconds for Level 1), and testing.
- Confirm generator transient performance (voltage dip, frequency stability) for motor loads and elevators.
- Optimize ATS placement to minimize transfer time and avoid load surges.
- Design ventilation for combustion air and cooling; include fire dampers and airflow direction control.
- Ensure fuel systems are:
  - Compliant with **NFPA 30**, **EN 12285**, or local fire codes.
  - Equipped with leak detection, spill containment, and refueling access.
- Provide access for delivery vehicles, maintenance, and acoustic upgrades.
- **Load Characterization:**
  - Perform a detailed load analysis using tools compliant with IEC or IEEE standards.
  - Identify **peak demand scenarios**, daily load cycles, and future expansion potential.
  - Distinguish **life-safety**, **essential**, and **nonessential** loads (see Section 3.4.4.1).
- **Generator Sizing and Load Stepping:**
  - Calculate both **baseline load** and **stepped load scenarios** (e.g., fire alarm activation, HVAC re-engagement, elevator recall).
  - Classify loads:
    - **Priority 1**: Fire alarms, emergency lighting, evacuation systems, and life safety.
    - **Priority 2**: Critical HVAC zones, data centers, and communication.
    - **Priority 3**: General lighting, admin loads.

- Size generator to handle **initial startup surge** and **long-term stepped load** without overload.
- Consider staged startup (sequential load reintroduction) using load-shedding contactors or controllers.

- **Fuel Autonomy Planning:**
  - Base fuel tank sizing on:
    - Load profile and rated kWh of generator.
    - **Code-mandated autonomy** (commonly 6–24 hours; hospitals may require 72 hours) (NFPA 110).
    - **Refueling logistics**, including site access and supplier availability.
  - Include **bunded/double-wall tanks**, fuel polishing systems, and level monitoring sensors.
  - Provide spill containment and venting as per **NFPA 30**, **EN 12285**, or local code.
- **Environmental and Room Design:**
  - Provide adequate **combustion and cooling airflow.**
  - Acoustic treatment to meet site noise ordinances.
  - Exhaust flue routing with fire rating and vertical discharge.
  - Fire-rated fuel lines, fire dampers, and emergency ventilation.
- **Compliance and Coordination:**
  - Adhere to **ISO 8528**, **NFPA 110**, **NFPA 30**, **EN12285**, and **local fire/electrical codes.**
  - Coordinate ATS ratings, response time, and controls.
  - Confirm generator transient performance (**voltage dip, frequency stability**) for motor loads and elevators.
- **Vendor Coordination:**
  - Engage manufacturers early to select models with correct derating for altitude and temperature.
  - Review maintenance intervals, spare parts availability, and warranty terms.
  - Ensure synchronization compatibility for parallel or redundant generator systems.

Generator Topology Options (Table 3.42)

Table 3.42 Generator Topology Configurations and Recommended Applications

Architecture	Description	Use Case
Single Generator	One unit sized for all emergency loads	Common in small to mid-sized facilities
N + 1 Redundant	Two or more generators, with one spare unit	Data centers, large hospitals, and critical infrastructure
Paralleling Generators	Multiple units operating in synchronization. Paralleling systems require synchronization controls, load-sharing governors, and protection coordination to ensure safe and stable operation	For scalable loads or staged startup loads
Separate Generators by Function	One for UPS and one for life safety	Use only where physical separation is required by code (e.g., NEC 700.10(B) or local equivalent)

Example Application

In a high-rise commercial building, the base design included a 1250 kVA generator sized to support full building loads. A VE review determined:

- Essential loads totaled only 620 kVA, with 8-hour autonomy needed per local fire authority
- A 2N system was not required based on building uptime classification
- Resulting system: One 800 kVA ESP-rated generator with a 6000-liter double-walled tank, centralized fuel polishing, and shared emergency distribution

Outcome: €84,000 in generator and fuel infrastructure savings; footprint reduced by 30%; and lowered emissions rating and fuel storage volume.

Key Takeaways:

- Generator systems must be sized and classified based on **true emergency load**, **use case**, and **code-mandated autonomy.**
- VE focuses on avoiding unnecessary generator capacity, redundancy, or tank oversizing.
- Shared generator design across systems enhances efficiency and simplifies control logic.
- Fuel system planning, noise mitigation, and room layout are critical early stage design items.

■ ***Sustainability Consideration: Biofuel Compatibility***

As part of long-term carbon reduction strategies, some clients or authorities may require that emergency generators be capable of operating on **biodiesel blends (e.g., B5*, B20, B100)** either immediately or in the future. This requires careful planning to ensure **compatibility, safety, and performance**. (**Even for B5, consult Original Equipment Manufacturer (OEM) documentation to confirm warranty coverage.)*

Table 3.43 Biofuel Compatibility Design Aspects for Emergency Generator Systems

Aspect	VE and Design Implication
Fuel Blend Type	Most Tier 3/4 generators can operate on **B5** (5% biodiesel) without modification. Blends above B20 typically require OEM approval.
Material Compatibility	Fuel system components (seals, hoses, gaskets) must be compatible with biofuel (especially B20–B100). Use fluoropolymer-based or nitrile seals.
Fuel Aging and Stability	Biofuels degrade faster than standard diesel. Systems must include **fuel polishing**, **filtration**, and **recirculation loops** to prevent microbial growth and phase separation.
Cold Flow and Storage	Biodiesel has higher cloud and pour points—may gel in cold climates, e.g., below 5 °C. Insulated, heated tanks and piping may be needed in northern zones.
Manufacturer Certification	Specify **"Biodiesel-compatible" or "Biofuel-certified"** generator models from OEMs. Require documentation that future upgrade or reprogramming is feasible without major engine modification.
Emissions Compliance	Biofuels reduce particulate emissions and lifecycle CO_2, supporting LEED/BREEAM points—but still must comply with **EU Stage V** or **EPA Tier 4 Final** emissions if applicable.
Applicable Standards	Reference ASTM D6751 (US biodiesel quality), EN 14214 (EU biodiesel standard), and NFPA 110 (fuel system design and testing) for compliance and specification alignment.

Key VE and Technical Considerations (Table 3.43)

Implementation Tip:

1) In specifications, include a clause such as:
   "The diesel generator system shall be capable of operating on a minimum of B20 biodiesel blend, and shall be constructed of fuel system components compatible with biodiesel up to B100. Fuel treatment and filtration equipment shall be designed to manage the increased degradation risk and ensure fuel quality during extended storage."
2) Biofuel systems should be tested quarterly for fuel quality and microbial contamination, per NFPA 110 Annex A.
3) Fuel polishing systems should include automated timers and monitoring for water, particulates, and microbial contamination to ensure fuel quality during extended storage.

3.4.4.5 DC Power Systems (110 V, 48 V, etc.)

System Overview

DC power systems are used to provide **stable, uninterruptible power** to low-voltage control, protection, and communication systems. Common voltage levels include **48 VDC** for telecom infrastructure and **110 VDC or 220 VDC** for control, SCADA, and switchgear applications (IEEE, 2020).

DC systems are particularly important in:

- Substations and power distribution networks
- Industrial automation and control panels
- Telecommunication base stations
- Critical BMS and relay protection systems
- Transportation and tunnel infrastructure

These systems typically include:

- Rectifiers/chargers
- Battery banks
- DC distribution panels (fused or breaker-based)
- Monitoring and alarm modules

Unlike UPS systems, DC systems provide power directly without conversion, resulting in high efficiency and reliability. Their autonomy is dictated by battery sizing, and they are often separate from the AC backup architecture (IEEE, 2020).

VE Objectives:

- Determine whether a **dedicated DC system** is truly required, or if the functionality can be integrated into UPS-based architectures.
- Right-size autonomy and battery capacity based on actual critical function timing.
- Minimize system complexity by avoiding redundant power paths (e.g., powering the same control device from both UPS and DC).
- Standardize DC panel ratings and protection devices to streamline maintenance and procurement.
- Avoid overspecification of rectifier current or battery backup durations beyond operational need (IEEE, 2020).

Typical Overdesign or Inefficiencies:

- Installing 110 VDC systems in environments where 48 VDC suffices.
- Providing 8–12 hours of autonomy when only 1–2 hours are required for switchover or orderly shutdown.
- Using oversized rectifiers rated for full load + future margin without documented expansion plan.
- Duplicating backup power (UPS + DC) for the same control components.
- Specifying custom battery enclosures instead of rack-mount or modular designs.

Targeted VE Strategies (Table 3.44)

Implementation Considerations:

Table 3.44 VE Strategies for DC Power System Design and Specification

Design Focus	VE Strategy
Autonomy Optimization	Size DC battery banks for actual system hold time (e.g., 1–4 hours), aligned with fault-clearing or generator startup intervals
Voltage Level Selection	Use 48 VDC for ICT and monitoring systems; 110/220 VDC for protection relays and industrial-grade control
Rectifier Sizing	Select rectifier based on average load + charging current; oversizing increases losses and cost (IEEE, 2020)
Protection Architecture	Use modular DC distribution with selective fusing or MCBs; standardize panel form factors across projects
Monitoring Integration	Ensure battery monitoring and charger diagnostics are connected to BMS or SCADA platforms for predictive maintenance (IEC, 2000)

- Select **battery chemistry** based on environment and cycle profile:
  - **VRLA** for standard indoor use.
  - **NiCd** for high-temperature or deep discharge resilience.
- Allow space for **safe ventilation**, battery replacement, and isolation switching.
- Provide **voltage alarms**, **earth fault detection**, and **remote communication interface (Modbus, SNMP).**
- Ensure clear **labeling of positive and negative rails**, with interlocks for maintenance safety.
- Coordinate power system grounding and bonding to avoid **DC loop instability** or ground loops.
- Ensure separation from AC cable trays and power panels to avoid EMI (IEC, 2000).

Example Application

In a medium-voltage substation, the original design included a 110 VDC system with 6-hour autonomy for switchgear protection and automation relays. A VE review adjusted:

- Autonomy reduced to 2 hours based on relay tripping window and generator startup time
- Charger derated by 25% after removing overspecified future loads
- Standardized modular rack battery replaced custom enclosure

Outcome: €18,500 saved in charger and battery costs, 30% reduction in room size, and improved maintainability.

Key Takeaways:

- DC systems are essential in control and protection applications, but must be sized based on real operation and code-driven autonomy.
- VE focuses on **rationalizing battery duration**, **rectifier sizing**, and **architecture standardization.**
- Coordination with grounding, panel layouts, and monitoring ensures long-term reliability and safety.
- Avoid design duplication—UPS and DC systems should complement, not replicate, backup coverage.

3.4.4.6 Power Transfer and Switching Mechanisms

System Overview

Power transfer systems are responsible for safely and efficiently switching critical loads between the **normal utility source**, **generator backup**, and, where applicable, **UPS or alternate feeders**. They ensure continuity of power supply with minimal delay, and often incorporate both **automatic** and **manual override functions**.

The two most common switching devices are:

- **ATS**—used to shift loads from utility to generator automatically when grid power fails (IEC, 2021e; UL, 2023).
- **STS**—used in high-reliability applications to switch loads between two independent live sources (e.g., UPS A and B) with millisecond response time (UL, 2023).

Power switching equipment may also include:

- Manual changeover panels
- Generator synchronizing panels
- Bypass arrangements for maintenance
- Load transfer logic (sequential, priority-based, or manual)

Proper selection and configuration are essential to maintain **safety**, **selectivity**, and **code compliance** under dynamic system conditions (IEC, 2021e).

VE Objectives:

- Select transfer mechanisms based on the criticality of the application and the expected power disturbance scenario.
- Avoid overspecifying STS unless loads are truly sensitive to transfer times.
- Ensure switchgear is modular and maintainable, avoiding custom hard-wired interlocks or proprietary logic.
- Reduce transfer time while minimizing inrush current or loss of synchronism during load connection.
- Integrate load prioritization in transfer logic to avoid overloading generators during initial switchover.

Typical Overdesign or Inefficiencies:

- STS applied universally even where 1–3 second ATS switching is acceptable.
- Oversized ATS units with unnecessary poles or switching capacity.

- Redundant transfer paths without documented requirement or failure mode justification.
- Use of manual changeover in systems that require automatic restoration (e.g., life-safety).
- No coordination with protection devices—resulting in load tripping or prolonged restoration.

Targeted VE Strategies (Table 3.45)

Table 3.45 Power Transfer Device Selection and Load Switching Optimization in Emergency Systems

Design Focus	VE Strategy
Device Selection	Use **ATS** for most commercial and life-safety systems; apply **STS** only where <20 ms transfer time is essential (e.g., Tier 3 UPS systems) (UL, 2023)
Sizing and Specification	Size ATS based on load and fault withstand rating; avoid overspecifying based on upstream breaker rating alone (IEC, 2021e)
Transfer Logic Design	Include time delays, retransfer logic, and inhibit conditions to reduce wear and nuisance transfers
Load Segmentation	Break loads into **critical**, **priority**, and **nonessential** blocks with staggered startup to minimize generator strain
Bypass and Testing Paths	Use ATS/STS units with **bypass isolation** to allow service without outage—cost-effective in high-uptime environments

Implementation Considerations:

- Ensure ATS complies with **IEC 60947-6-1** or **UL 1008**, and includes:
  - In-phase monitoring.
  - Time delays for transfer and retransfer.
  - Manual override (IEC, 2021e; UL, 2023).
- Confirm that ATS type (break-before-make vs. make-before-break) matches load tolerance and source behavior.
- Plan **generator warm-up** time and confirm fuel system readiness before initiating transfer.
- Coordinate transfer timing with **UPS hold time**—UPS should bridge the switching delay window.
- Provide **visual status indication** (normal/emergency/neutral) and **event logging** for diagnostics.
- Consider switchgear integration:
  - Line isolation.
  - Load busbar arrangement.
  - Arc flash mitigation via controlled re-energization sequence.

Example Application

In a municipal administrative building, the design specified STS for all IT and lighting panels. VE workshops showed that only server racks required true zero-interruption performance. The design was revised to:

- Use STS only for server and communication racks
- Apply ATS for all other essential systems (lighting, HVAC, elevators)
- Integrate priority load startup logic in ATS control to prevent generator overload

Outcome: €38,000 in savings on STS hardware, improved testing flexibility with ATS bypass options, and simplified installation across five floors.

Key Takeaways:

- Transfer devices must be selected based on load sensitivity and operational risk—not by default.
- ATS provides sufficient performance for most essential systems; STS is reserved for highly sensitive applications.
- Incorporate testing, bypass, and status monitoring into early design to streamline maintenance and fault isolation.
- Segment and stagger loads logically to prevent generator overload or UPS drain during switching events.
- VE in transfer systems reduces cost and increases maintainability without compromising uptime.

3.4.4.7 System Coordination and Protection in Emergency Power Systems

Overview and Cross-Reference

Protection coordination, cascading logic, and reliability modeling are critical to the safe and efficient operation of any electrical system. These principles have been discussed comprehensively in **Section 3.4.1.7**, which addresses their application within the main distribution architecture.

This subsection provides **focused guidance** on how these principles apply within **emergency power systems**—particularly those involving **UPS**, **generators**, **transfer switches**, and **DC backup networks**.

Emergency System-Specific VE Considerations (Table 3.46)

Table 3.46 Protection Coordination Considerations for Emergency Power Systems

Focus Area	VE Guidance
Generator Fault Behavior	Analyze the lower subtransient reactance of generators (X″d), which may limit available fault current. This impacts ATS selectivity and may allow the use of lower-rated protection devices downstream (IEEE, 2001).
UPS Coordination	Ensure UPS output protection curves are coordinated with downstream breakers. Avoid overrating output breakers that trip faster than necessary, as this increases arc flash exposure (IEEE, 2001).
STS Behavior	For STS, account for the challenge of coordinating two live sources with differing impedance and waveform. Select fast-acting, current-limiting protection where needed (IEEE, 2022b).
DC System Selectivity	Apply fused or MCB-based selective protection in 110/220 VDC distribution. Use modular DC boards with remote alarm capability (IEEE, 2001).
Bypass and Maintenance Modes	Ensure temporary bypass paths maintain protection selectivity, and do not introduce unprotected faults or overload scenarios. Document these paths clearly in O&M manuals (IEEE, 2001; NFPA, 2022).

Coordination with Broader System Studies:

- Use the results from the **main discrimination and arc flash study (Section 3.4.1.7)** to:
  - Validate selectivity of ATS/STS outputs during utility-to-generator switchover (IEEE, 2001).

- Confirm that emergency panels downstream of UPS maintain proper cascading behavior (IEEE, 2001).
- Ensure arc energy remains within PPE limits during transfer and backup operation (IEEE, 2022a; NFPA, 2022).

Key Takeaways:

- Emergency power introduces unique challenges in protection design due to varying source impedance, transfer logic, and load sensitivity.
- VE focuses on correct matching of protection curves, not overspecifying based on upstream equipment ratings.
- Refer to **Section 3.4.1.7** for foundational methodology; apply those principles in coordination with UPS/generator characteristics during detailed design.

3.4.4.8 Space Planning and Environmental Control

System Overview

Emergency power systems—including **generators**, **UPS units**, **battery banks**, **fuel tanks**, and **control gear**—require dedicated technical spaces with appropriate **ventilation**, **accessibility**, **structural support**, and **environmental control**. These systems are often large, heat-generating, and safety-critical, making spatial planning a key determinant of performance, safety, and cost.

Common installations include:

- Generator rooms or enclosures (indoor or rooftop)
- UPS rooms with battery storage
- Central battery systems for emergency lighting
- DC systems with protected battery cabinets
- Fuel storage rooms or external tanks
- Acoustic enclosures and fire-rated compartments

Poor spatial integration results in operational inefficiencies, retrofitting costs, code violations, and elevated OPEX for cooling and maintenance.

VE Objectives:

- Optimize spatial footprint by selecting **modular or compact form-factor** equipment.
- Avoid oversizing rooms due to early stage conservative assumptions.
- Ensure space allocation is proportional to **actual autonomy, cooling needs**, and **maintenance access.**
- Integrate fire, ventilation, acoustic, and fuel safety requirements early **into the architectural coordination**.
- Enable future expansion without major rework.

Typical Overdesign or Inefficiencies:

- Battery rooms sized for 30–60 minutes autonomy when only 5–10 minutes are needed.
- Dedicated UPS rooms separated by floor or core from critical loads, increasing cable losses and voltage drop.

- Generator room airflows designed with excessive redundancy, increasing duct sizes and shaft volume.
- Double containment rooms specified for small fuel tanks that could otherwise be externally housed.
- No provision for removal/replacement of large equipment, resulting in future demolition or rework (NFPA, 2022).

Targeted VE Strategies (Table 3.47)

Table 3.47 Space Planning and Environmental Control Strategies for Emergency Power Infrastructure

Design Focus	VE Strategy
Room Sizing	Right-size UPS/battery/fuel rooms based on selected autonomy and rack layout; avoid generic square-meter allowances
Thermal Management	Calculate actual heat rejection for UPS and batteries; use **passive cooling**, **air separation**, or **Li-ion UPS** to reduce HVAC loads (ASHRAE, 2021)
Access and Maintenance	Provide minimum working clearances per IEC/NEC; include overhead clearance for lifting batteries and side access for modular UPS modules (NFPA, 2023)
Equipment Footprint	Choose **modular, stackable, or integrated enclosures** (e.g., UPS + battery rack combo) to reduce overall room area
Generator Enclosure Options	Use **roof-mounted or containerized sets** when internal room footprint is constrained or costly
Routing and Cable Entry	Colocate UPS/generator near primary loads to reduce cable runs and trenching; avoid panel entry conflicts (e.g., top entry vs. ceiling tray height)
Acoustic and Emission Planning	Integrate silencers and baffles into enclosure design instead of constructing standalone acoustic chambers (NFPA, 2022)

Implementation Considerations:

- Confirm **minimum ventilation flowrates** and temperature rise limits for battery and UPS rooms (e.g., <25 °C for VRLA) (IEEE, 2020).
- Coordinate with **fire consultants** for required ratings of walls, doors, and fuel room construction (NFPA, 2022).
- Verify **fuel tank bunding volume**, access for refueling, and structural loading on slabs (NFPA, 2021).
- Ensure sufficient **space for maintenance**, including battery replacement, exhaust system inspection, and fuel filtration.
- Provide **double access paths** for major components when possible (UPS modules, battery racks, fuel tanks).
- Allow structural loads for battery systems—some may exceed **250–300 kg/m^2**, especially for NiCd or flooded cells (IEEE, 2020).
- Check ingress points and hoisting paths for containerized or roof-lifted generators.

- **Dedicated Technical Spaces**:
  - **Isolation**: House emergency systems in **physically separated, access-controlled technical rooms** to reduce the risk of accidental damage, unauthorized access, and to comply with fire zoning codes (NFPA, 2022).
  - **Security**: Limit access to UPS, generator, and fuel systems through lockable doors and monitoring, particularly in public or mixed-use buildings.
- **Proximity Planning**:
  - Strategically place generators and UPS rooms **close to the loads they serve** to:
  - Minimize voltage drop
  - Reduce cable size and cost
  - Shorten conduit routing and reduce installation time
  - For **fuel-fed generators**, plan location relative to external fuel filling access to reduce piping run, leak risk, and cost.
- **Generator Exhaust and Chimney Planning**:
  - Route generator exhausts **vertically and above roofline** where feasible, away from air intakes or occupied spaces.
  - Account for **backpressure limits** in genset exhaust specifications.
  - Confirm compliance with **local environmental regulations** for emissions and noise limits.
  - Consider **acoustic silencers and thermal insulation** for flues penetrating occupied spaces (NFPA, 2022).
- **Equipment Placement and Maintenance Strategy**:
  - Ensure placement allows for **full frontal access**, battery replacement clearances, and **UPS module slide-outs**.
  - Provide overhead lifting space or floor-level access hatches for battery rack or generator skid replacement.
  - Leave a **minimum of 800 mm** clearance for access behind and beside switchgear or battery racks, or as per manufacturer specs (NFPA, 2023).
- **Heat Dissipation and HVAC Coordination**:
  - Quantify actual heat output (in kW) from UPS, batteries, rectifiers, and DC chargers.
  - Use **passive air extraction** where battery heat output is <1 kW (ASHRAE, 2021).
  - Where possible, **separate thermal zones**: place batteries in thermally isolated racks or rooms from UPS inverters (especially for Li-ion).
  - Avoid overspecifying precision cooling if thermal loads are intermittent or low-density.

Example Application

A campus healthcare facility initially proposed a 60 m^2 battery room with 20-minute autonomy and VRLA batteries. Through VE and runtime analysis, autonomy was reduced to 7 minutes using Li-ion batteries. Resulting system used:

- Rack-mounted UPS with integrated Li-ion packs
- Downsized room to 24 m^2, no active cooling required
- Eliminated dedicated air-conditioning unit, reducing OPEX

Outcome: €41,000 in capex savings, €3500/year in HVAC energy savings, and freed up 36 m^2 for mechanical systems.

Key Takeaways:

- Space planning directly affects the **cost, maintainability, and thermal efficiency** of emergency systems.
- Right-sizing begins with autonomy evaluation and ends with coordinated layout, airflow, and structural integration.
- Modular and integrated systems offer substantial footprint reduction without compromising redundancy.
- Engage architectural, structural, and HVAC teams early to align spatial requirements with VE opportunities.

3.4.5 ICT and ELV Systems

System Overview

ICT and ELV systems form the digital and control backbone of modern facilities. These systems operate at voltage levels ≤50 V AC or ≤120 V DC (per IEC 60950-1) but are mission-critical for communication, security, life safety, and automation functions.

Although typically designed by specialist system integrators or ICT consultants, their infrastructure falls under the electrical engineering scope, particularly in relation to power supply, grounding/bonding, containment, coordination with other services, and emergency backup integration.

To ensure efficiency and scalability, these systems must be treated as VE assets, not as isolated subcontracts, while maintaining strict compliance with current telecommunications and electrical safety standards.

System Classifications

ICT Systems Include:

- Structured cabling networks (copper and fiber per ANSI/TIA-568-E)
- Data racks, patch panels, and network switches
- Server rooms, IDF/ODF cabinets, and communication hubs
- Voice and video communication systems
- PoE systems per IEEE 802.3at/bt

ELV Systems Typically Include:

- Fire detection and alarm systems (NFPA 72 compliant)
- Public address (PA) and voice alarm systems
- CCTV surveillance and digital video recording
- Access control and security alarm systems
- Intercom systems, master clock networks, and building automation interfaces

Design Coordination Note

While ICT and ELV system functionality is led by specialist teams, the infrastructure requirements (power, containment, grounding, spatial integration, and thermal management) must be tightly coordinated with the electrical and architectural disciplines.

Failure to align these during early design stages leads to rework, routing conflicts, and noncompliance with cable segregation or redundancy paths.

Standards Framework
All ICT/ELV system design shall comply with applicable current standards as mandated by the specific project location and governing regulatory authorities.

Commonly Referenced Standards (International—Verify Local Applicability):

Primary Standards:

- ANSI/TIA-568-E series (2020–2024) for telecommunications cabling
- ANSI/TIA-569-E for pathways and spaces
- ANSI/TIA-607-E for grounding and bonding
- IEC 61140 for protection against electric shock
- IEEE 802.3 for Ethernet and PoE standards
- NFPA 70 (NEC) Article 800 for communications circuits
- IEC 62368-1 for ICT equipment safety

Supporting Standards:

- EN 50575 (CPR) for cable fire performance
- IEC 60332 series for cable fire resistance
- IEC 61000 series for EMC requirements
- ASHRAE 90.1 for energy efficiency
- BICSI standards for installation practices

CRITICAL NOTE: Local codes and standards take absolute precedence over international references. Always consult the AHJ for the specific project location.

Critical Coordination Points:

- Telecommunications spaces sizing per TIA-569-E.
- Environmental controls (ASHRAE Class A1-A4 compliance).
- Grounding and bonding system integration per TIA-607-E.
- Cable separation requirements per applicable codes.
- Fire-stopping and pathway diversity per TIA-569-E.

Failure to align these during early design stages leads to rework, routing conflicts, and noncompliance with cable segregation, redundancy paths, and safety requirements.

3.4.5.1 Structured Cabling and Passive Infrastructure

System Overview
Structured cabling forms the physical layer of communication networks and should comply with applicable telecommunications cabling standards. The system includes:

Copper Cabling (Per ANSI/TIA-568-E Standards):

- Horizontal and vertical copper cabling (Cat.5e, Cat.6, Cat.6A, Cat.8).
- Maximum horizontal distance: 90 m permanent link + 10 m patch cords (TIA-568-E requirement).
- Total channel length: ≤100 m including all patch cords (TIA-568-E requirement).
- Individual patch cord limit: ≤5 m per TIA-568-E standards.

Note: Consult applicable local telecommunications standards for your jurisdiction.

Fiber Optic Cabling:

- Multimode: OM3, OM4, OM5 (50/125 μm).
- Single-mode: OS1, OS2 (9/125 μm).
- Laser-optimized 850 nm multimode recommended for backbone applications.

Passive Infrastructure:

- Patch panels and termination blocks.
- Passive enclosures, cabinets, and wall outlets.
- Cable trays, risers, and raceways per applicable pathway standards.
- Work area outlets (minimum two per workspace commonly recommended).

Standards Compliance Considerations

Pinout Requirements (Per ANSI/TIA-568-E Standards):

- T568A pinout recommended for horizontal cables (TIA-568-E specification).
- T568B acceptable (more commonly used in practice).
- All connectors should meet ANSI/TIA-568-E and IEC 60603-7 requirements.

Note: Consult applicable local telecommunications standards for connector requirements in your jurisdiction.

Performance Specifications (Per ANSI/TIA-568.2-E Standards):

- Category 5e: 100 MHz bandwidth (TIA-568.2-E specification).
- Category 6: 250 MHz bandwidth (TIA-568.2-E specification).
- Category 6A: 500 MHz bandwidth (TIA-568.2-E specification).
- Category 8: 2000 MHz bandwidth (TIA-568.2-E specification).
- Backward compatibility required per TIA-568.2-E standards.

Note: Consult applicable local standards for cable performance requirements in your jurisdiction.

This infrastructure supports a wide range of systems, including data networks, voice systems, CCTV, access control, and building automation. It typically follows the **TIA-568**, **ISO/IEC 11801**, and **EN 50173** standards.

VE Objectives:

- Eliminate unnecessary overspecification (e.g., Cat.6A where Cat.6 suffices).
- Right-size racks and patch panels based on actual port count + future growth.
- Standardize cable types across systems (data, security, BMS) where protocol allows.
- Optimize containment and cable routing to avoid wasted capacity and rework.
- Use modular solutions to simplify moves, adds, and changes (MACs) post-handover.

Typical Overdesign or Inefficiencies:

- Blanket specification of Cat.6A everywhere, even in low-bandwidth office or admin zones.
- Multiple independent cabling runs for systems that could share infrastructure (e.g., CCTV, access control, intercom).
- Oversized IDF rooms and underutilized racks based on outdated planning rules.
- Overlapping containment routes (e.g., separate tray for each system) causing ceiling congestion.
- Unused or unlabeled ports that require costly troubleshooting post-handover.
- Missing grounding/bonding system design per applicable standards.

Targeted VE Strategies (Table 3.48)

Table 3.48 Structured Cabling Infrastructure Optimization for ICT and ELV Systems

Design Focus	VE Strategy
Cable Category Selection	Use Cat.6 as default unless >10 Gbps is needed (e.g., in data centers); avoid unnecessary Cat.6A deployment
Fiber Type Optimization	Use OM3 for standard backbones; avoid OM4 unless required by link budget or legacy system interface
Rack and Patch Panel Planning	Plan port density per rack based on confirmed device counts + 20–30% growth margin; avoid excessive spare capacity
Containment Consolidation	Route ICT, CCTV, access, and BMS cabling through shared ELV trays (with proper segregation) to reduce cost and clutter
Modular Termination	Use preterminated or modular patch solutions in high-density zones for installation speed and easier MACs

Implementation Considerations:

- Follow TIA/EIA-568-B and ISO 11801 for cabling standards and color coding.
- Maintain cable bend radius: 4x cable diameter (installation), 1x diameter (patch cords) per TIA-568-E.
- Provide dual cable routes (A/B) in critical systems and server room backbones.
- Leave a minimum of 20–30% spare tray fill for future capacity.
- Terminate cables in standardized enclosures to reduce training and O&M complexity.
- Label all cables at both ends; follow structured numbering for racks, floors, and panels.

Distance Requirements (Per ANSI/TIA-568-E Standards):

- Horizontal cabling: ≤90 m permanent link.
- Work area patch cords: ≤5 m each.
- Equipment patch cords: ≤5 m each.
- Cross-connect jumpers: ≤20 m (intermediate/main).
- Equipment jumpers: ≤30 m maximum.
- Total channel length: ≤100 m including all patch cords.

Environmental Considerations:

- Cable jacket ratings per applicable electrical codes and fire performance standards.
- Fire-stopping per applicable fire codes.
- Temperature ratings appropriate for installation environment.

■ ***Grounding and Bonding Recommendations***

Recommended Requirements (Verify Local Codes):

- All metallic components should be bonded per applicable standards.
- Minimum six American Wire Gauge (IEC 2021) bonding conductors commonly recommended.

- Connection to telecommunications bonding system.
- Proper busbar sizing and installation per applicable requirements.

Equipment Bonding:

- Cable trays, racks, and enclosures bonded to appropriate bonding system.
- Shield termination for shielded cable systems.
- Patch panels bonded per manufacturer instructions.

Example Application

A commercial building originally called for Cat.6A cabling in all workspaces, with full-height 42 U racks per floor. VE analysis found that:

- Cat.6 was sufficient for 98% of devices based on actual application data
- Patch density per floor required only 18 U
- Shared ELV trays were coordinated to support access, CCTV, and Wi-Fi AP cabling

Outcome: €31,000 saved in cabling and passive hardware, ceiling space clearance improved by 150 mm, and improved maintainability due to reduced system fragmentation.

Key Takeaways:

- Structured cabling decisions impact both initial cost and long-term flexibility.
- Use data-driven port counts and bandwidth projections to right-size cable category and rack layout.
- Modular, standardized, and integrated cabling saves cost while enabling future expansion.
- Close coordination with ELV systems and ceiling/tray layout is critical to avoiding physical clashes and cost duplication.

3.4.5.2 Active ICT Equipment and Network Rooms

System Overview

Active ICT components serve as the backbone for **data transmission**, **voice services**, and **building control systems**. These systems should be designed for reliability, scalability, and standards compliance.

Active Equipment Categories:

- Ethernet switches (managed/unmanaged per IEEE 802.3).
- PoE switches (IEEE 802.3at: 25.5 W/port, IEEE 802.3bt: up to 71 W/port).
- Core and edge routers with appropriate throughput capacity.
- Network firewalls and security appliances.
- Media converters and distribution electronics.
- Server equipment (when applicable).

These devices are typically housed in:

- Main distribution frames (MDFs)
- Intermediate distribution frames (IDFs)

- Communication equipment rooms (CERs)
- Entrance facilities (EFs)

Proper planning of these spaces is essential for long-term reliability, efficient cooling, physical security, and scalability.

VE Objectives:

- Right-size network rooms and rack quantities based on actual equipment population + growth.
- Avoid allocating server-grade environments for passive switchgear or light-duty routing equipment.
- Optimize cooling strategy using passive airflow management and blanking panels before introducing precision HVAC.
- Standardize rack layouts and power provisioning to reduce procurement complexity.
- Coordinate room proximity to minimize cable lengths and access delays.

Typical Overdesign or Inefficiencies:

- Full-height 42 U racks specified where only 6–12 U of equipment is required.
- MDF/IDF rooms oversized based on old BICSI rules, without port density analysis.
- Precision cooling applied to light-load rooms with <1 kW of heat generation.
- Redundant power strips, cabling, and patch panels specified without documented uptime requirement.
- No airflow control—leading to recirculation, overheating, or HVAC inefficiencies.
- Missing grounding/bonding system integration per applicable standards.
- Environmental requirements not per applicable telecommunications space standards.

Targeted VE Strategies (Table 3.49)

Table 3.49 Active ICT Equipment and Network Room Design Strategies Under VE

Design Focus	VE Strategy
Rack Population	Size rack units based on confirmed switch counts and patching plans; avoid standard 42 U racks unless needed
Thermal Management	Use **blanking panels** in unused rack U-space to preserve front-to-rear airflow and avoid hot air recirculation
Power and UPS Support	Provide only essential power backup (5–10 minutes) for edge switches; full UPS only for core equipment
Room Type Classification	Distinguish between CERs, IDFs, and server rooms—apply thermal and security specs accordingly
Standardization	Use uniform rack layouts (power left, data right) and color coding for patching to reduce O&M complexity

Implementation Considerations:

- Follow **ANSI/TIA-569** and **BICSI 002** for room layout, spacing, and ventilation.
- For low-power IDFs, consider **passively cooled enclosures** without dedicated HVAC.

- Ensure dual power receptacles (A/B) only where equipment supports dual PSUs.
- Allow sufficient depth and side clearance for rear cabling and airflow.
- Install **environmental monitoring** (temperature, humidity, door access logs) for MDFs and CERs.
- Use **locking rack doors** and restrict access to network rooms via card access or monitored keypads.
- Label all patch cords, switches, and ports for efficient MAC (moves, adds, changes).

■ *Environmental Recommendations*

Environmental Requirements (Per ANSI/TIA-569-E Standards):

- Operating Temperature: 18–24 °C (64–75 °F) per TIA-569-E specifications.
- Relative Humidity: 45–55% (noncondensing) per TIA-569-E requirements.
- Equipment environmental class compliance per manufacturer specifications and ASHRAE guidelines.
- 24/7 monitoring recommended for critical spaces per TIA-569-E.

Note: Verify applicable local building and electrical codes for environmental requirements in your jurisdiction.

Space Sizing Recommendations:

- Distributor rooms sized per confirmed equipment + growth projections.
- Minimum 3 m × 2.2 m commonly recommended for basic IDFs.
- Height clearance: minimum 2.6 m typically required.
- Access clearance: 1 m front, 0.8 m rear recommended for equipment access.

■ *Power and UPS Considerations*

Power System Design:

- Dual power feeds (A/B) recommended for critical equipment with dual PSUs.
- Circuit protection per applicable electrical codes.
- Power density calculations should include efficiency losses (10–15% typical).

UPS Backup Recommendations:

- Essential power backup: 5–30 minutes as required by local authority or business requirements.
- Full UPS for core equipment per business continuity requirements.
- Battery monitoring and replacement scheduling recommended.

PoE Power Specifications (Per IEEE 802.3 Standards):

- IEEE 802.3at (PoE+): 25.5 W delivered per port (IEEE specification).
- IEEE 802.3bt (PoE++): 51 or 71 W delivered per port (IEEE specification).
- Include cable losses (typically 10–15% over 90 m based on IEEE calculations).
- Size PoE budget for 80% utilization maximum (recommended practice).

Note: Verify applicable electrical and telecommunications codes for PoE requirements in your jurisdiction.

Grounding and Bonding (Per ANSI/TIA-607-E Standards):

- All metallic equipment should be bonded to telecommunications bonding system.
- Racks and cabinets connected to appropriate bonding system.
- Minimum six AWG bonding conductors per TIA-607-E specifications.
- Switch and router chassis bonding per manufacturer instructions.
- Power strip and surge protector boding recommended.

Safety and Security:

- Fire suppression systems per applicable fire codes.
- Access control and monitoring recommended for equipment rooms.
- Emergency power-off (EPO) systems where required by applicable codes.
- Arc flash protection per applicable electrical safety standards.

Example Application

In a corporate tower, the initial design included full-size MDF rooms on every floor with dedicated air-conditioning. VE review determined:

- IDFs contained only 1–2 PoE switches per floor (<500 W heat)
- Blanking panels were not included, leading to anticipated recirculation
- Revised design implemented 18 U wall-mounted racks with integrated fans and passive air slots; blanking plates used for unoccupied rack space

Outcome: €24,000 saved in HVAC and room construction per four floors; improved energy performance and room space returned to tenant use.

Key Takeaways:

- Network rooms and equipment must be designed based on actual power, heat, and scalability—not legacy sizing rules.
- Simple interventions like **blanking panels** and **modular racks** offer tangible VE gains in energy and footprint.
- Coordinated power, access, and cooling designs support future scalability with minimal operational disruption.
- Proper labeling and standardization reduce O&M costs and prevent configuration errors.

3.4.5.3 CCTV, Access Control, and Security Systems

System Overview

Security systems provide protection for people, assets, and infrastructure through surveillance, access regulation, and real-time incident response. These systems are often **IP-based** and increasingly integrated with ICT and building management platforms.

Common components include:

- **CCTV systems**: IP cameras, network video recorders (NVRs), video management software (VMS), storage arrays
- **Access control**: Card readers, biometric scanners, turnstiles, electromagnetic locks
- **Intrusion detection**: Motion sensors, door contacts, security panels
- **Control rooms**: Monitoring stations, video walls, centralized servers

CCTV and access control systems are typically specified by security integrators, but their **power supply**, **cabling**, **containment**, and **room allocation** fall under electrical infrastructure.

VE Objectives:

- Avoid overcoverage and overspecification of cameras and access points based on risk zoning.
- Consolidate power, data, and control cables through shared pathways to minimize tray congestion.
- Standardize camera types and resolutions across the project.
- Align device count with centralized recording and server/storage capacity.
- Reuse infrastructure for multipurpose (e.g., surveillance + analytics, security + BMS badge access).

Typical Overdesign or Inefficiencies:

- Blanket CCTV coverage without reference to threat levels or critical asset locations.
- 4 K/12 MP cameras specified in low-risk areas or where 1080p suffices.
- Dedicated cable trays for CCTV and access when shared low-voltage containment is viable.
- Access control devices installed at all doors, including low-traffic interior rooms.
- Specified NVR or VMS servers with far greater capacity than needed (e.g., 180-day recording when 30 days required by regulation).

Targeted VE Strategies (Table 3.50)

Table 3.50 VE Approaches for CCTV and Access Control System Design

Design Focus	VE Strategy
Camera Placement	Use **risk-based layout**: reduce camera count in admin zones; increase in entries, elevators, and perimeter zones
Resolution and Storage	Match resolution and frame rate to application (e.g., 1080p @ 15fps standard); size storage for **actual retention policies**
Device Consolidation	Use **multisensor** or **panoramic cameras** to reduce camera count in wide areas
Power Strategy	Use **PoE switches** for CCTV and access control when compatible; avoid separate power supplies unless required
Reader Strategy	Use **read-in/free-exit** model for low-security doors to reduce device count; apply biometric or dual-authentication only where risk demands

Implementation Considerations:

- Coordinate headend/server rack location with IDFs to limit switch-to-camera cable length.
- Segregate security VLANs at switch level to prevent unauthorized access from other building systems.
- Provide UPS backup for NVRs, controllers, and critical camera feeds (e.g., emergency exits, perimeter).
- Review regulatory compliance for camera retention (e.g., 30–90 days per local authority).
- Ensure redundancy or failover for central controllers in large projects or mission-critical sites.
- Integrate access control with fire alarm systems for **fail-safe door release** in emergencies.
- Tag and document all camera viewpoints in coordination drawings for future maintenance.

Example Application

A commercial building initially specified 98 CCTV cameras and 160 access control points. A VE-led risk zoning approach adjusted:

- Reduced low-traffic interior doors to badge-free exit
- Replaced overlapping corridor views with multisensor units
- Aligned VMS storage with a 30-day policy using H.265 compression

Outcome: €52,000 saved in equipment, cabling, and installation; streamlined maintenance with 25% fewer endpoints.

Key Takeaways:

- Security systems should be **risk-zoned**, not uniformly applied across all areas.
- VE aims to right-size resolution, storage, and device count without compromising safety or compliance.
- Shared infrastructure, PoE use, and device consolidation reduce installation time and life-cycle cost.
- Integration with ICT and fire systems enhances response and interoperability.

3.4.5.4 Public Address and Voice Evacuation Systems

System Overview

PA and **VE** systems are essential for delivering audible messages across facilities, including routine notifications and emergency instructions. In many buildings—especially those classified under life safety codes—these systems are **legally required** and must be integrated with the **fire alarm system**.

Core components include:

- Central PA/VE controller and amplifiers
- Ceiling and wall-mounted speakers (zones)
- Microphones for manual announcements
- Fire-rated cabling and speaker loops
- Interfaces with fire alarm, access control, or BMS
- Storage for prerecorded emergency messages

Systems must comply with standards such as **EN 54-16**, **NFPA 72**, and relevant local fire authority codes.

VE Objectives:

- Optimize speaker layout based on intelligibility (STI), not just area coverage.
- Consolidate system architecture (PA + VE + intercom) where permitted.
- Reduce amplifier oversizing by applying diversity and zone priority.
- Use modular, scalable control systems that support phased implementation.
- Design with future partitioning or tenant changes in mind.

Typical Overdesign or Inefficiencies:

- Speaker density based on grid layout rather than acoustic analysis.
- Oversized amplifiers with no load analysis or standby justification.
- Separate systems for routine PA and evacuation (where hybrid is permitted).
- No zoning flexibility, requiring full-building broadcast for local announcements.
- Use of premium audio speakers in noncritical or utility zones.

Targeted VE Strategies (Table 3.51)

Table 3.51 Optimization of PA and Voice Evacuation Systems for Life Safety and Communication

Design Focus	VE Strategy
Speaker Layout	Use **acoustic simulation** to validate STI (≥0.5); reduce speaker count in low-occupancy areas
Amplifier Sizing	Size amplifiers per zone with load diversity; use standby amplifier for shared coverage instead of per-zone redundancy
System Integration	Combine **PA, VE, and intercom** on a unified IP-based platform where regulation allows
Cabling Approach	Use **looped speaker lines** for fault tolerance; avoid separate cable runs per speaker
Fire Interface Logic	Trigger prerecorded evacuation messages upon fire alarm signal; avoid duplicate logic programming in both systems

Implementation Considerations:

- Confirm system compliance with **life safety standards** (EN 54, NFPA).
- Separate speaker zones per **evacuation sequence, occupancy type**, or **floor zone.**
- Coordinate speaker installation height and angle with ceiling architecture and finishes.
- Provide **emergency paging stations** at security and fire command posts.
- Use **fire-rated cable and enclosure** for amplifiers and evacuation controllers.
- Ensure integration of PA/VE controller with fire alarm panel for VE logic.
- Provide UPS backup or 24 VDC supply with battery autonomy per code (e.g., 24–72 hours).

Example Application

A large university campus originally planned separate PA and fire VE systems. VE assessment recommended:

- Unified IP-based system with emergency messaging module
- Acoustic modeling to reduce ceiling speaker count by 20%
- Standby amplifier shared across four low-priority zones

Outcome: €67,000 saved in hardware and cabling; fewer devices improved commissioning speed and future zone flexibility.

Key Takeaways:

- PA/VE systems must balance life safety compliance with smart zoning and acoustic coverage.
- VE focuses on **speaker and amplifier count**, **system integration**, and **looped architecture.**
- Digital IP-based systems allow greater flexibility, zone control, and interoperability with BMS and security systems.
- Early integration with fire alarm logic is essential to avoid duplication and commissioning delays.

3.4.5.5 Integration, Power, and Containment Coordination

System Overview

While ICT and ELV systems are functionally distinct, their infrastructure—**power sources**, **containment routes**, and **mechanical coordination**—falls under the domain of electrical design. As these systems grow more complex and data-intensive, proper integration planning ensures long-term reliability, minimizes installation rework, and enhances VE outcomes.

Key infrastructure elements requiring coordination include:

- Cable containment (trays, ladders, and conduits)
- Segregation from power and interference sources
- Power provisioning (normal + UPS)
- Grounding and bonding
- Equipment access and thermal separation
- Labeling and routing documentation

VE Objectives:

- Eliminate redundant containment that runs across systems with similar cabling characteristics.
- Reduce EMI risk through appropriate segregation strategies.
- Optimize UPS deployment based on actual runtime need.
- Simplify system maintenance through structured routing and access design.
- Minimize spatial clashes and penetrations via early containment clash detection in BIM.

Typical Overdesign or Inefficiencies:

- Separate trays for each ELV system (e.g., CCTV, BMS, access control) despite same voltage class.
- Overuse of conduit or metallic trunking where shared ladder trays would suffice.
- UPS-backed sockets for all ELV devices, including noncritical panels.
- No predefinition of cable tray hierarchy, causing clashes and rerouting during construction.
- Power and data cabling co-routed in shared enclosures, risking EMI or code violations.

Targeted VE Strategies (Table 3.52)

Table 3.52 Integration and Containment Coordination Strategy for ICT and ELV Systems

Design Focus	VE Strategy
Containment Strategy	Use **shared ELV trays** (with separation barriers) for ICT, CCTV, access, and intercom systems of similar voltage class
Segregation Compliance	Maintain separation from low-voltage power (typically ≥300 mm or metal divider); document clearly in routing drawings
UPS Allocation	Prioritize UPS support for **server rooms, NVRs, central controllers**; use normal power for field devices with minimal risk
Labeling and Routing	Implement a standardized labeling and documentation protocol across all ELV systems for MAC efficiency
BIM Integration	Model all ELV and ICT containment in BIM with **predefined levels, access clearances**, and color codes to avoid overlap

Implementation Considerations:

- Maintain separation distances in compliance with **IEC 60364**, **NFPA 70**, or local codes.
- Coordinate with mechanical and architectural teams for ceiling space, access hatches, and tray fixings.
- Use **nonmetallic containment** only where permissible and flame-retardant ratings are met.
- Apply power diversity and surge protection at each IDF/MDF or equipment node.
- Ensure all containment is **accessible, maintainable**, and not located above critical patient care or sterile zones.
- Route fiber cables with extra bend radius space and termination enclosure allocation.

Example Application

In a mixed-use commercial complex, five different contractors were installing separate containment for each ELV system. A VE coordination workshop resulted in:

- Consolidation into shared horizontal trays for ELV systems
- Unified power strip allocation based on priority and runtime
- Color-coded BIM containment layers coordinated with ceiling HVAC and lighting

Outcome: €46,000 saved in supports, cable tray, and labor; improved ceiling serviceability and as-built clarity.

Key Takeaways:

- Early containment and power coordination reduces cost, ceiling congestion, and commissioning delays.
- VE strategies focus on **segregation, power hierarchy, and shared routing**, not just product selection.
- Integration across ICT/ELV systems must also include **labeling, access, and documentation protocols.**
- BIM coordination and field access planning are essential for long-term system performance and modification.

3.4.6 Fire Detection and Life Safety Systems

System Overview

Fire detection and alarm systems are among the most critical life safety systems in any building. These systems detect the early presence of smoke, heat, or flame and initiate alerts, evacuations, and system responses in accordance with safety codes. Their design is typically governed by **national fire codes** and international standards such as:

- **EN 54 series** (EU) (CEN 2021)
- **NFPA 72** (USA) (NFPA 2025a)
- **BS 5839-1** (UK) (BSI 2025a)
- **Local civil defense/fire authority regulations**

Fire alarm systems often integrate with:

- PA/VE systems
- Access control (for door release)

- HVAC (for smoke control shutdown)
- Sprinkler and fire suppression systems
- BMS for monitoring and reporting

Core components include:

- Fire alarm control panel (FACP)
- Addressable detectors (smoke, heat, multisensor)
- Manual call points (MCPs)
- Visual and audible alarms (beacons, sounders)
- Loop isolators, relays, interfaces, and power supplies

VE Objectives:

- Comply fully with codes while minimizing overspecification or redundant devices.
- Optimize device placement through proper fire zoning and ceiling height strategy.
- Leverage integrated platforms (PA, access, suppression) where permitted.
- Simplify maintenance by standardizing device types and mounting schemes.
- Ensure coordinated loop architecture for resilience without overspending on cabling.

Typical Overdesign or Inefficiencies:

- Excessive detector count due to lack of zone-based or area-based planning.
- Manual call points placed at all exits, including noncritical or staff-only areas.
- Redundant loops without cause (e.g., full A + B loops in small buildings).
- Specifying premium multisensor detectors where basic smoke or heat would suffice.
- Overuse of standalone interfaces for door control or HVAC shutdown, instead of shared relays.

Targeted VE Strategies (Table 3.53)

Table 3.53 VE Techniques for Fire Detection and Alarm System Design

Design Focus	VE Strategy
Device Selection	Use **smoke, heat, or multisensor** based on fire risk profile of the zone; avoid unnecessary use of multisensor in all areas
Loop Architecture	Plan for **single-ring loops** with isolators unless redundancy is required by local codes
Zoning Optimization	Design fire zones based on **compartmentation and evacuation paths**, not arbitrary floor plans
System Integration	Combine voice evacuation, fire alarm, and suppression signals on **shared interfaces** where allowed
Maintenance Simplification	Standardize detector heights and types to streamline access, testing, and replacements

Implementation Considerations:

- Confirm **AHJ** requirements for device placement, loop redundancy, and monitoring.
- Ensure system is **fail-safe** with supervised loops and end-of-line devices.
- Provide UL-listed rechargeable batteries with 25% correction factor for backup power (NFPA, 2025b), typically 24 hours standby plus alarm time duration per local code requirements.
- Coordinate manual call point height and beacon visibility with accessibility standards (e.g., ADA) (US Department of Justice, 2010).
- Use fire-rated cable complying with EN 50200 (CEN, 2015) (for cables ≤20 mm diameter, ≤2.5 mm^2 conductor size) with appropriate classification: standard fire cables require PH30/PH60 plus Annex E testing; enhanced fire cables require PH120 plus BS 8434-2 testing (BSI, 2008).
- Perform **false alarm risk analysis** to avoid oversensitivity in kitchens, mechanical rooms, etc.
- Interface fire alarm with **access control** for auto door release during evacuation.
- For connected systems, implement cybersecurity protocols per NFPA 72 Chapter 11 requirements (NFPA, 2025c), including security level assessment and regular security updates per ANSI standards.
- **False Ceiling Detectors**:
  - Only install detectors above false ceilings when required by **local codes**, **airflow patterns**, or where **high fire load (e.g., cabling, ducts)** exists in the plenum.
  - Avoid default installation above all ceilings—this adds device count and complicates access/testing.
- **Detector Technology Optimization**:
  - Use beam detectors only in large open areas (minimum 8 m ceiling height recommended) where point detectors are impractical, such as atriums and warehouses.
  - Avoid beam detectors in environments with high dust, vibrations, or movement that may trigger false alarms or misalignment.
  - Consider **aspirating smoke detectors (ASD)** for mission-critical or high-sensitivity spaces (e.g., data centers), but avoid overuse due to high cost and maintenance complexity.
- **Detector Spacing Requirements:**
  - Maximum detector spacing: 30 ft (9.1 m) for NFPA 72 jurisdictions (NFPA, 2025d), 7.5 m coverage radius per EN 54-14 standards (CEN, 2017c).
  - For ceilings above 40 ft (12.2 m) height, performance-based design analysis is required per NFPA 722025 (NFPA, 2025e).
  - Maintain minimum 0.5 m clearance from walls and obstructions per EN 54 requirements (CEN, 2017c).
- **Environmental Suitability**:
  - Select **heat detectors** for areas prone to smoke or steam (e.g., kitchens, plantrooms) to avoid nuisance alarms.
  - Use **multisensor detectors** only where required (e.g., variable ambient environments, critical spaces with fast detection needs).
  - Install smoke detectors listed for resistance to cooking nuisance sources within 20 ft (6.1 m) of cooking appliances per NFPA 722025 (NFPA, 2025f).

■ ***Additional System Requirements***

- **Remote Access and Monitoring:**
  - Implement restricted audible mode notification (RAMO) in areas where sudden loud noises could adversely impact occupants per NFPA 722025 (NFPA, 2025g).
  - For auxiliary service providers (ASPs), ensure communication path supervision every 90 seconds and UL 827 compliance (UL, 2019) with written notice to building owner, AHJ, and supervising station within 30 days.

Example Application

In a high-rise commercial office building, initial design showed 420 detectors and 96 call points. VE optimization:

- Re-zoned fire compartments to align with stair and lift cores
- Reduced detectors by 18% through strategic multisensor use and extended spacing
- Eliminated 16 unnecessary MCPs in internal back-of-house doors

Outcome: €29,000 saved in devices and cabling; simpler testing plan and faster system commissioning.

Key Takeaways:

- Fire detection systems must balance **regulatory compliance** with **design efficiency.**
- VE strategies include **device rationalization**, **zone-based planning**, and **interface consolidation.**
- Stay current with evolving standards: NFPA 72 and EN 54 series update on 3-year cycles, BS 5839-1:2025 superseded 2017 edition with new requirements for lift shaft detection and terminology updates.
- Early coordination with architectural layout, BMS, and public safety systems ensures minimal rework.
- Standardization and maintainability reduce lifecycle cost while improving safety reliability.
- Ensure AHJ approval for all VE decisions and maintain documentation for compliance verification.

3.4.7 Cable Management and Containment Systems

System Overview

Cable containment systems provide the structural support and physical pathways through which power and signal cables are routed throughout a building. They are essential for organizing infrastructure, protecting cables from physical damage, ensuring separation between systems, and enabling inspection or future modifications.

Primary containment systems include:

- **Cable trays** (perforated, ladder type)
- **Trunking and ducts** (metallic or PVC)
- **Conduits** (surface or embedded, rigid or flexible)

- **Underfloor raceways and service boxes**
- **Vertical risers and shafts**
- **Modular prefab containment (e.g., plug-and-play kits)**

These systems support all cable types—LV, MV (when armored or in ducts), ELV, fiber optics, and control wiring—and must be coordinated across architectural, mechanical, and structural trades.

VE Objectives:

- Optimize cable routing to reduce material usage and simplify installation.
- Use standardized, modular containment systems that support rapid installation and future expansion.
- Avoid unnecessary duplication of trays between systems with compatible voltage classes.
- Apply risk-based protection levels (e.g., conduit vs. tray) based on location and exposure.
- Minimize site labor by prefabricating or pre-assembling containment segments off-site.

Typical Overdesign or Inefficiencies:

- Separate trays for every low-voltage system (CCTV, access, BMS, etc.) instead of shared ELV trays.
- Oversized trays based on worst-case assumptions rather than actual cable schedules.
- Overuse of metallic trunking in secure areas where flexible conduit or tray suffices.
- Using embedment (cast-in conduit) in slabs where surface-mounted or raised-floor wiring is more adaptable.
- Full duplication of containment for future expansion without load forecast.

Targeted VE Strategies (Table 3.54)

Table 3.54 Cable Management and Containment System Optimization Through VE

Design Focus	VE Strategy
Tray Sizing	Size trays based on actual cable fill calculations + 20–30% future capacity margin; avoid oversizing for "just-in-case" scenarios
System Segregation	Use **shared trays with barriers or spatial separation** for compatible ELV systems (e.g., CCTV + BMS)
Routing Optimization	Plan shortest viable cable paths and avoid excessive bends or intersections; align with structural bays and ceiling grids
Prefab Containment	Use **preassembled tray sections** with factory-applied accessories (e.g., bends, tees, supports) to reduce site labor and errors
Coordination in BIM	Layer containment by voltage class and system, and verify clearances with HVAC, lighting, and architectural soffits early in the model

Implementation Considerations:

- Maintain physical separation per applicable standards (e.g., IEC 60364, NEC 300.3[C]):
  - LV vs. ELV: ≥300 mm or metallic barrier.
  - Power vs. data: separate trays or compartments.
- Support trays every **1.2–1.5 m** unless otherwise approved by manufacturer.

- Ensure **adequate access** for inspection, future pulls, and label visibility.
- Plan tray entries and exits into distribution boards or riser shafts **based on panel entry (top/bottom).**
- Avoid running trays through fire-rated walls without properly sealed firestop assemblies.
- Use **noncorrosive coatings or materials** in wet or exposed zones (e.g., GI + epoxy, SS304).

Example Application

In a high-rise commercial building, initial design showed that 12 dedicated ELV trays run across each floor level. VE workshops revealed:

- Only four voltage classes were required
- All CCTV, access, and BMS systems could share a common tray with metallic dividers
- Prefabricated trapeze supports were used instead of on-site welding

Outcome: €63,000 saved in tray materials and labor; 40% reduction in ceiling congestion and improved access for inspection.

Key Takeaways:

- Cable containment is a major cost and coordination driver—early VE input reduces rework and material waste.
- Modular and shared tray strategies deliver both cost and access benefits.
- Routing efficiency, bend minimization, and access provisions are critical for lifecycle maintenance.
- Coordination in BIM avoids ceiling clashes and overdesign, particularly with dense MEP layouts.

3.4.8 Energy Efficiency and Sustainable Design

System Overview

Energy efficiency measures in electrical systems directly influence both operational expenditure and sustainability credentials. By integrating performance-driven design strategies from the outset, electrical engineers can reduce energy losses, improve system utilization, and align with green building certification requirements and local energy codes. Sustainable design in VE does not solely focus on reducing energy use—it also includes power quality, integration of renewable sources, and lifecycle cost analysis.

This section identifies key opportunities to embed energy-conscious strategies within electrical infrastructure without undermining performance or reliability.

3.4.8.1 Energy Performance and Metering

VE Objectives:

- Ensure accurate load monitoring and control.
- Support energy dashboards and BMS integration.
- Enable postoccupancy optimization of system performance.

Typical Inefficiencies:

- Absence of submetering for major end users.
- Inflexible metering hierarchies that prevent benchmarking.
- Oversized systems with low load factors.

Targeted VE Strategies:

- Implement submetering at the level of tenant, lighting, HVAC, and plug loads.
- Use meters with communication capability (i.e., Modbus, BACnet) for BMS integration (IEEE, 2018a).
- Apply load profile analysis to optimize transformer and UPS sizing.

Implementation Considerations:

- Ensure CTs, pulse output devices, and network meters are specified early.
- Provide physical space and access for future expansion of metering infrastructure.
- Integrate energy data into facility dashboards for feedback-driven savings.

Example Application

A commercial office project introduced multilevel metering across lighting, HVAC, and tenant circuits, enabling postoccupancy energy audits and contract-based usage tracking. Payback was achieved within 2 years.

3.4.8.2 Power Factor Correction and Harmonic Mitigation

VE Objectives:

- Minimize utility penalties and transformer overloading.
- Improve real power delivery efficiency.
- Prevent unnecessary oversizing of upstream components.

Typical Inefficiencies:

- Reactive power penalties due to poor PF.
- Harmonic distortion leading to overheating or derating.
- Overdesigned correction systems that are rarely engaged.

Targeted VE Strategies:

- Design power factor correction (PFC) based on measured or simulated load profiles.
- Use automatic PFC panels with step controllers and detuned reactors.
- Evaluate active harmonic filters where nonlinear loads constitute a significant portion of total demand (IEEE, 2014).

Implementation Considerations:

- Coordinate placement of PFC units at MCCs or MDBs.
- Check resonance risk between capacitors and supply frequency.
- Ensure total harmonic distortion (THD) remains within IEEE 519 or local grid code limits.

VE Insight

Harmonics from VFDs and LED drivers can affect sensitive loads or UPS systems. Proactive filtering can reduce heat loss and cable oversizing, especially in IT-heavy environments.

3.4.8.3 Renewable Integration and Grid Interaction

VE Objectives:

- Offset utility consumption via on-site generation.
- Enable future-readiness for net-zero targets.
- Improve electrical autonomy during grid instability.

Typical Inefficiencies:

- Lack of infrastructure provision for future PV/BESS systems.
- Oversizing of generator or UPS systems where renewables could contribute.
- No provision for export limitation or feed-in optimization.

Targeted VE Strategies:

- Reserve roof/terrace space, containment, and spare breakers for future solar PV or battery storage systems.
- Use hybrid-ready inverters where appropriate.
- Design main switchboards with CTs and ATS interfaces to support net metering or load shaving.

Implementation Considerations:

- Evaluate building type and orientation for PV feasibility.
- Incorporate grid-tied and islanding logic into protection and synchronization studies.
- Comply with local grid interconnection standards (e.g., EN 50549, IEEE 1547) (CEN, IEEE).

Example Application

A hospital project routed spare 63 A feeders and EMT conduits to the roof in anticipation of future PV integration. This avoided core drilling and switchgear retrofits in later phases.

Key Takeaways:

- Energy efficiency is a core VE lever with strong ROI and sustainability alignment.
- Metering, power quality, and renewables must be integrated at the design stage.
- System flexibility and upgrade-readiness support long-term environmental and financial value.

3.5 Practical Design Support Methods

Section Overview

While much of this chapter focuses on system-specific strategies, successful VE implementation also depends on how designs are structured to accommodate real-world operation, evolution, and feedback. This section outlines two high-value methods that enhance maintainability and enable continuous improvement through project learning.

3.5.1 Design for Maintainability and Flexibility

VE Objectives:

- Reduce long-term operational costs through accessible and serviceable design.
- Ensure that systems can be adapted or expanded without major redesign.
- Improve safety and reduce downtime during maintenance or replacement.

Strategies:

- Use draw-out circuit breakers, modular UPS, and plug-in busducts in critical zones.
- Design dedicated maintenance clearances and access hatches in LV/MV rooms.
- Separate life safety and normal distribution paths for safe isolation and staged shutdowns.
- Reserve panel space and conduit routing for future expansion (minimum 20–25%).

Implementation Considerations:

- Coordinate with O&M teams during design phase reviews.
- Include maintenance access diagrams in the BIM coordination package.
- Specify standardized, replaceable components with short lead times.

Example Application

In a university data hall, locating UPS units in accessible plant corridors and specifying hot-swappable modules enabled 99.999% uptime, while reducing mean time to repair by over 60%.

3.5.2 Case-Based Reasoning and Learning Integration

VE Objectives:

- Leverage lessons from previous projects to avoid repeating costly errors.
- Integrate empirical outcomes into decision-making for current and future works.
- Establish feedback-driven improvement cycles in VE strategy.

Strategies:

- Maintain a log of VE opportunities evaluated, accepted, and rejected, including rationale.
- Use "VE memory" templates to capture product substitutions, coordination wins, or specification gaps.
- Conduct postoccupancy walkthroughs or FM interviews to capture operational feedback.
- Reference historical coordination problems (e.g., switchgear access, thermal buildup, rack airflow) during early design.

Implementation Considerations:

- Incorporate VE learnings into your office's design standards or specification library.
- Encourage consultants and contractors to share VE case studies during tendering or kickoff.
- Where practical, cross-reference past project submittals and vendor data.

Example Application

A logistics project referenced an earlier issue where cable tray congestion above MCCs delayed commissioning. In the current design, routing was revised in LOD 300 to maintain 300 mm vertical clearance, preventing rework and reducing ceiling coordination risk.

Key Takeaways:

- Maintainability and historical learning are often overlooked VE levers with high lifecycle value.
- Incorporating case-based insight into design processes improves decision-making, reduces risk, and strengthens institutional knowledge.

Note

Detailed methods for financial modeling, lifecycle costing, and comparative trade-off analysis are addressed in **Chapter 5 (Proactive VE)**, **Chapter 6 (Templates and Tools)**, and **Chapter 13 (Financial Strategies and Economic Impact)**. This section focuses specifically on support methods most relevant during electrical system implementation.

References

ANSI/TIA-568.2-E. (2020). *Balanced twisted-pair telecommunications cabling and components standard*. Telecommunications Industry Association.

ANSI/TIA-568-E. (2020). *Telecommunications cabling infrastructure*. Telecommunications Industry Association.

ANSI/TIA-569-E. (2019). *Telecommunications pathways and spaces*. Telecommunications Industry Association.

ANSI/TIA-607-E. (2024). *Generic telecommunications bonding and grounding (earthing) for customer premises*. Telecommunications Industry Association.

ASHRAE. (2022). *ASHRAE 90.1-2022 Energy Standard for Buildings Except Low-Rise Residential Buildings*. ASHRAE.

BICSI. (2024). *Telecommunications distribution methods manual*. Building Industry Consulting Service International.

Boothroyd, G., & Dewhurst, P. (1990). *Product Design for Assembly*. Boothroyd Dewhurst Inc.

BSI. (2008). *BS 8434-2:2003 - Methods of test for assessment of the fire integrity of electric cables*. BSI.

BSI. (2024). *BS 7671:2018+A3:2024 Requirements for Electrical Installations. IET Wiring Regulations*. British Standards Institution.

CEN. (2017a). *EN 15193-1:2017 Energy performance of buildings - Energy requirements for lighting - Part 1: Specifications, Module M9*. European Committee for Standardization.

CEN. (2017c). *EN 54-14:2017 - Fire detection and fire alarm systems. Part 14: Guidelines for drafting codes of practice for design, installation and use of fire detection and fire alarm systems*. European Committee for Standardization.

CEN. (2018). *EN 12285-1:2018 Workshop fabricated steel tanks - Part 1: Horizontal cylindrical single skin and double skin tanks for the underground storage of flammable and nonflammable water polluting liquids*. European Committee for Standardization.

CEN. (2019a). *EN 1838:2019 Lighting applications - Emergency lighting*. European Committee for Standardization.

CEN. (2019b). *EN 50549-1:2019 - Requirements for generating plants to be connected in parallel with distribution networks - Part 1: Connection to a LV distribution network - Generating plants up to and including Type A*. European Committee for Standardization.

CEN. (2021a). *EN 12464-1:2021 Light and lighting - Lighting of work places - Part 1: Indoor work places*. European Committee for Standardization.

CEN. (2021b). *EN 54 series - Fire detection and fire alarm systems*. European Committee for Standardization.

Clark, C., & Harris, B. (2024). *Valve Regulated Lead-Acid Battery Degradation Model for Industry Applications*. Springer.

European Commission. (2024). *Climate-friendly alternatives to F-gases -- Electrical equipment (SF6)*. Climate Action. https://climate.ec.europa.eu/eu-action/fluorinated-greenhouse-gases/climate-friendly-alternatives-f-gases/electrical-equipment-sf6_en

GE Grid Solutions. (2024) GE's g^3 - The SF_6 Insulating and Switching Gas. https://www.gevernova.com/grid-solutions/hvmv_equipment/catalog/sf6-free-hv-switchgear/

Guan, B., Cheng, Y., & Wu, X. (2025). A systematic data-based method for non-probabilistic convex modelling under dependent uncertainty variables. *Optimization and Engineering*, 26, 1–40.

IEC. (2000). *IEC 61000-4-29: Electromagnetic compatibility (EMC) -- Part 4-29: Testing and measurement techniques -- Voltage dips, short interruptions and voltage variations on DC input power port immunity tests*. International Electrotechnical Commission.

IEC. (2011a). *IEC 60076-1:2011 Power transformers - Part 1: General*. International Electrotechnical Commission.

IEC. (2011b). *IEC 60364-5-54:2011 Low-voltage electrical installations - Part 5-54: Selection and erection of electrical equipment - Earthing arrangements and protective conductors*. International Electrotechnical Commission.

IEC. (2011c). *IEC 61643-11:2011 Low-voltage surge protective devices - Part 11: Surge protective devices connected to low-voltage power systems - Requirements and test methods*. International Electrotechnical Commission.

IEC. (2016). *IEC 60909:2016 Short-circuit currents in three-phase a.c. systems*. International Electrotechnical Commission.

IEC. (2017a). *IEC 62271-1:2017 High-voltage switchgear and controlgear - Part 1: Common specifications for alternating current switchgear and controlgear*. International Electrotechnical Commission.

IEC. (2017b). *IEC 62561:2017 Lightning protection system components (LPSC)*. International Electrotechnical Commission.

IEC. (2018). *IEC 61727:2004+AMD1:2018 - Photovoltaic (PV) systems - Characteristics of the utility interface*. International Electrotechnical Commission.

IEC. (2019). *IEC 62386 series Digital addressable lighting interface (DALI)*. International Electrotechnical Commission.

IEC. (2020a). *IEC 60529:2020 Degrees of protection provided by enclosures (IP Code)*. International Electrotechnical Commission.

IEC. (2020b). *IEC 60598-2-22:2020 Luminaires - Part 2-22: Particular requirements - Luminaires for emergency lighting*. International Electrotechnical Commission.

IEC. (2020c). *IEC 60947-2:2020 Low-voltage switchgear and controlgear - Part 2: Circuit-breakers*. International Electrotechnical Commission.

IEC. (2021a). *IEC 60332 series Tests on electric and optical fibre cables under fire conditions*. International Electrotechnical Commission.

IEC. (2021b). *IEC 60502 series - Power cables with extruded insulation and their accessories for rated voltages from 1 kV (Um = 1,2 kV) up to 30 kV (Um = 36 kV)*. International Electrotechnical Commission.

IEC. (2021c). *IEC 60603-7 Connectors for electronic equipment - Part 7: Detail specification for 8-way, unshielded, free and fixed connectors.* International Electrotechnical Commission.

IEC. (2021d). *IEC 60947-1:2021 Low-voltage switchgear and controlgear - Part 1: General rules.* International Electrotechnical Commission.

IEC. (2021e). *IEC 60947-6-1: Low-voltage switchgear and controlgear -- Part 6-1: Multiple function equipment -- Transfer switching equipment.* International Electrotechnical Commission.

IEC. (2021f). *IEC 60950-1 Information technology equipment - Safety - Part 1: General requirements.* International Electrotechnical Commission.

IEC. (2021g). *IEC 61000 series Electromagnetic compatibility (EMC).* International Electrotechnical Commission.

IEC. (2021h). *IEC 61140 Protection against electric shock - Common aspects for installation and equipment.* International Electrotechnical Commission.

IEC. (2021i). *IEC 61439-1:2021 Low-voltage switchgear and controlgear assemblies - Part 1: General rules.* International Electrotechnical Commission.

IEC. (2021j). *IEC 61439-3:2021 Low-voltage switchgear and controlgear assemblies - Part 3: Distribution boards intended to be operated by ordinary persons (DBO).* International Electrotechnical Commission.

IEC. (2021k). *IEC 62271-1:2021 High-voltage switchgear and controlgear - Part 1: Common specifications for alternating current switchgear and controlgear.* International Electrotechnical Commission.

IEC. (2021l). *IEC 62271-100:2021 High-voltage switchgear and controlgear - Part 100: Alternating-current circuit-breakers.* International Electrotechnical Commission.

IEC. (2021m). *IEC 62271-200:2021 High-voltage switchgear and controlgear - Part 200: AC metal-enclosed switchgear and controlgear for rated voltages above 1 kV and up to and including 52 kV.* International Electrotechnical Commission.

IEC. (2021n). *IEC 62368-1 Audio/video, information and communication technology equipment - Part 1: Safety requirements.* International Electrotechnical Commission.

IEC. (2023). *IEC 60287 series - Electric cables - Calculation of the current rating.* International Electrotechnical Commission.

IEC. (2024a). *IEC 60364-4-44:2024 Low-voltage electrical installations - Part 4-44: Protection for safety - Protection against voltage disturbances and electromagnetic disturbances.* International Electrotechnical Commission.

IEC. (2024b). *IEC 62305-1:2024 Protection against lightning - Part 1: General principles.* International Electrotechnical Commission.

IEC. (2024c). *IEC 62305-2:2024 Protection against lightning - Part 2: Risk management.* International Electrotechnical Commission.

IEC. (2024d). *IEC 62305-3:2024 Protection against lightning - Part 3: Physical damage to structures and life hazard.* International Electrotechnical Commission.

IEC. (2024e). *IEC 62305-4:2024 Protection against lightning - Part 4: Electrical and electronic systems within structures.* International Electrotechnical Commission.

IEEE. (2001). *IEEE Std 242-2001: Recommended Practice for Protection and Coordination of Industrial and Commercial Power Systems (Buff Book).* Institute of Electrical and Electronics Engineers.

IEEE. (2013). *IEEE Std 80-2013 IEEE Guide for Safety in AC Substation Grounding.* Institute of Electrical and Electronics Engineers.

IEEE. (2014). *IEEE Std 519-2014: IEEE Recommended Practice and Requirements for Harmonic Control in Electric Power Systems.* Institute of Electrical and Electronics Engineers.

IEEE. (2018a). *IEEE 1584:2018 IEEE Guide for Performing Arc-Flash Hazard Calculations.* Institute of Electrical and Electronics Engineers.

IEEE. (2018b). *IEEE Std 1547-2018: IEEE Standard for Interconnection and Interoperability of Distributed Energy Resources with Associated Electric Power Systems Interfaces*. Institute of Electrical and Electronics Engineers.

IEEE. (2018c). *IEEE Std 2030.5-2018: IEEE Standard for Smart Energy Profile Application Protocol.* Institute of Electrical and Electronics Engineers.

IEEE. (2020). *IEEE Std 946-2020: Recommended Practice for the Design of DC Power Systems for Stationary Applications.* Institute of Electrical and Electronics Engineers.

IEEE. (2021). *IEEE Std 835-2021 - IEEE Standard Power Cable Ampacity Tables.* Institute of Electrical and Electronics Engineers.

IEEE. (2022a). *IEEE 3004.1: Recommended Practice for Power Systems Protection Coordination.* IEEE Standards Association.

IEEE. (2022b). *IEEE 3005.4: Recommended Practice for Emergency and Standby Power Systems for Industrial and Commercial Applications.* IEEE Standards Association.

IEEE. (2022c). *IEEE 802.3 (2022) IEEE standard for Ethernet.* Institute of Electrical and Electronics Engineers.

IEEE. (2023). *IEEE Standard for Performance Requirements for Uninterruptible Power Systems (UPS). IEEE Std 62040-3.* Institute of Electrical and Electronics Engineers.

ISO. (2018). *ISO 8528-1:2018 Reciprocating internal combustion engine driven alternating current generating sets --- Part 1: Application, ratings and performance.* International Organization for Standardization.

Kamrani, A. K., & Abouel Nasr, E. (2010). *Collaborative Engineering: Theory and Practice.* Springer. https://link.springer.com/chapter/10.1007/978-0-387-95863-7_6

NFPA. (2022). *NFPA 110: Standard for Emergency and Standby Power Systems.* National Fire Protection Association.

NFPA. (2023). *NFPA 70: National Electrical Code.* National Fire Protection Association.

NFPA. (2024a). *NFPA 30: Flammable and Combustible Liquids Code, 2024 Edition.* National Fire Protection Association.

NFPA. (2024b). *NFPA 70E:2024 Standard for Electrical Safety in the Workplace.* National Fire Protection Association.

NFPA. (2025b). *NFPA 72: National Fire Alarm and Signaling Code, 2025 Edition.* National Fire Protection Association.

Petrovic, S. (2020). Lead–acid batteries. In *Battery Technology Crash Course.* Springer.

Roy, C., & Das, D. K. (2025). Class topper optimizer for cost-efficient smart grid operation under renewable energy uncertainties. *Energy Efficiency*, 18(5), 47.

Siemens Energy. (2023). SF6 is no longer the gas of choice. Here's what could replace it. https://www.siemens-energy.com/global/en/home/stories/alternatives-for-sf6.html

U.S. Department of Justice. (2010). *2010 ADA Standards for Accessible Design.* DOJ.

UL. (2019). *UL 827: Standard for Central-Station Alarm Services* (13th ed.). Underwriters Laboratories.

UL. (2023). *UL 1008: Standard for Transfer Switch Equipment.* Underwriters Laboratories.

4

Practical Checklists for Electrical Value Engineering and Design Assurance

4.1 Introduction—Purpose and Use of the Electrical Design Review Checklists

This chapter presents a series of comprehensive checklists developed to support the systematic application of value engineering (VE) principles across all critical components of electrical system design. However, these tools are not limited to VE workshops alone. They are equally suited for peer reviews, quality assurance (QA/QC), and design validation processes, making them essential instruments for project governance, design efficiency, and technical excellence.

The structure of these checklists is directly derived from Chapter 3 of this guidebook, which defines the evaluation domains, design considerations, and subsystem classifications relevant to electrical VE. Each checklist corresponds to a specific section or subsection in Chapter 3, following the same upstream-to-downstream order, including power systems, emergency power, ELV, ICT, and supporting infrastructure.

4.1.1 Checklist Objectives

These checklists are structured to guide engineers, designers, and project managers through the evaluation of both strategic and system-level decisions, with the following core objectives:

- Identify opportunities for cost reduction without compromising system reliability or safety
- Ensure full compliance with applicable codes and regulatory standards
- Promote design efficiency, construction readiness, and spatial coordination across disciplines
- Encourage innovation through modularity, prefabrication, and smart system integration
- Support sustainability via energy-efficient technologies and right-sized infrastructure
- Facilitate continuous improvement through documentation, feedback, and knowledge transfer

4.1.2 How to Use the Checklists

These checklists can be used:

- Sequentially: to guide comprehensive system reviews from early concept to construction
- Selectively: by focusing on a specific subsystem, topic area, or milestone

Value Engineering for Electrical Systems in Construction, First Edition. Hazem Hussein.
 Published 2026 by John Wiley & Sons Ltd.
Companion website: www.wiley.com/go/hazem

- Iteratively: revisiting items at 30%, 60%, and IFC design stages for alignment and update
- Collaboratively: with cross-disciplinary teams, including electrical, MEP, ICT, fire, and architectural stakeholders
- Independently: for technical audits, peer reviews, or contractor submittal evaluation

Each item within a checklist includes a clear design intent or VE criterion, along with a structured space for validation, remarks, and action follow-up—making it easy to integrate into QA workflows and project documentation.

4.1.3 Checklist Management and Customization

To ensure maximum project relevance and impact:

- Treat checklists as living documents to be updated as project scope or requirements evolve
- Customize and expand checklist items based on project-specific risks, technologies, or regulatory constraints
- Maintain a record of decisions and rationale to support traceability and team alignment
- Implement a feedback loop post-project to refine checklist effectiveness and share lessons learned
- Use the checklist outputs to enhance handover packages, commissioning protocols, and lifecycle asset management

4.1.4 Legend and Review Marking

Symbol	Meaning
✓	Compliant/addressed as per design criteria
✗	Noncompliant or missing/requires action
N/A	Not applicable to current system or scope
Remarks	Used to document design decisions, clarifications, or follow-up actions

By adhering to this structure, these checklists serve as powerful tools not only for VE implementation but also for ensuring that each project achieves technical completeness, interdisciplinary alignment, and design integrity from concept through commissioning.

4.1.5 Availability of Editable Soft Copies

For practical application, editable soft copies of all checklists presented in this chapter are made available as part of the supporting document package. These working files are formatted in standard spreadsheet and document formats, allowing project teams to directly incorporate them into design reviews, peer reviews, technical audits, and construction phase QA/QC processes.

The soft copies may be downloaded from: www.wiley.com/go/hazem

Users are encouraged to maintain proper version control when customizing or expanding these checklists to suit project-specific requirements.

4.1.6 Closing Remarks—Embedding Checklist Practice into Project Delivery

The systematic use of these checklists transforms VE from a one-time exercise into a continuous quality assurance and design improvement process. Project teams are encouraged to **integrate the relevant checklists into their internal workflows**, apply them across design phases, and adapt them to suit project-specific requirements, delivery models, and stakeholder roles.

By documenting observations, verifying compliance, and tracking follow-up actions, these tools not only **support better design outcomes** but also establish a clear audit trail for **technical decisions, coordination efforts, and VE interventions**. When used consistently, they help prevent scope gaps, reduce late-stage redesign, and improve the constructability and maintainability of electrical systems.

As electrical systems continue to evolve in complexity and integration, these checklists offer a scalable, adaptable method for achieving **design discipline, stakeholder alignment, and long-term performance assurance**. Their application should be iterative, collaborative, and continuously refined through real project experience.

4.1.7 Reference Standards and Applicability Note

The checklists presented in this chapter have been developed based on industry-standard engineering practices and structured to align with commonly adopted international and regional standards. While not intended as regulatory instruments, they reflect key principles found in the following references:

- **IEC 60364**: Low-voltage electrical installations
- **IEC 61537**: Cable tray and ladder systems
- **IEC 61439**: Low-voltage switchgear and controlgear assemblies
- **EN 50174 series**: Cabling installation standards for ICT and ELV systems
- **BS 7671**: Requirements for electrical installations (IET Wiring Regulations)
- **BS 5839-1**: Fire detection and fire alarm systems for buildings
- **NEN 1010**: Electrical installations in buildings (the Netherlands)
- **NFPA 70/NFPA 72**: National Electrical Code and Fire Alarm Code (the United States)
- **IEEE 446**: Recommended practice for emergency and standby power systems

These checklists are designed to support compliance and quality, but they must be applied in conjunction with the **specific statutory codes, employer's requirements, and technical specifications applicable to each project**. Users should ensure that all checklist applications are cross-referenced with the governing local regulations and authority requirements.

5

Proactive Value Engineering in Electrical Systems: Principles, Implementation, and Project Integration

5.1 Introduction

In electrical system design for construction projects, conventional value engineering (VE) is often deployed reactively—typically after the design is completed or cost overruns are identified. While this traditional approach can generate short-term cost reductions, it frequently overlooks opportunities for early design optimization, system integration, and innovation.

Proactive value engineering (PVE) introduces a fundamental paradigm shift. Rather than serving as a late-stage corrective mechanism, PVE embeds value optimization from the conceptual design phase onward. It emphasizes early stakeholder collaboration, functional analysis, and design modularity to achieve solutions that maximize long-term performance, cost-efficiency, and sustainability.

This chapter introduces the principles, methodologies, and implementation strategies of PVE within the context of electrical engineering for construction. It builds on the general VE framework presented in Chapter 1 and highlights how proactive integration of VE principles can lead to improved outcomes in both design and project execution. A comparative overview of PVE and traditional VE is presented in Table 5.1.

5.2 The Need for PVE

The increasing complexity of electrical system design—driven by rapid technological evolution, regulatory changes, and stringent sustainability targets—necessitates a strategic and anticipatory design approach. In this context, PVE serves not only as a process framework but also as a guiding mindset that integrates value optimization from the earliest design stages.

Unlike reactive VE, which is typically introduced after design finalization or cost escalation, PVE aligns with the evolving demands of modern construction by prioritizing:

- Lifecycle performance over short-term cost savings
- Adaptability to emerging technologies and smart infrastructure
- Compliance with sustainability and energy efficiency objectives
- Resilience in power continuity and system redundancy

Value Engineering for Electrical Systems in Construction, First Edition. Hazem Hussein.

Companion website: www.wiley.com/go/hazem

Table 5.1 Comparison Between Traditional and PVE

Aspect	Traditional VE	PVE
Timing	Applied during cost-cutting or postdesign phase	Integrated from concept development
Focus	Primarily cost reduction	Value maximization: performance, sustainability, and lifecycle cost
Stakeholder Involvement	Often limited to engineers and cost consultants	Inclusive: designers, clients, operations, and sustainability advisors
Design Flexibility	Limited, due to fixed designs	High, alternatives evaluated early
Risk of Rework	High, due to late changes	Low, due to early alignment and scenario testing
Use of Reference Designs	Typically overlooked	Prominently used for benchmarking, standardization, and speed

This approach directly supports global industry trends, including carbon reduction commitments, digital construction methodologies (e.g., BIM integration), and client-driven expectations for operational flexibility.

By embedding flexibility, efficiency, and sustainability from the outset, PVE ensures electrical systems are not only cost-effective but also future-ready and aligned with long-term performance goals (Table 5.2).

Table 5.2 PVE Timeline in Electrical Projects

Project Phase	VE Opportunity (% of Total Impact)	Potential Cost Savings (% of MEP Value)	Relative Implementation Cost (Baseline = Concept Phase)
Concept Design	100	15–25	1.0 × baseline
Design Development	60	8–15	1.5 × baseline
Construction	25	3–8	3.0 × baseline
Operational Phase	10	1–3	5.0 × baseline

Source: Adapted from Amuda-Yusuf et al. (2015), SAVE International guidelines, and ISO 21500/56002 principles on project value delivery. Values are indicative and reflect typical patterns observed in MEP-heavy infrastructure projects.

Notes:

- VE Opportunity reflects the proportion of total achievable value engineering benefit that can be realized in each phase.
- Potential Savings are indicative ranges based on electrical scope within total MEP (mechanical, electrical, plumbing) packages.
- Implementation Cost refers to the relative effort, time, and financial resources required to apply VE strategies at each stage. Late-stage VE incurs higher costs due to rework, procurement constraints, and construction limitations.

5.3 Potential Cost Savings and Implementation Costs

PVE, when applied from the early stages of electrical system design, offers substantial financial benefits. Industry benchmarks indicate that effective VE implementation can yield cost savings between 5% and 15% of the total cost of building services installations (Amuda-Yusuf et al., 2015), without compromising technical performance, system reliability, or functional requirements. (Table 5.3).

These savings arise from early identification of value alternatives, elimination of over-specification, and optimized component selection. In contrast to its financial impact, the cost of implementing VE is relatively modest, typically ranging between **0.5% and 1%** of total building services cost (Amuda-Yusuf et al., 2015). This investment enables significant returns through reduced capital expenditure, minimized rework, and enhanced design coordination (Figure 5.1).

Table 5.3 Summary of VE Cost-Benefit Metrics in Electrical Design

VE Metric	Industry Range	Remarks
Estimated Cost Savings from VE	10–25%	Dependent on project complexity and early intervention
Typical Cost of VE Implementation	0.5–1.0% of MEP cost	Includes team workshops, redesign effort, analysis
Cost-Benefit Ratio (Typical)	10:1–25:1	Indicates significant return on VE investment

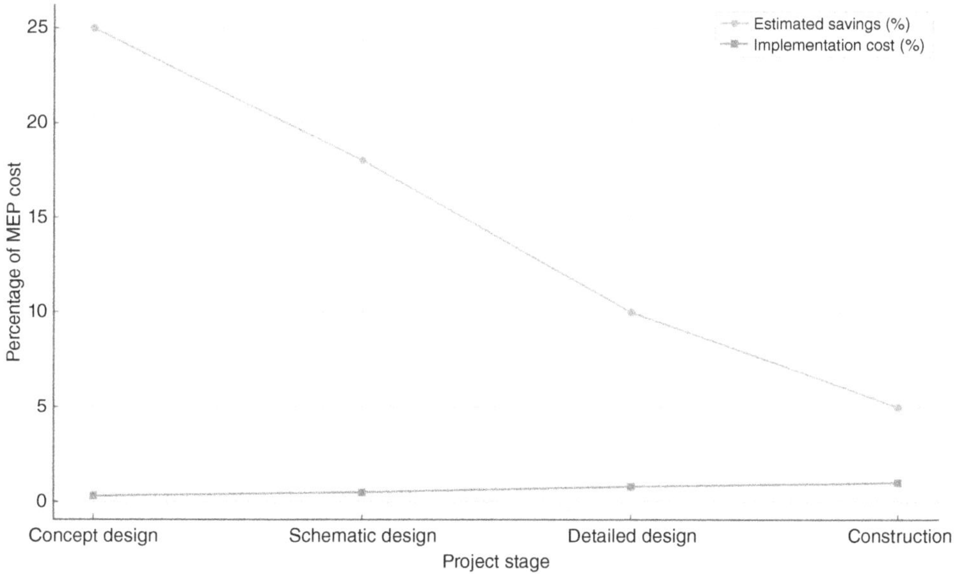

Figure 5.1 VE Cost vs. Saving Across Project Lifecycle

Beyond direct financial benefits, the PVE process also enhances qualitative project outcomes. These include improved interdisciplinary communication, more informed stakeholder engagement, and a greater emphasis on innovation during design development.

5.4 Causes of Unnecessary Cost in Electrical Services

In the integration of PVE into electrical systems design, it is essential to understand that unnecessary costs often arise not from major design flaws but from a combination of systemic inefficiencies, traditional practices, and limited collaboration. The identification of these factors is the first step toward achieving value-driven solutions. The main causes of unnecessary cost include the following:

- **Lack of Innovation**: Opting for the first workable solution without exploring more cost-effective or innovative alternatives.
- **Insufficient Information**: Decisions made with incomplete or outdated information can lead to inefficiencies and higher costs.
- **Rigid Adherence to Tradition**: Engineers and designers might stick to familiar methods without considering newer, more efficient options.
- **Habitual Practices**: Old habits and traditional practices can hinder the adoption of more efficient, modern solutions.
- **Resistance to Consultation**: A reluctance to seek or accept advice from other experts can prevent the identification of more value-driven solutions.
- **Outdated Specifications**: Specifications that haven't been updated to reflect current best practices or technologies can lead to over-engineering or the use of more expensive materials and methods than necessary.
- **Poor Communication**: Ineffective communication among team members can result in missed opportunities for cost savings or the selection of suboptimal solutions (Amuda-Yusuf et al., 2015).

To overcome these challenges, the following solutions can be implemented:

- **Foster a Culture of Continuous Improvement**: Encourage a mindset where innovation and efficiency are prioritized, and traditional methods are regularly reevaluated.
- **Improve Information Sharing**: Utilize digital tools and platforms to ensure that all team members have access to the latest, most accurate information.
- **Promote Interdisciplinary Collaboration**: Encourage teams to work closely together, sharing knowledge and perspectives that can lead to more innovative solutions.
- **Emphasize Training and Education**: Regular training sessions can help update team members on the latest technologies, methods, and best practices.
- **Adopt Modern Specifications and Standards**: Regularly review and update specifications to ensure they reflect current best practices and allow for the use of new materials and methods.
- **Improve Communication Channels:** Implement tools and practices that enhance communication among team members, ensuring that ideas and concerns can be shared freely and effectively (Amuda-Yusuf et al., 2015).

5.5 Strategies for Implementing PVE in Electrical Design

Implementing PVE within electrical design demands strategic integration of value-focused techniques from project inception to detailed design. While the principles of early engagement and iterative analysis are broadly applicable across disciplines, their application to electrical systems must address the unique complexities of power distribution, redundancy, safety, and lifecycle performance. The following strategies, when applied specifically to electrical engineering tasks, facilitate the delivery of robust, efficient, and scalable power solutions aligned with long-term project value.

a) Early Collaboration

Engage electrical engineers from the conceptual stage to shape space allocation, infrastructure hierarchy, and system topologies. Cross-disciplinary collaboration is critical to integrating core power needs with architectural and mechanical systems from the outset.

> **Example:** In a hospital project, early electrical input ensured that emergency switchboards were colocated with mechanical equipment rooms to reduce feeder lengths for life-safety systems, improving both voltage regulation and fault discrimination.

b) Real-Time Function-Cost Analysis

Electrical systems are composed of numerous functionally equivalent alternatives with vastly different cost, performance, and space profiles. Real-time function-cost analysis during design development enables informed decision-making on high-impact items.

> **Example:** During the selection of an Low voltage (LV) distribution strategy, comparing radial cabling versus plug-in busducts revealed that although busducts had a higher unit cost, they reduced installation labor and supported future load changes—yielding better lifecycle value.

c) Dynamic Brainstorming and Solution Generation

Facilitate structured workshops to ideate alternative approaches for key electrical systems. These sessions should challenge standard specifications and encourage cross-pollination of ideas among disciplines.

> **Example:** A brainstorming session led to the adoption of a centralized earthing bar system in lieu of multiple local earth rods, reducing conductor duplication and simplifying testing and maintenance across the facility.

d) Iterative Technical Evaluation

Electrical design evolves rapidly with technology trends and procurement realities. Iterative reviews ensure that new solutions—such as arc-resistant switchgear, smart metering panels, or high-efficiency transformers—are continuously evaluated and integrated if they offer value improvements.

> **Example:** A design initially specified conventional Motor control centers (MCCs) but was revised mid-design to include intelligent MCCs with embedded network monitoring after a comparative review showed a 20% reduction in commissioning and fault diagnostic time.

e) Client Feedback Loop

Ongoing dialogue with the client is essential to verify that electrical system decisions—such as backup durations, automation level, or metering capabilities—continue to reflect their operational priorities and financial expectations.

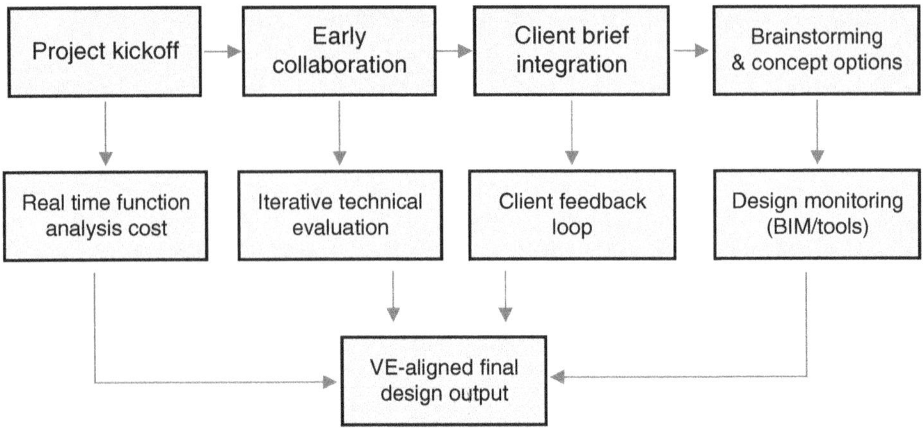

Figure 5.2 PVE Integration Flowchart in Electrical Design Workflow

> **Example:** Feedback from the facility operations team led to the inclusion of sectionalized busbar systems with load-break isolators in the main LV switchboard to support safe live maintenance, improving uptime without significant cost impact.

f) **Continuous Monitoring and Digital Validation**

Leverage digital design tools such as BIM and electrical analysis software to validate the impact of evolving design decisions in real time. This enables proactive identification of spatial conflicts, under/oversizing, and coordination issues.

> **Example:** BIM-based clash detection identified a cable ladder routing conflict with HVAC ductwork, which was resolved by vertically staggering the containment—a change that avoided both site rework and schedule delays.

Together, these strategies form the operational basis of a PVE methodology that is not only focused on design optimization but also promotes resilience, constructability, and lifecycle value (Figure 5.2).

5.6 Implementing PVE in Project Management

To fully realize the benefits of PVE, its principles must be embedded not only in the design process but also in the broader framework of project management. This requires an intentional shift in mindset—from focusing purely on scope and schedule delivery to systematically integrating value considerations throughout all phases of project execution.

Effective PVE implementation within the project management plan (PMP) can be achieved through the following structured practices:

a) **Integration of PVE from the Initiation Phase**: Define value drivers and performance goals during project setup. Include VE targets in the project charter and align them with client priorities, sustainability objectives, and long-term operational goals.
b) **Early Stakeholder Engagement**: Involve key internal and external stakeholders—consultants, end users, facilities management, and compliance authorities—from the early planning stages. Their input is critical for identifying functional needs and cost-effective solutions.

c) **Incorporation of Function-Cost Analysis into Planning Workflows**: Embed continuous function-cost analysis checkpoints into the project schedule. Ensure that system-level decisions are regularly evaluated for value optimization, especially during design freeze milestones.
d) **Allocation of VE-Focused Review Milestones**: Schedule formal PVE review workshops aligned with critical path activities (e.g., schematic design, detailed design submission). Assign responsible parties for tracking VE opportunities and documenting value outcomes.
e) **Alignment of Procurement Strategy with Value Objectives**: Include value-based selection criteria in tendering and procurement processes. Prioritize components, systems, and vendors that demonstrate technical performance, adaptability, and cost-effectiveness over the full asset life cycle.
f) **Establishment of a Feedback and Adaptation Loop**: Monitor value outcomes during implementation and commissioning phases. Capture feedback from site teams and end users to refine design assumptions, improve specifications, and inform future projects.

Projects that embed PVE into their project management framework consistently demonstrate improved performance. Research has documented significant benefits, including reduced lifecycle costs, shortened time-to-operation, improved user satisfaction, and enhanced design resilience (Rachwan et al., 2016). Studies indicate that systematic VE implementation can deliver substantial cost reductions, with studies showing savings typically ranging from 8% to 15% across various construction sectors (Chen et al., 2022). Case studies across various sectors, including educational facilities and infrastructure projects, have demonstrated significant cost reductions, with some projects achieving savings of 20–30% through systematic value engineering implementation (Rachwan et al., 2016; Youssef et al., 2023).

This integrated approach also aligns with established international standards, specifically ISO 21500 principles on project management guidance and ISO 56002 guidelines for innovation management systems (ISO, 2019, 2021), reinforcing PVE's role not only as a design tool but also as a strategic driver of engineering project success.

5.7 Challenges and Solutions

Adopting PVE is not without its challenges, from resistance to change within organizations to the complexities of integrating new technologies.

The adoption of PVE in electrical design and project delivery is not without challenges. While its benefits are increasingly evident, implementation can be hindered by organizational, technical, and procedural barriers. Recognizing these obstacles and establishing mechanisms to overcome them is essential for achieving consistent value outcomes.

The key challenges associated with implementing PVE, along with practical solutions (Figure 5.3)., are summarized as follows:

a) **Resistance to Change**: Many organizations remain rooted in traditional design workflows and VE methods that are applied late in the process. A shift to proactive integration requires changes in roles, timing, and responsibilities, which can meet cultural or managerial resistance.
b) **Mindset Inertia**: Teams accustomed to reactive decision-making and scope-based management may struggle to adopt a value-centric approach. This includes difficulty in transitioning from cost control to lifecycle performance thinking.
c) **Learning Curve for New Tools and Methods**: Implementing PVE often involves advanced software platforms (e.g., real-time collaboration tools, BIM-integrated VE dashboards) and systematic functional analysis methods. Teams may face an initial productivity dip due to unfamiliarity.

Figure 5.3 Common Barriers to PVE Adoption and Corresponding Interventions

d) **Compatibility with Existing Systems**: New design methods or VE frameworks may not easily align with legacy tools, specifications, or workflows. Integrating digital PVE practices into existing project structures can be technically complex and resource-intensive.
e) **Limited Access to Expert Input**: Projects may lack internal resources or technical leadership to evaluate alternative systems and facilitate early VE workshops, especially in smaller firms or low-budget projects.
f) **Inadequate Incentives**: Without a contractual or procedural driver, consultants may not prioritize early VE efforts. Deliverables are often judged on completeness rather than lifecycle value, which disincentivizes proactive optimization.

To address these challenges, the following solutions can be implemented:

i) **Training and Capacity-Building**: Invest in continuous learning programs to upskill staff on PVE concepts, digital collaboration tools, and function-cost analysis. Sharing real-world success stories helps teams understand the practical benefits of PVE.
ii) **Leadership Support and Vision Alignment**: Executive endorsement of PVE principles and alignment with organizational value goals fosters top-down support. Leaders must communicate that value optimization is a strategic priority.
iii) **Phased Implementation Approach**: Introduce PVE incrementally, focusing first on high-impact systems or stages (e.g., power distribution, space planning). Gradual adoption allows teams to build experience and confidence before full-scale deployment.

iv) **Engagement with External Expertise**: Bring in independent advisors or technical consultants with experience in PVE to guide workshops, design reviews, or lifecycle assessments. External support strengthens quality assurance and provides an objective perspective.
v) **Incentivization and Recognition**: Establish reward systems or evaluation criteria that recognize design teams for achieving value-based improvements, not just cost or compliance targets. This reinforces a culture of innovation and performance.
vi) **Structured Feedback Mechanisms**: Use post-project reviews and performance benchmarking to evaluate the effectiveness of implemented VE strategies. Lessons learned should be captured and disseminated to refine future practice (Amuda-Yusuf et al., 2015).

Together, these measures form a robust framework for embedding PVE as a sustained practice within engineering project teams. They support not only the technical deployment of value engineering tools but also the organizational and cultural shift necessary for long-term adoption (

5.8 Leveraging Reference Designs for Efficient Project Implementation

In modern engineering practice, the development of bespoke solutions for every project can be time-consuming, costly, and inconsistent. Leveraging reference designs—tested, validated, and repeatable design frameworks—offers a highly efficient pathway for delivering electrical systems that meet performance, regulatory, and operational standards while aligning with value engineering principles.

Reference designs provide a foundation for both new construction and upgrade projects. Their application contributes to PVE by reducing design time, minimizing errors, improving coordination, and streamlining procurement. These designs may encompass entire systems (e.g., data center electrical architecture), subsystems (e.g., medium-voltage distribution or emergency power), or room-based configurations (e.g., typical substations or lighting layouts for parking garages).

The following value drivers are associated with reference design adoption:

a) **Cost Savings**: Reusing standardized, proven designs reduces engineering hours, material cost uncertainty, and procurement inefficiencies. Established solutions help avoid overdesign and reduce contingency allowances, particularly for high-volume or repeatable project types.
b) **Risk Reduction**: Reference designs offer validated performance histories, minimizing the likelihood of unexpected design flaws or coordination issues. Lessons learned from previous applications inform enhancements and help avoid technical rework.
c) **Time Efficiency**: Reference designs allow projects to progress more rapidly through design and review stages, enabling early procurement, streamlined authority approvals, and faster transition to construction phases.
d) **Quality and Consistency**: Standard designs maintain a uniform technical standard across multiple projects or sites. This supports streamlined quality control and reduces ambiguity in execution. Detailed design packages (e.g., one-line diagrams, bills of materials) provide clarity and repeatability.
e) **Scalability and Modularity**: Well-structured reference designs can be adjusted for different load profiles or spatial constraints without starting from first principles. This is particularly useful for phased developments and system expansions.

f) **Knowledge Transfer**: Reference documents serve as valuable resources for onboarding new team members, training junior engineers, and aligning stakeholder understanding. They promote consistent application of engineering principles across distributed teams.
g) **Sustainability Alignment**: Optimized, pre-evaluated designs often support energy efficiency, material reduction, and integration of low-carbon technologies. Their reuse reduces design waste and accelerates compliance with green certification frameworks.

To be effective, reference designs should be well-documented and maintained through regular updates. They typically include two categories of content:

i) **Graphical Design Documents**: These include system one-line diagrams (for power, earthing, lighting, ICT, etc.), typical floor layouts for technical spaces (e.g., electrical rooms, corridors, control areas), and equipment arrangement drawings with spatial clearances and cable routing pathways. These visual elements help ensure constructability and alignment with site constraints.
ii) **Written Specifications and Design Reports**: These define system performance criteria, material selections, construction methods, and compliance references. Proper customization is essential to ensure they remain aligned with project-specific conditions, local codes, and client standards.

The use of reference designs does not eliminate the need for engineering judgment. Rather, it provides a structured starting point, allowing teams to focus engineering effort on areas of true complexity or uniqueness. To ensure optimal value delivery:

- Reference documents must be reviewed and adapted to suit project-specific scope, spatial conditions, and regulatory requirements.
- Blind copying must be avoided, as uncritical replication can lead to misalignment with project objectives or site limitations.
- All reference-based documentation must be verified for compliance, constructability, and coordination with other disciplines.
- Regular review cycles should be established to ensure reference content remains current with technological and regulatory developments.

When used correctly, reference designs enhance the PVE process by introducing speed, confidence, and repeatability into engineering delivery. They are particularly valuable in sectors such as mission-critical facilities, infrastructure, and institutional campuses—where design variation is limited, but reliability and standardization are essential (Figure 5.4).

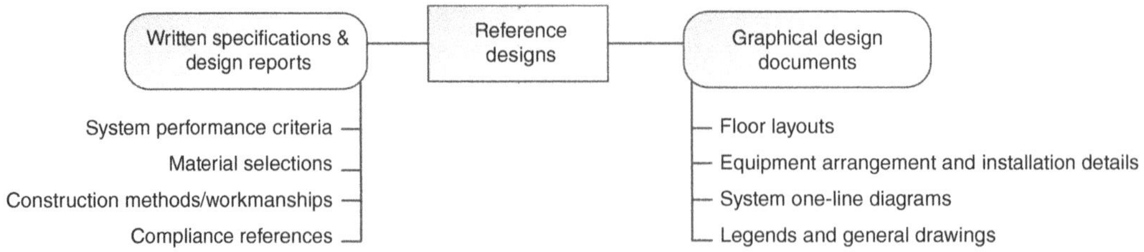

Figure 5.4 Typical Components of an Electrical Reference Design Package

5.9 Future of PVE in Electrical Design

PVE is poised to play an increasingly influential role in the future of electrical system design. As the construction industry evolves to meet the demands of digitalization, decarbonization, and system resilience, PVE provides a structured framework for integrating value-based decision-making throughout the project lifecycle.

Several key trends are expected to shape the future development and application of PVE in electrical engineering:

a) **Integration of Digital Design Intelligence**: Advancements in design automation, artificial intelligence (AI), and digital twins will support faster evaluation of design alternatives and real-time value analysis. Intelligent systems will enable designers to simulate multiple scenarios and dynamically optimize cost, performance, and sustainability outcomes.
b) **Expansion of Lifecycle Value Modeling**: PVE will increasingly extend beyond capital expenditure (CAPEX) optimization to include operational expenditure (OPEX), carbon footprint, and end-of-life considerations. Lifecycle performance metrics will be embedded into early-stage decision frameworks and procurement specifications.
c) **Standardization of Modular Reference Designs**: Growing adoption of prefabrication and modular construction methods will further encourage the development of scalable reference designs. These will support rapid deployment, reduce design cycle times, and ensure consistency across regional and international project portfolios.
d) **Enhanced Stakeholder Cocreation Platforms**: Emerging collaboration environments, powered by cloud-based tools and immersive technologies (e.g., VR/AR), will enable broader and earlier engagement of stakeholders in value-based design discussions. This democratization of design input will improve alignment and reduce change orders.
e) **Regulatory Emphasis on Value-Based Delivery**: Public procurement frameworks and green certification programs are increasingly incorporating value criteria into evaluation models. This will reinforce the need for PVE approaches that demonstrate lifecycle performance, user-centricity, and sustainability compliance.
f) **Integration with Circular Economy Principles**: Future PVE strategies will align with circular economy goals by emphasizing system adaptability, reuse of components, and reduction of environmental impact through low-carbon design practices and material optimization.

Looking forward, PVE will evolve from a project enhancement tool to a **strategic enabler of resilient, efficient, and sustainable infrastructure**. Its proactive, systems-oriented mindset aligns with global priorities such as the United Nations Sustainable Development Goals (SDGs), net-zero commitments, and digital transformation initiatives across the built environment.

To remain relevant and effective, engineering teams and project stakeholders must continuously adapt their processes, tools, and skillsets to align with this evolving value paradigm. Integrating PVE into education, procurement frameworks, and professional standards will be key to unlocking its full potential across the sector.

5.10 Case Study: PVE in Data Centre Power Distribution

5.10.1 Project Context

A Tier III enterprise data center was planned with an initial IT load of 1.2 MW and future scalability to 2.4 MW. Traditional design methodology would involve a fixed dual-path power architecture with a central UPS system and static mechanical layouts. However, the client sought to optimize capital investment and enhance adaptability to future technology changes.

5.10.2 PVE Application

From the concept stage, a multidisciplinary team adopted PVE strategies to evaluate multiple power distribution topologies. The process included real-time function-cost analysis, reference design benchmarking, and modular scalability assessment.

Key decisions made during PVE workshops included:

a) **Modular UPS Deployment**: Instead of a single centralized UPS, the team proposed distributed, scalable UPS units sized to match initial IT loads, reducing up-front capital costs by 22% and improving maintainability.
b) **Busbar Trunking System**: A plug-in busbar system replaced traditional cabling in technical spaces. Though slightly higher in material cost, it reduced installation time by 30% and allowed future IT cabinet additions without major rework.
c) **Space Reservation Strategy**: Equipment layouts were planned with future expansion zones, based on the reference design library from previous sites. This allowed deferred capital spending on switchgear and cabling while preserving upgrade flexibility.
d) **Integrated Maintenance Bypass**: Pre-engineered maintenance bypass panels were included to eliminate the need for temporary rerouting during future equipment replacements, improving uptime compliance and safety.

5.10.3 Outcome

By proactively applying VE principles, the project achieved the following results:

- 17% reduction in initial electrical capital cost
- Avoidance of $150,000 in projected upgrade rework
- Enhanced maintainability and lifecycle resilience
- Shorter commissioning timeline through reference-based layouts

The PVE-led approach also facilitated smoother coordination with mechanical and ICT disciplines, helping deliver a right-sized, scalable power infrastructure without compromising Tier III compliance or future growth capacity.

This example illustrates the measurable value of PVE when applied early in high-performance infrastructure projects, especially where design flexibility, lifecycle cost, and uptime are critical success factors.

References

Amuda-Yusuf, G., Sanni, A., & Olowa, T. (2015). Appraisal of application of value engineering methodology in mechanical and electrical services installations. *Jurnal Teknologi (Sciences & Engineering)*, 77(14), 85–93.

Chen, Z., Li, H., Ross, A., Khalfan, M. M. A., & Madden, P. (2022). A decade of value engineering in construction projects. *Advances in Civil Engineering*, 2022, 2324277.

ISO. (2019). *ISO 56002:2019 Innovation management — Innovation management system — Guidance.* International Organization for Standardization.

ISO. (2021). *ISO 21500:2021 Project, programme and portfolio management — Context and concepts.* International Organization for Standardization.

Rachwan, R., Abotaleb, I., & Elgazouli, M. (2016). The influence of value engineering and sustainability considerations on the project value. *Procedia Environmental Sciences*, 34, 431–438.

Youssef, M., AlDeep, S. M. H., & Olwan, M. M. (2023). Value engineering: Case study of Libyan educational buildings. *Alexandria Engineering Journal*, 14(8), 102106.

6

Practical Sheets and Templates for Effective Value Engineering of Electrical Works

6.1 Value Engineering Forms and Templates—Introduction

In this essential section of the book, we present a comprehensive collection of specialized forms and templates designed to systematize and enhance the electrical value engineering (VE) process. These carefully structured documents provide a framework that guides practitioners through each critical phase of VE—from project initiation through implementation, tracking, and capturing lessons learned. Each form represents a specific methodology developed through years of industry experience to ensure thorough analysis, clear documentation, and effective decision-making.

6.1.1 The VE Document System

The following 11 forms create an integrated system for managing the complete VE lifecycle:

1) **VE Project Initiation Form**

   This foundational document establishes the scope, objectives, and parameters of the VE study. It clearly defines what systems will be analyzed, identifies key stakeholders, outlines expected outcomes, and sets the timeline for the VE process. By formalizing these elements at the outset, the project team ensures alignment on goals and expectations before detailed analysis begins.

2) **Data Collection Sheet for VE Exercise**

   Comprehensive data gathering is essential for informed decision-making. This structured form ensures systematic collection of project specifications, design criteria, performance requirements, cost constraints, and other critical information. By organizing this data methodically, VE teams establish a solid factual foundation for their analysis.

3) **Electrical System Cost Breakdown Template**

   This detailed template disaggregates electrical system costs into their component parts, allowing for precise identification of high-cost elements that may present value improvement opportunities. By visualizing where project dollars are allocated, teams can strategically focus their VE efforts on areas with the greatest cost impact potential.

Value Engineering for Electrical Systems in Construction, First Edition. Hazem Hussein.

Companion website: www.wiley.com/go/hazem

4) **Function Analysis Worksheet**
   At the heart of the VE methodology, this worksheet facilitates the systematic analysis of project functions using verb-noun pairs and function classifications. This disciplined approach helps teams distinguish between essential primary functions and supporting secondary functions, revealing opportunities for cost reduction without compromising required performance.
5) **VE Idea Generation Log**
   This form documents the creative ideation phase, capturing alternative approaches generated through brainstorming and other creativity techniques. By recording ideas systematically with preliminary evaluation notes, teams establish a comprehensive inventory of potential solutions for detailed analysis.
6) **VE Idea Evaluation Matrix**
   This analytical tool enables objective comparison of VE proposals against multiple weighted criteria. The structured matrix format allows teams to quantitatively assess each alternative's impact on cost, performance, schedule, reliability, and other critical factors, facilitating data-driven selection of the most valuable solutions.
7) **Electrical Load and Capacity Analysis Form**
   This technical worksheet documents the detailed analysis of electrical demand requirements, system capacities, and utilization factors. It helps teams identify overdesigned elements and right-sizing opportunities while ensuring that all modified systems will still meet actual project needs reliably.
8) **Energy Efficiency Impact Assessment**
   This comprehensive assessment quantifies the operational impact of VE proposals on energy consumption and long-term costs. By calculating energy savings, utility incentives, payback periods, and lifecycle costs, this form helps justify decisions that might have higher initial costs but superior long-term value.
9) **VE Summary Report Template**
   This executive-level document compiles all findings and recommendations into a clear, comprehensive format for stakeholder review. It presents the methodology, analysis results, and implementation requirements in a structured manner that facilitates informed decision-making at all organizational levels.
10) **VE Implementation Tracking Log**
    This management tool monitors the execution of approved VE changes through design, procurement, and construction phases. By tracking implementation status, issues encountered, and actual savings achieved, this log ensures that projected benefits are realized and documented.
11) **VE Lessons Learned Register**
    This knowledge management document captures experience gained during the VE process for application to future projects. By systematically recording successful strategies, challenges encountered, and recommendations, organizations create a valuable resource that improves the efficiency and effectiveness of subsequent VE efforts.

6.1.2 Flexible Application Based on Project Needs

It is important to note that these forms are intentionally detailed and comprehensive to address complex VE exercises. However, not every project will require the use of all 11 forms. Practitioners should select and adapt these tools based on:

- Project size and complexity
- Available time and resources for the VE exercise
- Specific goals of the VE study
- Client requirements and expectations
- Project phase when VE is conducted

For smaller projects or targeted VE exercises, a streamlined approach using selected forms may be more appropriate. The key is to maintain the structured methodology while scaling the documentation to fit the specific context. Even when using a subset of these forms, maintaining consistent terminology and formats will help build organizational expertise over time.

6.1.3 Maximizing Value Through Structured Documentation

When used appropriately, these forms guide practitioners through each critical phase of the VE process while creating a comprehensive documentation trail. This structured approach offers several advantages:

- **Consistency**: Establishes a standardized methodology that can be applied across multiple projects
- **Thoroughness**: Ensures all critical aspects of VE analysis are addressed systematically
- **Clarity**: Improves communication among team members and stakeholders
- **Accountability**: Creates clear documentation of decisions and responsibilities
- **Knowledge Transfer**: Facilitates organizational learning and continuous improvement
- **Efficiency**: Reduces time spent developing analytical frameworks for each new project

6.1.4 Digital Access to VE Resources

To enhance your experience and accessibility, we have provided a link and QR codes to access a soft copy of these resources. This digital format allows for easy customization and integration into your specific project workflows. Simply scan the QR code below or visit our dedicated resource portal to download editable versions of all forms and templates presented in this section.

Resource Portal: www.wiley.com/go/hazem

The digital versions of these forms include:

- Editable file formats (Excel templates)
- PDF Format

While these forms provide powerful structure, they should be adapted as needed to address specific project requirements. The templates serve as a starting framework that can be modified based on project complexity, client preferences, and organizational standards. When properly implemented, this documentation system transforms VE from an ad hoc activity into a rigorous, replicable process that consistently delivers maximum value to electrical construction projects.

7

Applying the SAVE Methodology and Global Best Practices to Electrical Works

7.1 Introduction: Value Engineering in a Global Context

Value engineering (VE) is a globally adopted methodology that seeks to maximize project value by optimizing the relationship between function and cost (Kelly et al., 2015; SAVE International, 2020). While the core principles of VE—function analysis, structured creativity, evaluation, and implementation—remain consistent, the ways in which these principles are institutionalized, standardized, and applied vary significantly across different countries and sectors (CEN, 2020; Ellis and Woodward, 2001).

In the United States, the SAVE International's value methodology (VM) has become the dominant framework, particularly for infrastructure, public buildings, and federally funded projects (SAVE International, 2020). However, across the globe, other regions have developed frameworks tailored to their policy environments, industry structures, and performance goals. For example:

- Japan integrates VE with **kaizen** and advanced prefabrication methods (SJVE, 2017).
- Europe applies VE under EN 12973, emphasizing value management and lifecycle performance (CEN, 2020; Value for Europe, 2020).
- Singapore and the United Kingdom embed VE into government procurement, BIM processes, and sustainability frameworks (BCA, 2022; HM Treasury, 2018).
- Germany and Nordic countries stress energy performance and digital integration in VE application (NVMN, 2020; VDI, 2021).

Despite these variations, most countries align with the same foundational objectives: to deliver optimal function at minimal cost, while meeting performance, safety, and stakeholder requirements (Kelly et al., 2015; Woodward, 1997).

This chapter introduces the application of SAVE's VM to electrical works in construction. It outlines how the principles of function analysis, creativity, evaluation, and implementation can be adapted for practical use in electrical disciplines, complementing the VE strategies developed in earlier chapters.

The objective is not to replace the book's earlier framework, but to connect it to an internationally validated methodology, thereby equipping electrical professionals with tools recognized across industries, sectors, and geographies (SAVE International, 2020).

Value Engineering for Electrical Systems in Construction, First Edition. Hazem Hussein.

Companion website: www.wiley.com/go/hazem

Table 7.1 Relationship Between This Book's Framework and SAVE's VM

This Book's VE Framework	SAVE's VM
Specific to Electrical Works	Applicable to all engineering disciplines
Based on System Design and Coordination	Based on function analysis and structured phases
Emphasizes Design, Layout, Procurement	Emphasizes function, creativity, and evaluation
Field-Tested in Construction Contexts	Standardized and internationally recognized
Practical and Component-Level	Procedural and team-level

7.1.1 Positioning SAVE's Methodology Within This Book

In previous chapters, this book has introduced a practical framework for applying VE to electrical works in construction. That framework emphasized areas such as load estimation, system topology, modular design, spatial coordination, and component-level optimization—each grounded in project realities and electrical engineering practice.

The VM developed by SAVE International, in contrast, is a discipline-independent framework. It is widely adopted in sectors such as civil engineering, defense, healthcare, and manufacturing (SAVE International, 2020). Rather than focusing on specific systems, SAVE's methodology provides a universal, structured process for defining and optimizing value based on function (Miles, 1972).

SAVE's VM offers procedural depth and analytical structure, while the book's earlier VE framework provides domain-specific content, criteria, and techniques for electrical systems. These two are complementary (Table 7.1).

By understanding both perspectives, readers gain a more **holistic toolkit**. When projects require formal VE workshops, cross-discipline collaboration, or international certification, the SAVE Job Plan can be directly applied. For day-to-day value-driven engineering decisions, the electrical-focused strategies developed in prior chapters remain fully applicable.

◆ Note to the Reader

If you have already reviewed the VE methodology presented in Chapter 2, you will find both similarities and important distinctions here. Chapter 2 outlined a VE process tailored to electrical works, deeply rooted in engineering workflows, component selection, and cost optimization. In contrast, this chapter presents SAVE International's VM, a discipline-agnostic, structured framework used worldwide in formal VE studies. It reinforces the principles of function-driven evaluation, creative ideation, and transparent evaluation, and complements the hands-on techniques introduced earlier.

Suggested Roadmap for Chapter 7

This chapter introduces the VM as defined by SAVE International and explains how it can be adapted for electrical works in construction projects. To guide the reader through the structure, the chapter is organized in a logical sequence reflecting both process and practice:

- **Section 7.2** outlines the **adaptation of the SAVE methodology** to the realities of electrical systems, including technical baselining, function-led design, creative ideation, and evaluation.

- **Section 7.3** presents the **standard SAVE Job Plan**, customized step-by-step for electrical works, with practical examples at each phase.
- **Section 7.4** provides a **focused deep dive on function analysis**, expanding upon earlier sections and introducing tools such as the Function Analysis System Technique (FAST) diagram.
- **Subsequent sections (7.5–7.7)** explore individual VE techniques—**creative problem-solving, evaluation strategies, and implementation follow-through**—in greater detail, building on the job plan.
- **The chapter concludes (7.8)** with a synthesis of how structured VM enhances long-term performance, cost control, and innovation in electrical works.

Throughout this chapter, care is taken to avoid repetition by linking each section to its practical purpose in the VE process. Readers are encouraged to cross-reference with earlier chapters of the book, particularly Chapter 2, which addresses VE integration in electrical design phases.

7.1.2 How VE Is Applied Around the World

VE is a globally recognized methodology for achieving optimal value by balancing function, performance, and cost (Kelly et al., 2015; SAVE International, 2020). However, the way VE is institutionalized, standardized, and executed varies significantly across countries and regions. This diversity reflects different regulatory systems, professional cultures, public procurement models, and industry structures.

The following overview illustrates how VE is practiced internationally—highlighting the national authorities, frameworks, and distinctive characteristics in each region. This global context is not intended to replace the SAVE International methodology used throughout this chapter but to provide additional perspective for engineers working across borders or in multijurisdictional environments (Tables 7.2 and 7.3).

7.1.2.1 Interpretation and Strategic Insight

While SAVE's six-phase methodology offers a robust, codified approach used widely across North America and internationally (SAVE International, 2020), other regions bring valuable variations that reflect:

- Cultural preferences (e.g., kaizen in Japan) (SJVE, 2017).
- Procurement mechanisms (United Kingdom's public value mandates) (HM Treasury, 2018).
- Sustainability imperatives (Nordic LCA alignment) (NVMN, 2020).
- Urban efficiency models (Singapore's space and energy optimization) (BCA, 2022).

These variations suggest that while the principles of VE are globally convergent, the methodological packaging and institutional drivers differ (Kelly et al., 2015). This context allows engineers to adapt the SAVE VM where appropriate, while respecting local frameworks and regulations.

7.2 Adaptation to Electrical Works in the Context of VE

The successful application of VE in electrical works requires more than adopting generic tools or methodologies. It demands a careful alignment of VE principles with the technical, regulatory, spatial, and operational realities of electrical systems in construction projects (SAVE International,

Table 7.2 Global VE Frameworks by Region

Region/ Country	VE Authority	Core Standards or Method	Distinct Features
USA	SAVE, FHWA, GSA	SAVE VM, Circular A-131	Formalized workshops, VECPs in design and construction
UK	IVM, HM Treasury	Green Book, IVM BoK	Integrated with BIM, public procurement mandates
EU	CEN	EN 12973:2020	Lifecycle costing and sustainability embedded
Japan	SJVE	Kaizen-integrated VE	Continuous improvement, modular/ prefabricated systems
China	CECA, MOHURD	National VE Standards	Focused on efficiency and public sector cost control
Germany	VDI	VDI 2800	Emphasis on energy performance and digital integration
India	CPWD, NHAI, INVEST	Government-mandated VE	Common in infrastructure and (Public–Private Partnership) PPP optimization
Canada	PSPC, CSVA	TAC VE guidelines	Aligned with national codes and lifecycle design
Australia	AIQS, TfNSW	National VE frameworks	Sustainability, stakeholder inclusion
Singapore	BCA	Public VE mandate	Integrated with Green Mark scheme and urban infrastructure
Middle East	Ashghal, SCE, Others	Project-based standards	Infrastructure-focused, VE blended with sustainability
Nordic Countries	NVMN, Trafikverket	Circular economy–aligned VE	Strong Life Cycle Assessment (LCA)and lifecycle-based performance criteria
Russia	Minstroy, GOST R	Adapted SAVE/ISO-based frameworks	Cold-climate adaptations, standardization via state systems

Data sources for all countries referenced: BCA (2022), CECA (2020), CPWD (2018), CSVA (n.d.), HM Treasury (2018), INVEST India (n.d.), Minstroy (2020), SAVE International (2020), SJVE (2017), TAC (2021), Value for Europe (2020), VDI (2021).

2020). SAVE International's VM provides a structured foundation; however, its practical deployment must be contextualized to the nature and demands of electrical works.

Electrical systems are inherently interdependent. Power distribution, lighting, grounding, life safety systems, emergency power, and ICT/ELV infrastructure all interact with the architectural envelope, mechanical systems, and civil utilities. Each decision, from conductor sizing to switchgear placement, must be evaluated not only for technical adequacy but also for functional contribution, cost efficiency, maintainability, and interface coordination. Adapting the VM to this environment requires integrating its structured phases—particularly function analysis, creative ideation, evaluation, and implementation—with the engineering constraints and deliverables of construction projects.

The following subsections describe how each component of the SAVE VM can be tailored to electrical works in construction, maintaining fidelity to both engineering rigor and value optimization.

Table 7.3 Comparative VE Diagnostic Matrix by Country/Region

Question	US USA	GB UK	EU EU	JP Japan	CN China	SG Singapore	IN India	CA Canada	Others (Nordics, Russia, ME)
1. National VE body/standard?	YES (SAVE, FHWA, NIBS)	YES (IVM, HM Treasury)	YES (CEN, VDI)	YES (SJVE)	YES (CECA, MOHURD)	YES (BCA)	YES (CPWD, NHAI)	YES (CSVA, TAC)	YES (Minstroy, SCE, NVMN)
2. Defined methodology/job plan?	YES (SAVE 6-phase)	YES (Green Book, IVM BoK)	YES (EN 12973, VDI 2800)	YES (Kaizen-aligned)	YES (national stds)	YES (BCA VE phases)	YES (SAVE-based hybrid)	YES (TAC VE model)	YES (varies)
3. Function analysis & LCC central?	YES	YES	YES	YES	Partial	YES	YES	YES	Mixed
4. Integrated into public procurement?	YES	YES	YES	YES	YES	YES	YES	YES	Partial
5. Applied across project phases?	Design + VECPs	Full lifecycle	Design + LCA	Early to late	Mostly design	Design + commissioning	Design + construction	Full cycle	Growing trend
6. Certification/ training system?	YES (CVS®, AVS®)	YES (PVM/ TVM)	YES (EN, VDI)	YES (SJVE)	Emerging	YES (Gov support)	Expanding	YES	Partial
7. Common project types?	Infrastructure, public	Public, healthcare	Industry, infrastructure	Buildings, products	Infrastructure, housing	Urban/civil	Infra, metro	Buildings, infra	Infra, energy, defense
8. Terminology consistency?	VE	VM + VE	VM (EN), VE (practice)	VE (Kaizen-linked)	VE	VE	VE	VE	VE, VM—varies
9. Creativity techniques embedded?	YES (TRIZ, brainstorming)	YES	YES	YES (Kaizen)	Limited	YES	Partial	YES	Variable

This diagnostic matrix maps the degree to which VE is formalized, applied, and integrated across nine key questions. It shows that while frameworks vary, most countries emphasize structured value analysis, functional logic, and lifecycle orientation.

Data compiled from BCA (2022), CECA (2020), Ellis and Woodward (2001), HM Treasury (2018), Kelly et al. (2015), Minstroy (2020), SAVE International (2020), SJVE (2017), Value for Europe (2020), VDI (2021).

7.2.1 Function Analysis in Electrical Works

7.2.1.1 Defining and Structuring Electrical Functions

Every electrical system and component serves one or more primary functions, which must be explicitly defined to enable proper evaluation. Functions should be described using action-oriented, measurable terms, such as:

- Distribute power to floor-level panels
- Illuminate occupied zones within emergency egress timeframes
- Detect fault currents within 30 milliseconds
- Transmit high-integrity data to control nodes

By clearly articulating basic functions (essential for performance) and secondary functions (enhancing convenience or efficiency), the VE team can prioritize elements that deliver maximum value.

7.2.1.2 Identifying Functional Dependencies

Electrical functions rarely operate in isolation. Distribution equipment depends on upstream reliability, cable routing depends on containment capacity and spatial conflicts, and control systems depend on interoperability with BMS or fire systems. Early function mapping—often via tools such as a **FAST diagram**—reveals these interdependencies and highlights opportunities to consolidate or simplify system functions (Miles, 1972) Figure 7.1).

7.2.2 Information Gathering and Technical Baselines

7.2.2.1 Establishing Technical and Regulatory Inputs

A robust VE exercise begins with gathering all baseline data: load assessments, project standards, equipment schedules, spatial allocations, interface diagrams, and national/regional codes

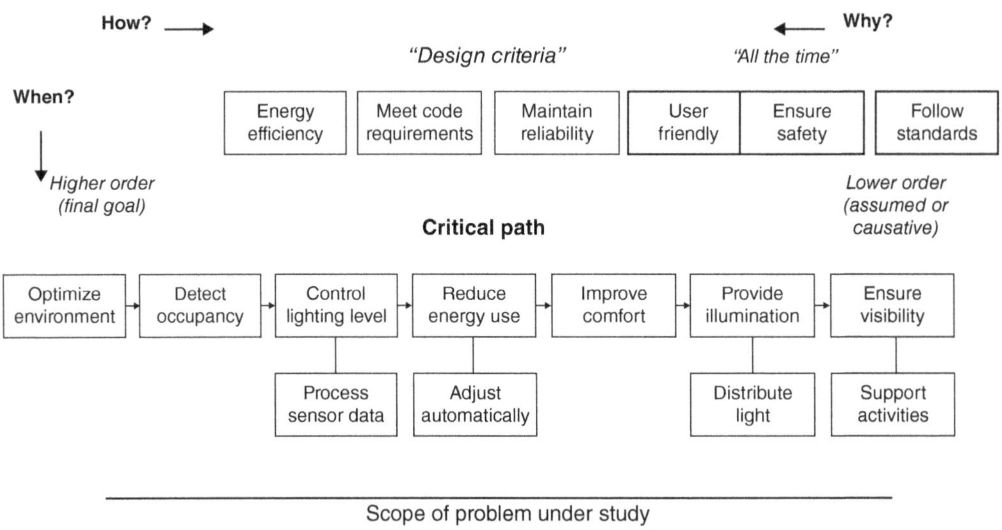

Figure 7.1 FAST Diagram for Lighting Control System

(e.g., IEC 60364, NEN standards, NEC, BS 7671) (IEC, 2015; NFPA, 2023). This foundation ensures that any proposed changes align with:

- Safety and performance criteria
- Legal and code compliance
- Project constraints (budget, phasing, and lead times)

7.2.2.2 Cost Structure Mapping

In electrical works, cost drivers include not only equipment but also installation labor, containment complexity, spatial constraints, and coordination with other trades. Cost mapping should identify:

- **High-cost components** with questionable functional contribution
- **Redundant systems** (e.g., duplicated metering or over-specified lighting control zones)
- **Lifecycle impacts**, such as maintenance burden or energy inefficiency

A sample **cost-function matrix** can help identify candidates for value improvement.

7.2.3 Creative Ideation for Electrical Systems

7.2.3.1 Facilitating Systematic Innovation

Using structured ideation techniques—such as brainstorming, morphological analysis, and function-based innovation prompts—the VE team can explore alternatives across several dimensions (Altshuller, 1996; SAVE International, 2020):

- Can fewer components serve multiple functions?
- Can prefabrication reduce cost or spatial demand?
- Can smart controls replace redundant infrastructure?

Creative sessions should be grounded in functional requirements, not just cost-cutting, to ensure that essential system integrity is preserved or improved.

7.2.3.2 Example Focus Areas for Ideation

- Lighting control consolidation using DALI or KNX
- Busbar vs. cable distribution for vertical risers
- Alternative containment routing to mitigate core congestion
- Hybrid AC/DC distribution for data-intensive facilities

7.2.4 Evaluation and Technical Feasibility Assessment

7.2.4.1 Structured Evaluation Criteria

Once alternatives are generated, they must be evaluated across multiple criteria:

- Functional performance
- Compliance with electrical standards
- Capital and operational cost
- Spatial feasibility
- Installation complexity
- Long-term maintainability

A **weighted evaluation matrix** helps stakeholders make transparent, justifiable decisions.

7.2.4.2 Example Evaluation Case

In a commercial facility, consider replacing copper with aluminum conductors:

- **Functional Impact**: Acceptable with derating
- **Cost Impact**: Lower material cost but increased termination complexity
- **Space**: Slightly larger containment required
- **Maintainability**: Requires trained personnel

This structured comparison prevents cost-driven decisions from compromising functionality or compliance (Table 7.4).

Table 7.4 Evaluation Matrix: Distribution System Alternatives

Criteria	Busbar Trunking	Cable Tray + Cables
Initial Cost	3	4
Space Efficiency	5	3
Ease of Installation	4	3
Maintainability	4	3
Compliance with Standards	5	5
Lifecycle Cost	5	4

Scoring Key (suggested for use in your documentation):

- **1 = Poor**
- **2 = Fair**
- **3 = Acceptable**
- **4 = Good**
- **5 = Excellent**

7.2.5 Implementation Planning and Integration

7.2.5.1 Actionable Engineering Strategy

Proposed VE options must be translated into tangible implementation actions:

- Revised schematics and schedules
- Updated bill of quantities (BOQs)
- Installation sequence modifications
- Coordination with other disciplines (e.g., HVAC, architecture)

This phase also identifies procurement and phasing impacts. For example, shifting from on-site cable termination to preassembled busduct segments may affect delivery timelines and access strategies.

7.2.5.2 Stakeholder Alignment

Clear communication is critical. The VE team should document:

- Functional justification for changes

- Expected value outcomes (cost, space, and performance)
- Compliance validation
- Change management plan

7.2.6 Documentation and Knowledge Transfer

The final output of the adapted VM process must include:

- A detailed VE report, including function diagrams, cost analyses, and implementation impacts.
- Lessons learned and VE strategies applicable to future phases or similar projects.
- Knowledge capture for internal engineering standards or design guides.

By closing the loop between analysis and implementation, VE becomes a continuous improvement process, not a one-time workshop.

Key Takeaways:

- Adapting SAVE's VM to electrical works requires grounding each phase in engineering context and constructability realities.
- Function-driven thinking, rather than component-driven selection, enables more efficient, compliant, and value-optimized solutions.
- Creative alternatives must be evaluated not only on cost but also on technical, spatial, and operational criteria.
- Successful VE outcomes are those that can be implemented, maintained, and justified across stakeholders.

While previous chapters focused on targeted VE strategies specific to electrical works—including load classification, modular distribution, procurement optimization, and coordination practices—this chapter introduces a structured process that can be used to **facilitate formal VE efforts** across disciplines.

The VM promoted by SAVE International is not a replacement for system-level engineering insight. Rather, it is a complementary framework that supports the **systematic execution** of value studies through function analysis, structured ideation, and transparent evaluation.

When applied together, the **practical depth of engineering-specific VE** and the **structured rigor of SAVE's methodology** form a comprehensive toolkit for maximizing the performance, constructability, and long-term value of electrical works in construction.

7.3 The Job Plan: Customized for Electrical Works

The **VM Job Plan**, as defined by SAVE International, provides a structured, phased framework for conducting VE studies (SAVE International, 2020). While the standard six-phase structure—information, function analysis, creative, evaluation, development, and implementation—can be applied across disciplines, its effectiveness in electrical works depends on proper adaptation to the technical, regulatory, and integration demands of construction projects.

This section reinterprets each job plan phase through the lens of electrical works, providing system-specific examples and identifying practical deliverables at each stage. It is intended for use by electrical engineers, project managers, and VE facilitators working within the built environment.

7.3.1 Information Phase

Objective: Establish a comprehensive understanding of the electrical scope, constraints, stakeholder expectations, and regulatory framework.

Key Activities:

- Review of current design schematics, single-line diagrams, load schedules, and containment layouts.
- Collection of design criteria, performance requirements, and redundancy classifications (e.g., tier ratings).
- Compilation of project-specific cost data (equipment, labor, and commissioning).
- Analysis of space allocations, interdiscipline coordination issues, and installation logistics.
- Identification of critical interfaces (e.g., between MV and LV systems, ICT and ELV systems).

Example: In a data center project, the VE team compiles power usage effectiveness (PUE) targets, redundancy goals (N + 1 or 2N), existing MEP coordination drawings, and UPS vendor proposals. This informs all subsequent phases of the study.

7.3.2 Function Analysis Phase

Objective: Deconstruct each system or component into its primary and secondary functions and assess its contribution to system value.

Key Activities:

- Identification of essential functions using verb–noun language (e.g., *Distribute power*, *Detect faults*, and *Isolate circuits*).
- Categorization of functions as basic (must be preserved), secondary, or supportive.
- Application of FAST diagrams to visualize interrelationships (Miles, 1972; SAVE International, 2020).
- Identification of functional overlaps, gaps, or redundancy across the system architecture.

Example: In analyzing a lighting control system, functions such as *Illuminate space*, *Adjust brightness*, and *Detect occupancy* are mapped. The team identifies that some functions are duplicated across multiple control layers (manual switches, sensors, and BMS), presenting an opportunity to streamline and consolidate.

7.3.3 Creative Phase

Objective: Generate alternative approaches to deliver identified functions more effectively, efficiently, or at lower lifecycle cost.

Key Activities:

- Facilitated brainstorming sessions (using methods like SCAMPER, mind mapping, or morphological analysis) (Altshuller, 1996).
- Generation of alternative designs, materials, layouts, or technologies.
- Exploration of prefabrication, modularization, or smart system integration.
- Engagement of cross-disciplinary expertise to foster innovation.

Example: The VE team proposes replacing traditional ladder tray and field-wired cabling with modular busbar trunking systems for vertical distribution in a hospital tower. The idea emerges from identifying congestion in core risers and high labor costs for traditional routing.

7.3.4 Evaluation Phase

Objective: Assess the technical, financial, and implementation viability of generated alternatives.

Key Activities:

- Technical feasibility analysis (load capacity, thermal performance, harmonics, and fault current ratings).
- Compliance checks against applicable codes (IEC, NEC, BS7671, local regulations)
- Spatial integration reviews (e.g., clash detection with BIM tools).
- Development of evaluation matrices (scoring alternatives against value drivers: cost, maintainability, redundancy, and installability).

Example: Three UPS configurations (centralized, decentralized, and distributed) are compared using a weighted evaluation matrix considering space, scalability, single point of failure, and capital cost. The decentralized approach scores highest in value and maintainability.

7.3.5 Development Phase

Objective: Convert selected alternatives into actionable engineering recommendations, ready for integration into design and procurement.

Key Activities:

- Detailed definition of chosen alternatives, including equipment specifications, layouts, and BOQs.
- Development of design modifications and revised schematics.
- Preparation of procurement and delivery strategy aligned with VE outcomes.
- Quantification of expected value gains (cost savings, space optimization, and energy reduction).

Example: Following the VE review, the team refines a proposal to downsize the sub-main cables by shifting to a ring-fed LV distribution system. They generate revised cable schedules, update containment drawings, and prepare a revised technical submittal for client review.

7.3.6 Implementation Phase

Objective: Ensure that approved VE recommendations are integrated into project execution and tracked for compliance and performance.

Key Activities:

- Coordination with the project team to incorporate changes into IFC (issued for construction) drawings.
- Adjustment of the construction program and procurement lead times as required.
- On-site QA/QC to validate execution against revised design.
- Documentation of implemented changes and performance benchmarks for postimplementation review (Table 7.5).

Table 7.5 Summary Table: VE Job Plan Adapted to Electrical Works

Job Plan Phase	Electrical Works Application
Information	Design review, load profile analysis, and regulatory compliance mapping
Function Analysis	FAST diagrams, function-cost mapping for components and systems
Creative	Generation of alternative topologies, materials, and control strategies
Evaluation	Scoring against compliance, cost, performance, and installation feasibility
Development	Engineering refinement, design modification, and procurement integration
Implementation	Execution tracking, QA/QC, and integration into project delivery and commissioning

Example: During construction of a utility substation, the VE-approved shift to aluminum busbars requires coordination with suppliers for specialized lugs and terminals. The project team updates the installation plan and includes additional QA steps for torque checks and contact surface preparation.

Key Takeaways

- The SAVE Job Plan, when tailored to electrical works, becomes a powerful process framework for structured value optimization (Figure 7.2).
- Each phase must translate function-based insights into practical engineering decisions, grounded in compliance, constructability, and stakeholder alignment.
- Success depends on cross-functional collaboration, clear documentation, and traceable implementation throughout the design and build process

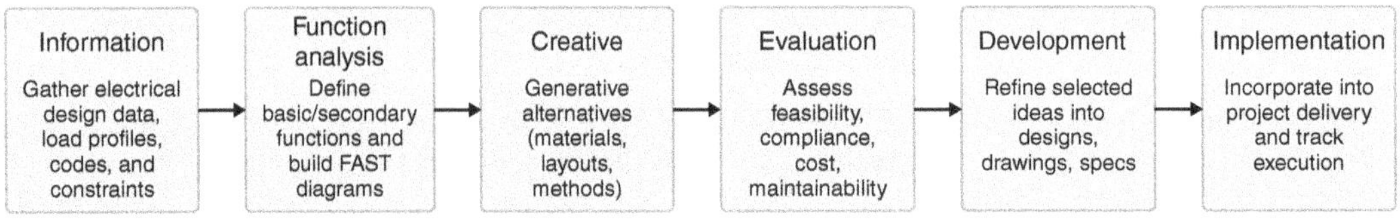

Figure 7.2 VM Job Plan for Electrical Works

7.4 Function Analysis in VE for Electrical Works

Function analysis lies at the core of the VM and serves as the intellectual foundation for identifying value improvement opportunities in electrical works. In VE, "function" refers to the specific action that a system, component, or process must perform to fulfill the project's objectives. Function analysis enables engineers to separate what a system must do from how it is currently designed to do it, opening the door for innovation, cost optimization, and simplification.

In electrical systems—where multiple layers of distribution, control, safety, and integration exist—function analysis is particularly powerful for identifying redundancy, overspecification, or inefficiencies. This section outlines how to perform function analysis effectively in the context of construction-related electrical works and introduces the use of the **FAST** diagram as a structured tool (Miles, 1972; SAVE International, 2020).

7.4.1 Purpose and Application of Function Analysis

Function analysis is used to:

- Define and clarify the essential and supporting functions of electrical systems
- Identify overlapping or redundant functions
- Enable comparative analysis of design alternatives based on their functional merit
- Serve as a foundation for brainstorming and creative phases
- Ensure that design decisions are **function-led**, not solely component- or specification-driven (Table 7.6)

Table 7.6 Types of Functions

Function Type	Description	Electrical Example
Basic Function	Essential purpose that must be preserved	*Distribute power* and *Detect faults*
Secondary Function	Adds convenience or performance enhancement	*Improve access* and *Enhance monitoring*
Supportive Function	Facilitates system operation but is not core	*Mount enclosure* and *Label circuits*

7.4.2 Practical Implementation of Function Analysis

7.4.2.1 Step 1: Identify Functional Scope

Begin by selecting a target system (e.g., LV switchgear, UPS, lighting control) and listing its major components. For each, ask:

- What does this item do?
- What value does that function provide?
- Is this function duplicated elsewhere?

7.4.2.2 Step 2: Use Verb–Noun Structure

Define functions using concise verb–noun pairs. For example:

- *Transmit signal* (for instrumentation cable)
- *Activate lighting* (for control switch)
- *Store energy* (for UPS battery system)

Avoid naming the component itself; focus on what it **does**, not what it **is**.

7.4.2.3 Step 3: Classify Functions

Use the earlier table to categorize each function. Functions that do not directly contribute to performance or safety may be candidates for elimination or consolidation.

7.4.3 FAST Diagram Application

The **FAST** diagram provides a visual map of function logic, answering two core questions:

- How is this function achieved? (→ to the left)
- Why is this function needed? (→ to the right)

FAST diagrams are particularly useful when evaluating:

- Multicomponent systems (e.g., lighting or emergency power)
- Control architectures (e.g., BMS integration)
- Layered distribution paths (e.g., MV–LV–final circuits)

7.4.3.1 Example: FAST Diagram—Lighting Control System

The diagram below illustrates a simplified FAST diagram for a lighting control system:

Detect Occupancy → Control Lighting Level → Reduce Energy Use → Improve Comfort → Provide Illumination

- **How?** is read left to right (each function supports the next)
- **Why?** is read right to left (each function exists to enable the next)
- *Refer to visual diagram provided in Section 7.2 for graphical representation.*

7.4.4 Integrating Function Analysis into the VE Process

Function analysis is not an isolated step. It feeds directly into:

- **Creative Ideation**: clarifies where innovation can target underperforming or overengineered functions
- **Evaluation Phase**: provides a functional baseline against which to score alternatives
- **Development and Implementation**: ensures that proposed modifications continue to fulfill required functions (Table 7.7)

In this example, function analysis may reveal that the generator has been oversized or that autonomy requirements could be met more efficiently with remote fuel support.

Key Takeaways:

- Function analysis enables objective, functional thinking in electrical system design and evaluation.
- FAST diagrams are a powerful tool for mapping and communicating function logic.

Table 7.7 Case Example: Emergency Generator System

Component	Basic Function	Secondary Function	Optimization Opportunity
Generator Set	*Supply backup power*	*Stabilize voltage*(Through integrated controls)	Evaluate sizing against actual load profile
Fuel Tank	*Store fuel*	*Enable runtime autonomy*	Analyze autonomy vs. refueling logistics
Control Panel	*Start generator*	*Display status*	Consider digital HMI integration

- By defining functions clearly, VE efforts become targeted, measurable, and nonredundant.
- Function analysis supports the transition from component-driven design to value-driven engineering.

7.5 Creative Problem-Solving in Electrical Works

The creative phase of the VM is designed to break free from conventional thinking and generate a wide range of viable alternatives for delivering essential functions (SAVE International, 2020). In electrical works, this phase is particularly crucial because design decisions are often constrained by standards, space limitations, predefined vendor solutions, and risk-averse specifications. Effective creative problem-solving within the VE process requires a structured approach—one that maintains technical feasibility while promoting innovation, cost reduction, and functional efficiency.

This section explores creativity techniques suitable for electrical engineering applications and offers guidance on how to structure ideation sessions, engage stakeholders, and translate creative insights into tangible design alternatives.

7.5.1 Purpose of the Creative Phase in Electrical VE

In electrical works, the goal of the creative phase is to:

- Discover new ways to fulfill defined system functions.
- Develop alternatives that offer improved value without compromising safety or compliance.
- **Challenge assumptions** embedded in existing specifications or design traditions.
- **Broaden the solution space** before prematurely committing to known methods.

Unlike informal brainstorming, this phase is part of a structured job plan, and creativity is driven by clearly defined function statements generated in the prior function analysis phase.

7.5.2 Techniques for Creative Idea Generation

The following creativity techniques are most relevant to VE efforts in electrical engineering contexts (Table 7.8).

Table 7.8 Techniques for Creative Idea Generation and Their Application in Electrical Works

Technique	Application in Electrical Works
Brainstorming	Rapid generation of system alternatives (e.g., different lighting controls, cable routing strategies)
Mind Mapping	Visual exploration of subsystem relationships (e.g., between distribution paths, containment, and spatial zones)
Morphological Analysis	Matrix-based generation of component combinations (e.g., busbar + smart breaker + centralized control)
SCAMPER (Eberle, 1977)	Structured prompts: Substitute, Combine, Adapt, Modify, Put to use, Eliminate, Reverse
TRIZ (Theory of Inventive Problem Solving) (Altshuller, 1996)	Systematic contradiction analysis (e.g., reduce fault risk without increasing redundancy cost)
Reverse Engineering	Start from functional output and ask: *What's the simplest, leanest way to achieve this?*

7.5.3 Structuring Creative Sessions

For creativity to be effective in engineering projects, it must be organized and purpose-driven. Consider the following structure:

1) Pre-Session Preparation
   - Distribute relevant function statements and evaluation criteria.
   - Provide baseline information: current design, known constraints, and performance targets.
   - Invite cross-disciplinary participants (e.g., electrical, mechanical, IT, construction).
2) Facilitated Ideation
   - Use a trained facilitator to guide the session through divergence (many ideas) and convergence (idea grouping).
   - Record all ideas without critique during initial brainstorming.
   - Use visual tools (e.g., whiteboards, digital pinboards, morphological tables).
3) Idea Capture and Categorization
   - Sort ideas into categories:
     - Immediately feasible
     - Requires development
     - Long-term possibilities
   - Flag ideas requiring technical simulation or supplier consultation.

7.5.4 Practical Examples of Creative Alternatives (Table 7.9)

Table 7.9 Practical Examples of Creative Alternatives

System	Standard Approach	Creative Alternative
Lighting Distribution	Conventional wiring with individual control zones	Use of DALI-based digital addressable control with sensor integration
UPS System	Centralized static UPS	Decentralized modular UPS units per critical load area
Busduct Routing	Overhead containment in congested ceilings	Underslab prefabricated containment with removable covers
Generator Fuel Storage	One large diesel tank adjacent to building	Smaller distributed tanks with integrated fuel polishing
Cable Sizing	Oversized for perceived "futureproofing"	Adjusted using actual diversity factors and monitored loads

7.5.5 Digital and Collaborative Tools

In modern VE practice, creativity is increasingly supported by **collaborative digital tools** (Table 7.10).

Table 7.10 Digital and Collaborative Tools and Their Function

Tool	Function
Virtual Whiteboarding Tools	Virtual whiteboards for mind mapping and brainstorming
Morphological Analysis Tools	Generate and filter combinations efficiently
BIM Tools with VE Overlays	Evaluate spatial impact of creative alternatives visually
Cost/Performance Simulators	Model impact of creative ideas on lifecycle outcomes

Key Takeaways:

- Creativity in VE is not random—it is targeted at improving function delivery and reducing unnecessary cost, complexity, or material use.
- Structured creativity sessions allow electrical professionals to unlock new possibilities that are technically feasible and economically justifiable.
- Techniques like brainstorming, TRIZ, and morphological analysis must be applied with clear functional intent and documented outcomes.
- The most effective creative ideas are those that can move quickly into the evaluation and development phases with realistic paths to implementation.

7.6 Evaluation and Development Strategies in VE for Electrical Works

Following the generation of creative alternatives, the VM advances into the evaluation and development phases (SAVE International, 2020). These stages involve critical review, comparative analysis, and refinement of shortlisted ideas to determine which provide the highest value when considering technical performance, cost, feasibility, and compliance.

In electrical works, these phases are particularly sensitive because safety, reliability, and regulatory compliance cannot be compromised. As such, the evaluation process must be both rigorous and multidisciplinary, ensuring all ideas are technically sound and practically implementable within the constraints of the construction project.

7.6.1 Strategic Evaluation of VE Alternatives

The objective of the evaluation phase is to assess each alternative based on a set of predefined, project-relevant criteria. This structured analysis supports informed, transparent, and auditable decision-making (Tables 7.11 and 7.12).

7.6.2 Qualitative and Quantitative Review Techniques

7.6.2.1 Qualitative Techniques

- **Expert Review Panels**: Electrical engineers, safety officers, and maintenance leads review alternatives for real-world applicability.
- **Functional Trade-Off Analysis**: Determine what is gained or lost when moving from one alternative to another.
- **Stakeholder Feedback Sessions**: Gain user/operator insights for operational reliability and maintainability.

Table 7.11 Evaluation Criteria for Electrical Works

Category	Key Considerations
Technical Feasibility	Load capacity, harmonics, thermal performance, selectivity
Safety and Compliance	Adherence to NEC, IEC, BS 7671, NEN, or other local codes
Cost Efficiency	Capital cost, installation cost, operational cost, lifecycle cost
Maintainability	Accessibility, spare parts, maintainability index (MTTR, MTBF)[a]
Constructability	Coordination with structure and other trades, modularization potential
Flexibility and Scalability	Capacity for expansion, adaptability to load changes or future phases

[a] MTTR (Mean Time to Repair): The average time required to diagnose, access, repair, and restore a failed component or system back to operational status. Lower MTTR values indicate faster repairability and reduced downtime. MTBF (Mean Time Between Failures): The predicted elapsed time between inherent failures of a system during operation. Higher MTBF values indicate greater system reliability and longer intervals between failures.

Table 7.12 Example: Evaluation Matrix for Generator Sizing Alternatives

Criteria	600 kVA Generator	2 × 300 kVA in Parallel
Initial Cost	3	4
Redundancy/Fault Tolerance	2	5
Fuel Efficiency (Load %)	4	3
Space Requirement	5	3
Ease of Maintenance	3	4
Compliance & Standards	5	5

Score: 1 = Poor; 5 = Excellent.

7.6.2.2 Quantitative Tools

- **Lifecycle Cost Analysis (LCCA)** Woodward (1997): Assess cost impact over expected system life.
- **Load Flow and Short-Circuit Studies**: Validate system performance using specialized electrical power system analysis software.
- **Constructability Modeling**: Use BIM to simulate physical routing and installation feasibility.
- **Sensitivity Analysis**: Evaluate how fluctuating variables (e.g., energy price, material cost) affect value outcomes.

7.6.3 Refinement and Development of Chosen Alternatives

Once the best alternative is identified, it must be **technically matured and documented** to a level that allows for design integration, procurement, and construction planning.

7.6.3.1 Development Deliverables

- Updated schematics and single-line diagrams
- Revised load schedules and circuit protection plans

- Modified equipment specifications and submittal documentation
- Adjusted construction schedules and phasing plans
- Risk register entries for VE-affected systems

7.6.3.2 Case Example: Lighting Redesign Using Digital Addressable Lighting Interface (DALI) Controls

- **Original Concept**: Standard on/off relay-based lighting circuits with occupancy switches.
- **VE Proposal**: Use DALI addressable system to reduce cabling, increase control flexibility.
- **Evaluation Outcome**: Higher initial cost offset by zone flexibility and lifecycle energy savings.
- **Development Outputs**:
  - New wiring schematics with DALI addressing
  - Control panel redesign with programmable gateway
  - Revised BOQs with fewer cores per circuit

7.6.4 Integration with Project Management and Procurement

As part of the development phase, VE-aligned alternatives must be embedded into the project ecosystem, including:

- **Procurement**: Align specifications with supplier offerings and delivery schedules
- **Cost Control**: Track projected vs. actual savings
- **Quality Assurance**: Update test procedures and commissioning plans to reflect new designs
- **Change Management**: Ensure updates are reflected in all disciplines' IFC packages and project records

7.6.5 Applying Pareto Analysis to Focus the VM

Within the VM, Pareto analysis serves as a strategic tool for prioritizing efforts during the information and evaluation phases. Based on the well-established Pareto principle (Pareto, 1896)—commonly known as the **80/20 rule**—this technique posits that a majority of project costs or risks are often concentrated in a small number of systems or components (Juran, 1951).

7.6.5.1 Purpose in VE

Pareto analysis helps teams:

- Identify the **few high-impact systems** that drive the majority of electrical CAPEX
- Target those systems for **function analysis, creative ideation, and evaluation**
- Avoid dispersing VE efforts on low-impact items with minimal value return

7.6.5.2 Application in Electrical Works

During VE workshops or project reviews, teams can generate **Pareto charts** from:

- Preliminary or detailed cost breakdowns
- System reliability/failure data
- Energy consumption or operational cost profiles

7.6.5.3 Example Insight

In a typical medium- to large-scale construction project, components such as emergency generators, UPS systems, MV switchgear, and transformers may represent over 60–70% of the electrical budget. Focusing the VE process on these items first ensures the highest return on analysis and redesign efforts (Table 7.13).

7.6.5.4 Benefits

- Enables **resource-efficient VE workshops**
- Facilitates **quantifiable targeting** of engineering time and creativity
- Increases **client confidence** by aligning VE efforts with tangible impact areas (

This chart illustrates a typical electrical CAPEX distribution across major systems in mid- to large-scale construction projects. It shows that a small number of systems (e.g., emergency generators, UPS, MV switchgear) account for the majority of electrical cost.

Table 7.13 Use Case in VM Phases

VM Phase	Pareto Application
Information	Identify top-cost or critical systems for analysis
Function Analysis	Focus effort on highest-value functions
Creative	Direct ideation toward top 3–5 components
Evaluation	Apply deeper technical scrutiny to top alternatives
Development	Prioritize refinement of high-value opportunities
Implementation	Schedule and manage rollout for largest savings first

The Pareto chart presented in this chapter (see Figure 7.3) illustrates how this prioritization can be visualized and acted upon within a structured VE process.

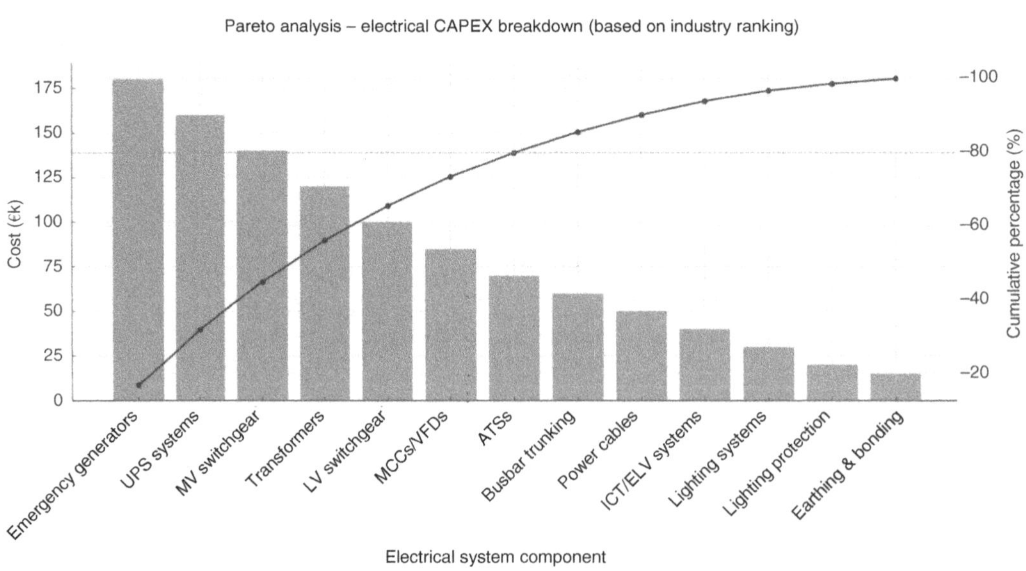

Figure 7.3 Pareto Analysis—Electrical CAPEX Breakdown (Representative Example)

Actual distributions may vary by project type, scope, and design strategy, but the principle of cost concentration remains consistent. This justifies the use of Pareto analysis to focus VE efforts on the most impactful systems.

Key Takeaways:

- Evaluation and development are not about discarding ideas—they are about refining the **most valuable ones into implementable, compliant, and cost-justified solutions**.
- Electrical works require a **multicriteria, cross-disciplinary evaluation strategy** to ensure no aspect of safety, performance, or coordination is overlooked.
- Structured tools such as evaluation matrices, LCCA, and constructability reviews bring transparency and rigor to decision-making.
- Development ensures that good ideas do not die in discussion—they are **engineered, documented, and delivered**.

7.7 Implementation and Follow-Up in VE for Electrical Works

The final phase of the VM **is implementation**—the point at which approved VE recommendations are formally adopted into the project and transitioned from concept into construction and operation. However, effective implementation in electrical works requires more than issuing revised drawings. It demands deliberate coordination, stakeholder alignment, technical validation, and postexecution review.

This section outlines how to successfully deliver VE changes in the field, ensure performance verification, and maintain a feedback loop for continuous improvement across projects.

7.7.1 Strategic Implementation in Electrical Projects

Implementation in electrical works often intersects with multiple systems, subcontractors, and trades. Therefore, execution must be methodical and documented to prevent disruption or scope misalignment (Table 7.14).

7.7.2 Practical Example: Cable Containment Optimization

VE Outcome: Shift from heavy-duty ladder tray to light-duty perforated tray in corridor ceilings due to lower cable density.

Table 7.14 Key Implementation Activities

Activity	Purpose
Update Construction Documents	Ensure all IFC drawings and BOQs reflect VE changes
Communicate Changes to Site Team	Prevent confusion and rework
Coordinate with Other Disciplines	Maintain inter-system compatibility (e.g., HVAC, ELV, civil)
Adjust Procurement Timelines	Accommodate lead times or vendor shifts
Revise Testing and Commissioning Plans	Ensure performance validation reflects new designs
Monitor Field Installation	Confirm compliance with revised specifications

Table 7.15 Recommended Follow-Up Actions

Action	Purpose
Conduct VE Performance Audit	Validate outcomes and document lessons
Update As-Built Documentation	Ensure O&M manuals, schematics, and logs reflect actual install
Host Lessons Learned Workshop	Capture feedback from all stakeholders
Integrate Into Design Standards	Apply successful VE strategies to future projects

Implementation Steps:

- BOQ updated to reflect reduced material quantity.
- Drawings revised for tray profile, spacing, and support details.
- Installer briefing conducted to explain changes and new handling guidelines.
- Inspection checklist modified to verify installation against revised specifications.

7.7.3 Follow-Up and Performance Monitoring

Implementation must be followed by a structured **postexecution review** to determine whether the VE solution delivered the intended value.

7.7.3.1 Postimplementation Metrics

- **Cost Savings Realized** (vs. projected)
- **Installation Time** (actual vs. planned)
- **Defect Reports** or commissioning issues arising from the VE change
- **System Performance** under operational loads
- **Client/User Feedback** on operability or maintainability (Table 7.15)

7.7.4 Ensuring Traceability and Accountability

To institutionalize VE practices, each implemented recommendation must be:

- **Traceable** back to its functional rationale
- **Documented** with its original function analysis, evaluation, and design decision trail
- **Auditable** as part of the project's QA/QC or postoccupancy review

Use a **VE implementation log** to track each change through its lifecycle:

VE Ref.	Description	Date Approved	Issued for Construction	Monitored Outcome	Verified Savings
VE-EL-003	Busbar trunking for riser A	March 15, 2025	Rev 3.2—March 18, 2025	Installed; QA passed	€14,800 CAPEX saved

Key Takeaways:

- Implementation is the most visible and risk-sensitive phase of the VE process; it must be executed with discipline and clarity.

- Success hinges on clear communication, integration into project control systems, and active site coordination.
- Follow-up ensures that value gains are measured, validated, and leveraged across the organization or future phases.
- A well-executed implementation process strengthens the credibility of VE and builds trust among clients, consultants, and contractors.

7.8 Conclusion: The Impact of VM on Electrical Works

The structured application of the VM, as developed by SAVE International, offers a transformative approach to achieving higher value outcomes in electrical works across construction projects. By focusing on function-led thinking, systematic analysis, and structured creativity, the methodology enables teams to optimize technical solutions while maintaining compliance, safety, and operational integrity.

This chapter has demonstrated how the SAVE VM process—through its six-phased job plan—can be effectively adapted to the specific needs and constraints of electrical systems. When applied correctly, the methodology bridges the gap between theoretical VE principles and the real-world complexity of power distribution, control systems, emergency power, and ELV infrastructure.

7.8.1 Promoting Function-Centric, Value-Led Engineering

At its core, the VM challenges engineers to ask:

- What is the function of this system or component?
- Is this the most efficient, cost-effective, and reliable way to achieve that function?
- What alternatives could achieve equal or better performance with reduced cost, complexity, or footprint?

This shift in perspective—from design-driven to function-driven—empowers teams to make strategic engineering decisions that are justified not only by specification, but also by demonstrable value.

7.8.2 Driving Creativity and Innovation

Through structured creativity techniques such as brainstorming, morphological analysis, and TRIZ, the VM creates a controlled environment where innovative alternatives can be explored without compromising technical rigor. This is particularly valuable in electrical works, where system design is often constrained by:

- Regulatory requirements
- Multitrade spatial conflicts
- Legacy design practices
- Risk aversion in critical systems

The creative phase of the VM ensures that innovation is not an afterthought, but an integrated part of the value-creation process.

Table 7.16 Summary of Strategic Benefits

Impact Area	VM Contribution
Cost Optimization	Function-cost alignment and lifecycle cost visibility
Space and Coordination	System simplification and routing alternatives
Performance and Compliance	Function preservation with standards alignment
Innovation	Structured ideation and alternative generation

7.8.3 Supporting Stakeholder Alignment and Transparency

The use of tools such as FAST diagrams, evaluation matrices, and implementation logs improves communication among stakeholders—engineers, designers, clients, contractors, and facility managers—by making trade-offs and value decisions transparent and traceable.

Whether the objective is reducing capital expenditure, improving installation efficiency, or enhancing system maintainability, the methodology fosters cross-functional understanding and buy-in, particularly in multidisciplinary construction projects.

7.8.3.1 Enabling Continuous Improvement

VE is not a one-time exercise. The VM reinforces a continuous improvement mindset, where successful strategies are:

- Captured in lessons learned sessions
- Reflected in design standards
- Propagated into future projects

For organizations working across multiple construction programs, this leads to systematic uplift in design quality, cost control, and delivery efficiency (Table 7.16).

References

Altshuller, G. S. (1996). *The innovation algorithm: TRIZ, systematic innovation and technical creativity.* Technical Innovation Center.

Ashghal (Qatar Public Works Authority). (2019). *Project value engineering guidelines.* Ashghal.

Building and Construction Authority (BCA). (2022). Design for maintainability guidelines. https://www1.bca.gov.sg

Canadian Society of Value Analysis (CSVA). (n.d.). CSVA training and practice guidelines. https://www.scav-csva.org

CECA. (2020). *China engineering cost association: Technical guidance on value engineering.* CECA Press.

CEN. (2020). *EN 12973:2020 Value management.* European Committee for Standardization.

Central Public Works Department (CPWD). (2018). *Guidelines on value engineering for public works.* Government of India.

Eberle, B. (1977). *SCAMPER: Games for imagination development.* DOK Publishers.

Ellis, R. D., & Woodward, D. G. (2001). International applications of value engineering. *Journal of Construction Engineering and Management*, **127**(6), 447–456.

GSA (U.S. General Services Administration). (2020). *Public buildings service value engineering program guide*. GSA.
HM Treasury. (2018). *The green book: Central government guidance on appraisal and evaluation*. Her Majesty's Treasury.
IEC. (2015). *International electrotechnical vocabulary and IEC 60364 standards*. International Electrotechnical Commission.
Institution of Value Management (IVM). (2020). *Value management practice guide*. IVM.
INVEST India. (n.d.). Certified VE programs. http://www.investindia.org
Juran, J. M. (1951). *Quality control handbook*. McGraw-Hill.
Kelly, J., Male, S., & Graham, D. (2015). *Value management of construction projects* (2nd ed.). Wiley-Blackwell.
Miles, L. D. (1972). *Techniques of value analysis and engineering* (3rd ed.). McGraw-Hill.
Minstroy (Russian Ministry of Construction). (2020). *VE integration guidelines for national infrastructure projects*. State Construction Standards.
MOHURD (Ministry of Housing and Urban-Rural Development, China). (2019). *Technical guide for construction value engineering*. MOHURD.
NFPA. (2023). *NFPA 70: National electrical code*. National Fire Protection Association.
Pareto, V. (1896). *Cours d'économie politique*. F. Rouge.
Public Services and Procurement Canada (PSPC). (2021). *Value engineering in federal construction projects*. PSPC.
SAVE International. (2020). *Value methodology standard and body of knowledge (VMBoK)*. SAVE International.
SAVE International. (n.d.). About SAVE. https://www.value-eng.org
Society of Japanese Value Engineering (SJVE). (2017). *Value engineering learning brochure*. SJVE.
TAC (Transportation Association of Canada). (2021). *Value engineering guidelines for infrastructure projects*. TAC.
Value for Europe. (2020). European training and certification system in value management. https://www.valueforeurope.com
VDI. (2021). *VDI 2800 – Value analysis series*. Verein Deutscher Ingenieure.
Woodward, D. G. (1997). Life cycle costing—Theory, information acquisition and application. *International Journal of Project Management*, **15**(6), 335–344.

8

Digital and Technological Drivers of Value Engineering in Electrical Systems

8.1 Introduction

The design and delivery of electrical systems are undergoing a profound transformation. As the demands on buildings and infrastructure evolve—driven by electrification, decarbonization, digitalization, and increased system complexity—traditional design methodologies have reached their limits. In this context, value engineering (VE) must also evolve, extending beyond cost reduction to encompass a broader mandate: delivering functional, resilient, and future-proof electrical solutions.

Today's engineering practice is shaped by technologies that were largely absent from earlier generations of VE analysis. Digital tools such as Building Information Modeling (BIM) now underpin multi-disciplinary coordination and parametric design evaluation. Data-driven platforms including Artificial Intelligence (AI), Internet of Things (IoT) networks, and digital twins are reshaping system operation, monitoring, and control. Simultaneously, advancements in materials science—from high-conductivity nanomaterials to next-generation photovoltaics (PVs)—are redefining the physical and performance limits of electrical infrastructure.

For engineers and VE practitioners, these developments open new avenues for system optimization. Intelligent modeling environments enable the early comparison of design alternatives with quantifiable impacts on energy use, reliability, spatial requirements, and life-cycle cost. Embedded analytics facilitate predictive maintenance and operational tuning. Emerging technologies introduce flexibility in deployment, greater integration with renewable energy sources, and potential reductions in material usage and embodied carbon.

This chapter sets out a structured approach to applying VE in this new digital and technological landscape. It examines the tools and trends shaping current practice, highlights practical use cases for VE integration, and addresses key implementation challenges—including digital interoperability, skills alignment, and data assurance. Throughout, the focus remains on supporting engineers in making informed, value-based decisions that align with project objectives, operational requirements, and long-term sustainability goals.

Ultimately, by embracing digitalization, VE becomes a proactive, performance-focused discipline—one that ensures electrical systems are not only economically optimized but also technologically resilient and future-ready.

Value Engineering for Electrical Systems in Construction, First Edition. Hazem Hussein.

Companion website: www.wiley.com/go/hazem

8.2 BIM-Enabled Workflows for VE

BIM represents a fundamental shift in the way electrical systems are designed, analyzed, and coordinated across the project life cycle. At its core, BIM is a collaborative digital process that enables the creation, management, and interrogation of three-dimensional models containing both geometric and nongeometric data. When strategically aligned with VE principles, BIM allows engineers to optimize system functionality, reduce waste, improve stakeholder communication, and enhance decision-making across every design phase.

Electrical engineers can now use BIM tools not only to produce schematic layouts and schedules but also to test design alternatives, simulate system performance, and verify compliance with spatial, functional, and regulatory requirements. The following BIM-driven capabilities support value-oriented electrical engineering.

8.2.1 Clash Detection and Multidisciplinary Coordination

One of the most critical uses of BIM is automated clash detection. This feature identifies physical interferences between electrical components (e.g., cable trays, distribution boards) and other building systems, such as mechanical ductwork or structural beams. Tools such as Autodesk® Navisworks®, Revizto®, and Solibri™ Model Checker highlight these conflicts during early design stages, enabling preemptive resolution and avoiding costly on-site modifications.

VE Impact: Detecting and resolving clashes before construction reduces waste, accelerates timelines, and ensures that space allocation for electrical systems is functionally optimized.

8.2.2 Load Calculations and Energy Modeling

Modern BIM platforms integrate electrical load analysis and energy simulation tools. Engineers can model real-time power demand, simulate lighting and HVAC loads, and optimize cable and switchgear sizing. Software such as IES VE or Revit Systems Analysis enables fine-tuning of power distribution layouts in response to operational or regulatory requirements.

VE Impact: Enables early-stage evaluation of multiple design scenarios to balance performance, reliability, and cost—core to life-cycle optimization.

8.2.3 Lighting Design and Photometric Integration

BIM environments support integration of photometric data (IES/LDT files), allowing lighting simulations to evaluate illuminance levels, glare indices, and spatial distribution patterns. Designers can assess lighting layouts against international workplace standards such as EN 12464-1 or WELL Building criteria.

VE Impact: Optimizes luminaire placement and specification to meet functional requirements with fewer components—reducing material, energy use, and maintenance.

8.2.4 Automated Documentation and Drawing Production

Electrical BIM models automatically generate coordinated single-line diagrams, schedules, cable lists, containment layouts, and power distribution schematics. Any model changes update associated documentation, ensuring version control and reducing manual drafting errors.

VE Impact: Enhances design clarity, speeds up approvals, and reduces rework by ensuring documentation is accurate and up to date throughout all design stages.

8.2.5 Embedded VE Tools and Analysis Methods

By integrating VE methodologies such as Function Analysis System Technique (FAST) and Life-Cycle Cost Analysis (LCCA) into BIM workflows, engineers can evaluate design alternatives based on function, cost, and environmental impact. For example, electrical spaces can be analyzed using LCCA to compare high-efficiency transformers with standard options across installation, energy use, and maintenance intervals.

VE Impact: Aligns digital design workflows with the formal VE process, enabling structured, defensible decision-making.

8.3 Intelligent and Adaptive Electrical Systems

The evolution of intelligent and adaptive electrical systems is reshaping how functionality, reliability, and efficiency are delivered and maintained throughout the building life cycle. These systems employ advanced technologies such as AI, Machine Learning (ML), the IoT, and digital twin modeling to create responsive environments capable of optimizing themselves based on real-time conditions and historical trends.

When integrated within VE frameworks, these technologies allow for the quantification and optimization of both capital and operational expenditures. Electrical infrastructure becomes not only a static provision of power but also an active participant in energy management, environmental control, and predictive asset maintenance.

8.3.1 Artificial Intelligence and Machine Learning

AI refers to the simulation of human decision-making processes by computational systems, while ML is a subset of AI that uses algorithms to learn from data patterns and make predictions or automate tasks (Chen et al., 2023). In the context of electrical systems, AI and ML are applied in areas such as:

- **Energy Optimization**: AI platforms can dynamically adjust lighting, HVAC loads, and distribution equipment settings to minimize energy consumption while maintaining service levels (Agarwal et al., 2024).
- **Fault Prediction and Condition Monitoring**: ML models trained on historical data can forecast component failures (e.g., in switchgear, transformers, or uninterruptible power supply [UPS] units), enabling proactive intervention and reducing downtime (Sharma et al., 2022).
- **Design Optimization**: AI-assisted design tools can generate and evaluate thousands of layout configurations, identifying options that meet functional and cost criteria under VE analysis.

VE Implications: AI/ML capabilities allow engineers to model long-term performance outcomes during early design stages, aligning proposals with both life-cycle cost goals and operational resilience strategies.

Figure 8.1 summarizes the comparative value impact of conventional and intelligent systems across capital expenditure, operational expenditure, and system resilience. Table 8.1 provides detailed numeric comparisons and VE insights for these dimensions.

Interpretation:

- **CAPEX** for intelligent systems is moderately higher (+10%) due to the inclusion of AI engines, IoT sensors, and digital twin platforms.

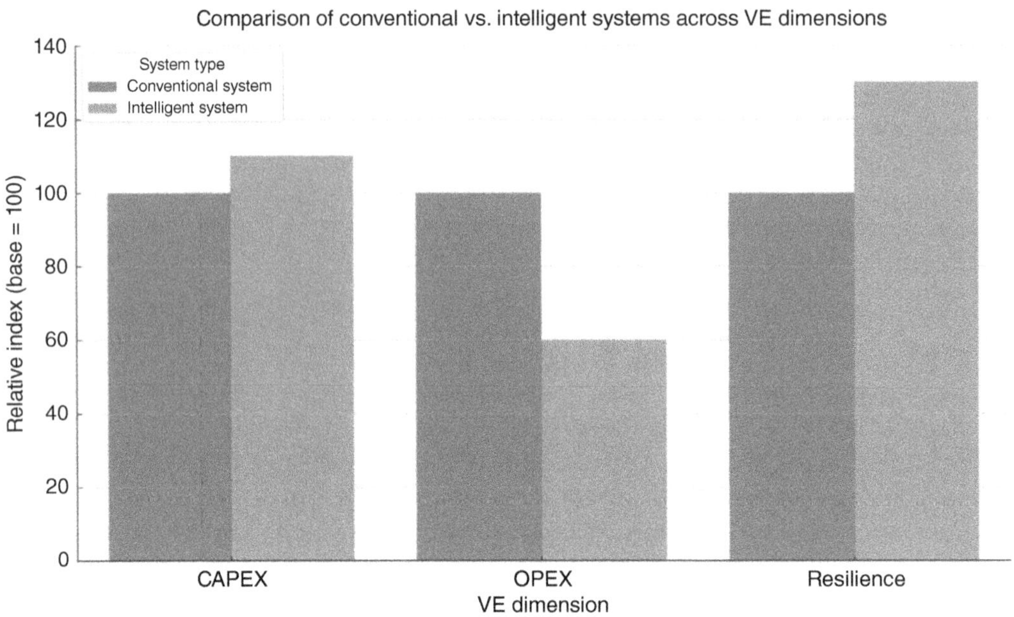

Figure 8.1 Comparison of Conventional vs. Intelligent Systems Across VE Dimensions: CAPEX, OPEX, and Resilience

Table 8.1 Comparison of Conventional vs. Intelligent Systems Across VE Dimensions

VE Dimension	Conventional System (Baseline = 100)	Intelligent System (Index)	VE Insight
CAPEX	100	110	Slightly higher due to added sensors, platforms, and analytics infrastructure.
OPEX	100	60	Significantly lower through automation, energy savings, and predictive maintenance.
Resilience	100	130	Substantially enhanced via real-time monitoring, self-healing systems, and fault tolerance.

- **OPEX** is reduced significantly (−40%) through predictive analytics, energy optimization, and reduced manual intervention.
- **System resilience** is markedly improved (+30%) via real-time diagnostics, automated failover protocols, and adaptive response capabilities.

8.3.2 Internet of Things

The IoT refers to a distributed network of interconnected sensors, devices, and systems that can monitor, transmit, and react to environmental and operational conditions (Kumar et al., 2024). Within electrical engineering, IoT implementation typically includes:

- **Smart meters and submetering** for real-time power usage data.
- **Environmental sensors** for temperature, occupancy, and air quality influencing lighting and power usage.

- **Automated control systems** such as dimmable ballasts, variable speed drives (VSDs), and networked distribution boards.

VE Implications: By integrating IoT, buildings become self-regulating environments where energy consumption, system loading, and power quality are monitored and adjusted continuously—enabling VE teams to include dynamic system efficiency and performance stability within their value analysis.

8.3.3 Digital Twins and Performance Analytics

Digital twins in the context of electrical are real-time, virtual replicas of physical electrical systems created using operational data, 3D models, and sensor feedback (Gharaibeh et al., 2024). These digital environments simulate asset behavior, allowing engineers to:

- Visualize equipment status, loading profiles, and fault conditions.
- Conduct scenario planning for load shedding, redundancy testing, or maintenance strategies.
- Benchmark system performance against baseline expectations and industry standards (Liu et al., 2023).

VE Implications: Digital twins serve as a tool for both proactive design evaluation and continuous commissioning. This supports ongoing VE activities post-handover, enabling system adjustments aligned with user behavior and operational priorities. The system architecture for intelligent electrical designs is illustrated in Figure 8.2, showing the layered integration of sensing, electrical distribution, and digital optimization elements.

Figure 8.2 Architecture of an Intelligent Electrical System Integrating Smart Sensors, Electrical Infrastructure, and Data-Driven Platforms for Performance Analysis and Optimization

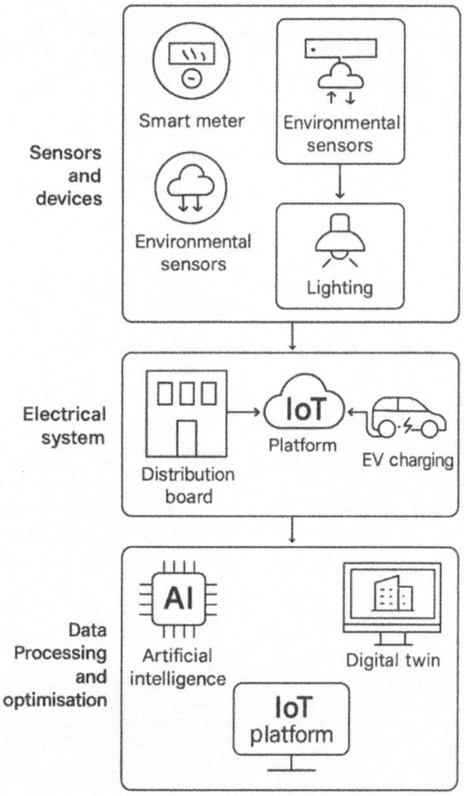

8.4 Electrification, Decarbonization, and Distributed Energy Integration

The global decarbonization agenda and the accelerating shift toward net-zero carbon targets are reshaping the design priorities of electrical systems in the built environment. Electrification—the replacement of fossil-fuel-based systems with electricity-driven technologies—has become a principal strategy in reducing carbon emissions across buildings, infrastructure, and industry. This is supported by technological advancements, policy mandates, and the increasing viability of distributed energy resources.

VE plays a crucial role in integrating electrification and decarbonization goals into the electrical design framework. The evaluation of cost, function, and long-term performance must now include emissions reduction, energy flexibility, and future energy market compatibility. This section examines the key vectors of transformation and their implications for VE decision-making.

8.4.1 Renewable Energy Systems and Building-Integrated Photovoltaics

The integration of renewable energy systems—particularly PV arrays—into electrical infrastructure enables buildings to become partial energy producers. Building-integrated photovoltaics (BIPV go further by embedding PV modules within architectural elements such as façades, canopies, or skylights (Shukla et al., 2017; Yang & Zou, 2016). BIPV offers dual-purpose functionality, reducing façade material requirements while generating on-site electricity.

VE Implications:

- Capital expenditure must be assessed against long-term energy yield and avoided grid costs.
- Life-cycle analysis can reveal favorable return on investment (ROI) when feed-in tariffs, carbon incentives, or building certification schemes (e.g., BREEAM, LEED) are considered.
- Integration constraints such as inverter selection, string configuration, and spatial orientation should be evaluated during early VE stages. Figure 8.3 illustrates the cumulative ROI over

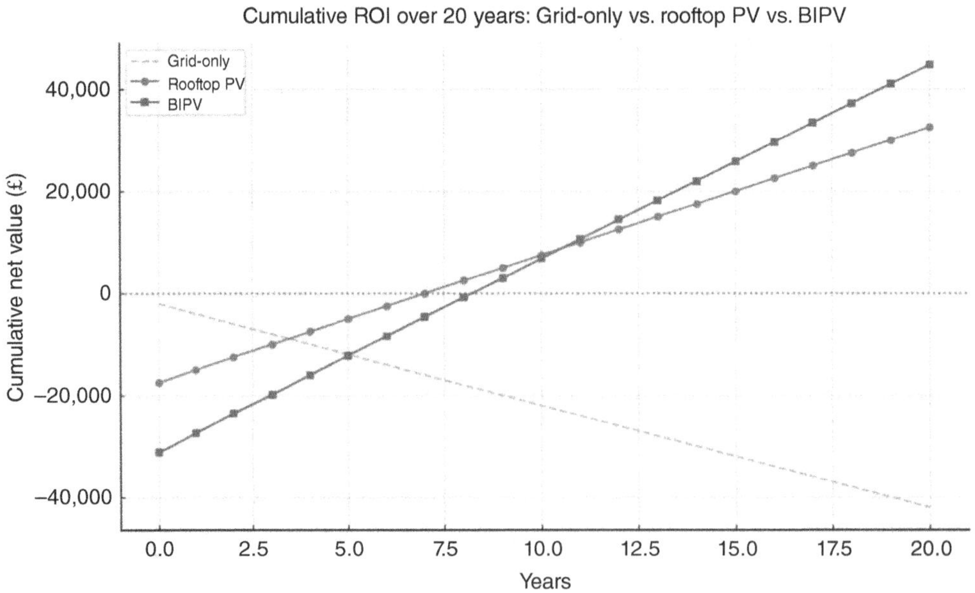

Figure 8.3 Cumulative ROI, Grid-only, Rooftop PV, and BIPV

Table 8.2 Cumulative ROI Over 20 Years: Grid-Only vs. Rooftop PV vs. BIPV[a]

Scenario	Initial Investment	Annual Net Savings	Break-Even Point	Key Insight
Grid-Only	£0	−£2000/year	N/A	High and continuous operating cost with no asset ownership
Rooftop PV	−£20,000	£2500/year	Year 8	Recoverable within mid-term horizon; moderate savings thereafter
BIPV	−£35,000	£3800/year	Year 10	Higher upfront cost but better long-term performance and dual architectural use

[a] The values used are for conceptual VE illustration only, not prescriptive modeling.

20 years for three scenarios: Grid-only, rooftop PV, and BIPV. Table 8.2 provides the underlying comparative data.

The chart above illustrates the **cumulative ROI** over 20 years for three scenarios: **grid-only**, **rooftop PV**, and **BIPV**.

Note: The comparative ROI curves presented in Figure 8.3 and Table 8.2 are based on simplified, illustrative financial models and should not be considered universally applicable. Actual performance and ROI for BIPV and rooftop PV systems are highly dependent on project-specific factors, including:

- Local solar irradiance and climate conditions
- System design, orientation, and tilt
- Building geometry and envelope availability
- Incentive structures, feed-in tariffs, and electricity tariffs
- CAPEX variation based on integration complexity and material costs
- Maintenance costs, inverter replacement cycles, and degradation rates

While BIPV may offer dual functionality (i.e., power generation and façade material substitution), this advantage must be weighed against higher initial design and integration costs. Engineers should conduct detailed LCCA and energy modeling for each project to determine the optimal PV solution.

8.4.2 Electric Vehicle Charging Infrastructure

As electric vehicles (EVs) become mainstream, buildings—particularly residential complexes, commercial sites, and transport hubs—require scalable EV charging infrastructure. This involves careful consideration of electrical capacity, load management, and redundancy (Sadeghian et al., 2022).

VE Implications:

- Load impact assessments must inform the sizing of main distribution boards, transformers, and cabling.

- Smart charging solutions (e.g., demand response, scheduled charging) reduce peak demand and infrastructure oversizing (NREL, 2024).
- VE assessments can compare centralized fast-charging stations with distributed lower-power chargers from a cost-performance perspective.

8.4.3 Bi-directional Energy Flow and Prosumer Systems

The emergence of prosumer models—where buildings both consume and generate energy—introduces bidirectional power flows into the electrical grid. Distributed energy resources such as rooftop PV, small wind turbines, and local storage require grid-compatible interconnection and metering

VE Implications:

- Bidirectional metering and export capability must be considered in distribution board layouts and utility interface points.
- Electrical protection schemes need to address reverse power flow, anti-islanding, and dynamic voltage support.
- VE can support cost-benefit optimization of storage plus export vs. self-consumption-only systems (Table 8.3).

Table 8.3 VE Comparison Matrix: Conventional vs. Electrified Buildings

VE Dimension	Conventional Building	Electrified Building
CAPEX	£0 (baseline)	+10–30% (with PV/EV systems)
OPEX	High (heating, cooling, vehicles)	Low (efficient systems, EVs, renewables)
Carbon Emissions	High (fossil fuel-based systems)	Low (electrification + renewable sources)
Grid Reliance	High—fully dependent	Medium–low (with storage and building-integrated photovoltaics)

Usage Note: This table summarizes the high-level VE considerations associated with transitioning from conventional to electrified building strategies. The electrified column reflects scenarios incorporating renewable generation (e.g., BIPV), EVs, and local storage integration.

8.5 Energy Storage and Resilience Planning

Energy storage systems (ESS) are increasingly recognized as critical enablers of flexible, reliable, and sustainable electrical infrastructure. As electrification and renewable integration expand, the ability to store excess generation, manage demand fluctuations, and provide backup power becomes central to long-term system resilience. In VE analysis, energy storage is no longer treated as an auxiliary feature, but as a core element of distributed energy system design.

This section examines key storage technologies, deployment strategies, and VE-related evaluation criteria relevant to contemporary and future-ready electrical engineering practice.

8.5.1 Battery Technologies

Battery-based storage systems are the most commonly adopted energy storage solutions in buildings and small- to medium-scale infrastructure (Olabi et al., 2022). Options include:

- **Lithium-ion (Li-ion)**: Widely used due to high energy density, cycle efficiency (>90%) (Xu et al., 2023), and declining costs. Applications include peak shaving, self-consumption optimization, and backup power.
- **Solid-State Batteries**: Emerging technology with improved safety (nonflammable), higher cycle life, and energy capacity. Recent developments promise cycle lives extending to over 500 cycles (Huang et al., 2025). Still under development for widespread use.
- **Redox Flow Batteries**: Suitable for longer-duration storage. These systems offer scalability and minimal degradation over time but have lower energy density (Yuan et al. 2019).

VE Implications:

- Capital and replacement cost must be weighed against operational savings, demand charge mitigation, and grid independence.
- Technology selection should consider duty cycle, ambient operating conditions, available space, and compliance with applicable battery installation standards (e.g., IEC 62619, UL 9540)

8.5.2 Long-Duration and Distributed Storage Solutions

Beyond batteries, long-duration energy storage (LDES) and distributed storage technologies offer further opportunities for resilience and decarbonization:

- **Thermal Energy Storage (TES)**: Stores energy as heat or cold (e.g., in chilled water or phase-change materials). Often used in HVAC-dominant facilities.
- **Hydrogen-based Storage**: Converts electricity to hydrogen via electrolysis and stores it for reconversion to electricity or other energy uses.
- **Supercapacitors**: Provide short-duration, high-power storage for momentary load smoothing or bridging transitions during switching events.

VE Implications:

- LDES enhances system flexibility during outages or prolonged low-generation periods.
- Distributed storage, located closer to the load, reduces transmission losses and defers infrastructure upgrades.
- Cost-benefit evaluations should include secondary services (e.g., frequency regulation, demand response participation).

8.5.3 Energy Storage as a Service (ESaaS)

Under the ESaaS model, energy storage assets are installed, owned, and maintained by a third party. The building owner pays a fixed or performance-based fee for the availability or use of the storage system.

Table 8.4 VE Evaluation Matrix for Storage Technologies[a]

Technology	CAPEX	OPEX	Life Cycle	Energy Density	Duration Suitability	Safety/Risk Profile	Best Use Case
Lithium-Ion	Medium	Low	3000–6000[b]	High	Short–medium	Moderate	General building storage, peak shaving
Solid-state	High	Low	10,000+[c]	Very high	Short–medium	High (early stage)	Future-ready resilient backup (R&D phase)
Redox Flow	High	Medium	10,000+[d]	Low	Medium–long	Low	Large-scale renewable integration
Thermal Storage	Low–medium	Low	15+ years	N/A (thermal)	Medium–long	Low	HVAC-driven buildings
Hydrogen Storage	High	Medium–high	10–20 years	Low	Long	High (gas handling)	Industrial-scale long-term storage
Supercapacitor	Medium	Very low	1,000,000+	Very low	Very short	Very low	Power buffering and voltage stability

[a] Table data compiled from multiple sources, including manufacturer specifications and academic literature.
[b] Based on standard operating conditions and depth of discharge.
[c] Projected values for commercial solid-state batteries.
[d] Typical for vanadium redox flow batteries under standard conditions.

VE Implications:

- ESaaS allows storage deployment without upfront capital expenditure, improving project cash flow and financing flexibility.
- Contracts must be carefully evaluated for performance guarantees, maintenance liabilities, and service continuity.
- VE assessments must compare life-cycle cost and operational risk between owned vs. service-based models (Table 8.4).

8.6 Augmented and Virtual Reality in Electrical Systems: Design, Construction, and Life-Cycle Applications

Augmented reality (AR) and virtual reality (VR) are immersive technologies that are transforming how electrical systems are visualized, reviewed, and maintained. These tools bridge the gap between abstract digital models and physical construction environments, enabling improved communication, error detection, and life-cycle decision-making. When integrated into VE practices, AR and VR provide new capabilities to optimize designs, reduce construction risk, and enhance operational support.

While still emerging in many engineering contexts, AR and VR have become increasingly viable due to improvements in hardware (e.g., wearable headsets, mobile tablets) and software platforms that support BIM integration, real-time simulation, and field-based interaction (Du et al., 2018; Saredakis et al., 2020). The utilization of these technologies in the architecture, engineering, and construction (AEC) sector enhances preventative safety, productivity, and quality (Getuli et al., 2020).

8.6.1 Design Review and Coordination

In the design stage, VR enables stakeholders to experience electrical layouts, component clearances, and spatial coordination within a fully navigable 3D environment. This is especially valuable in dense plant rooms, riser cores, or ceiling voids, where spatial constraints and multidiscipline interfaces are critical.

AR overlays BIM data onto the physical construction space via tablets or headsets, enabling side-by-side comparison between the intended design and the actual site condition. This assists in:

- Detecting potential constructability issues before installation
- Verifying spatial allowances for cable containment, lighting, and distribution boards
- Reviewing access and maintenance clearances in complex assemblies

VE Implications:

- Reduces design errors by visualizing real-world scale and clash zones.
- Improves stakeholder engagement and decision-making during design workshops.
- Enables cost avoidance through early detection of coordination issues.

8.6.2 Construction and Installation Support

On-site AR can guide electricians and installers by projecting installation pathways, cable tray routes, and equipment positions directly onto building surfaces. Research demonstrates that mixed reality enables faster conduit construction with fewer errors compared to traditional 2D plans (Chalhoub & Ayer, 2018). VR-based training simulators also provide safe, repeatable environments to rehearse installation procedures or hazardous scenarios.

VE Implications:

- Improves labor productivity by reducing interpretation errors from 2D drawings.
- Reduces installation rework and site variations.
- Enhances accuracy of "as-built" documentation by verifying work against the design model (Figure 8.4).

8.6.3 Operations and Maintenance

Once in operation, AR tools can display live sensor data, maintenance instructions, and component health indicators overlaid on actual equipment. This enables maintenance personnel to carry out diagnostics and procedures without consulting separate manuals. The integration of AR with digital twin systems facilitates predictive maintenance and operational efficiency by enabling real-time monitoring and analysis of electrical systems (Aivaliotis et al., 2019; Saniuk et al., 2020).

VE Implications:

- Minimizes downtime by speeding up fault diagnosis and part identification.
- Enhances safety by providing visual step-by-step guidance for complex procedures.
- Supports integration with digital twins for continuous commissioning and asset life-cycle tracking (Figure 8.5).

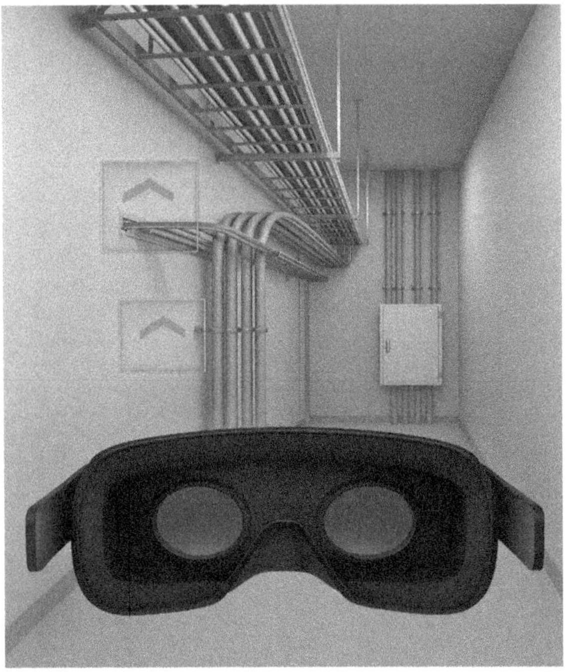

Figure 8.4 Virtual Reality (VR) Interface Used for Immersive Walkthrough of an Electrical Riser Corridor. The View Illustrates Spatial Coordination of Cable Trays, Conduits, and Distribution Boards Within a Controlled Digital Environment, Enabling Early-Stage Clash Detection, Constructability Review, and Stakeholder Engagement in Electrical Design

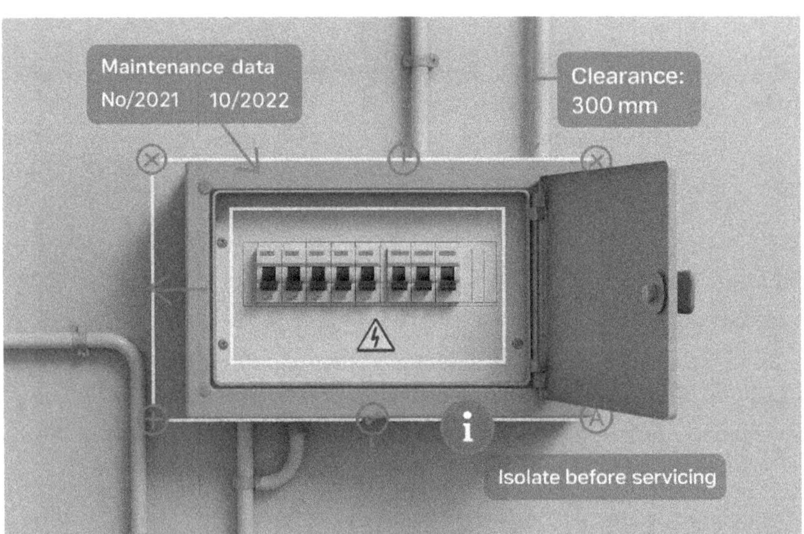

Figure 8.5 Augmented Reality (AR) Overlay Used During Electrical Distribution Board Inspection. Key Data Such as Service History, Spatial Clearances, and Isolation Instructions are Visualized in Real Time to Support Maintenance Safety, Compliance, and Verification

8.7 Wireless Power Transmission and Embedded Systems

Wireless power transmission (WPT) represents a paradigm shift in how electrical energy can be delivered and utilized across a variety of building and infrastructure contexts. By eliminating physical conductors for power transfer, WPT technologies offer novel possibilities in environments where cabling is impractical, aesthetically constrained, or where mobility and flexibility are essential. In parallel, embedded systems—such as wireless sensors and microcontrollers—are increasingly used for decentralized intelligence and control.

While WPT is still in early stages of commercial maturity for large-scale deployment, its relevance in specialized applications is growing. VE assessments must consider both the transformative potential and technical limitations of these technologies, especially where safety, efficiency, and cost justification are key (Figure 8.6).

8.7.1 Principles and Technologies of Wireless Power Transmission

WPT operates on various physical principles depending on application range and power level:

- **Inductive Coupling (Short-Range)**: Common in consumer electronics and EV charging pads; requires precise coil alignment. Efficiency typically ranges from 85% to 95% for properly aligned systems (Ahmad et al., 2021).
- **Magnetic Resonance Coupling (Mid-Range)**: Enables more flexible spatial positioning with higher efficiency over moderate distances. Research demonstrates power transfer efficiency of approximately 60–80% at distances of 10–50 cm (Kurs et al., 2007; Zhang & Donaldson, 2021).
- **Radiative Transmission (RF or Microwave-based)**: Used in low-power, long-distance applications such as remote sensors or aerospace systems.

VE Implications:

- High initial cost and system complexity must be weighed against maintenance savings, space flexibility, and safety improvements.
- Efficiency losses (typically 10–30% lower than wired systems) must be included in life-cycle cost assessments.
- Interference with metallic structures or critical equipment may limit applicability in dense urban or industrial environments.

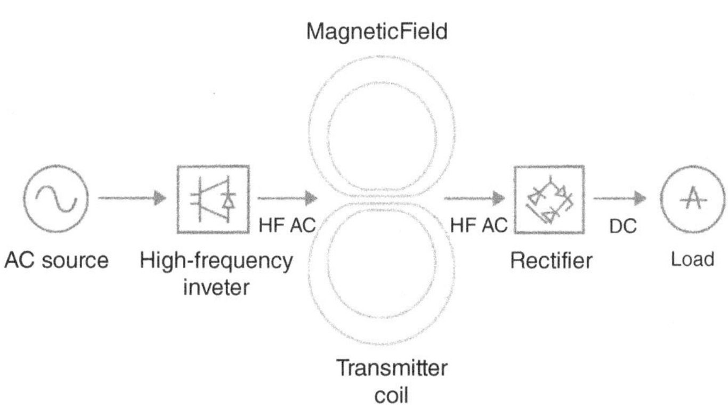

Figure 8.6 Basic WPT Concept Illustration

8.7.2 Embedded Systems and Wireless Sensor Networks

Embedded systems, including low-power microcontrollers, wireless sensor nodes, and system-on-chip devices, provide decentralized control and monitoring capabilities. These are often powered by small batteries, energy harvesting, or, in some cases, low-wattage WPT.

Applications Include:

- Smart lighting systems with wireless dimming and occupancy sensors.
- Environmental sensing for HVAC control and IAQ (indoor air quality) management.
- Structural health monitoring embedded into building fabric.

VE Implications:

- Reduce wiring, conduit, and containment requirements—particularly in retrofit or prefabricated assemblies.
- Enable real-time data acquisition for performance tuning and predictive maintenance.
- Support modular, upgradeable system architectures that defer or phase capital investment (Figure 8.7).

This schematic compares three core WPT methods:

1) **Inductive Coupling**
   - **Range**: <10 cm
   - **Efficiency**: High (85–95%)
   - **Application**: Consumer devices, electric toothbrushes, near-field EV pads
   - **Key Limitation**: Requires tight alignment of transmitter and receiver coils
2) **Magnetic Resonance Coupling**
   - **Range**: 10–50 cm
   - **Efficiency**: Medium (60–80%)

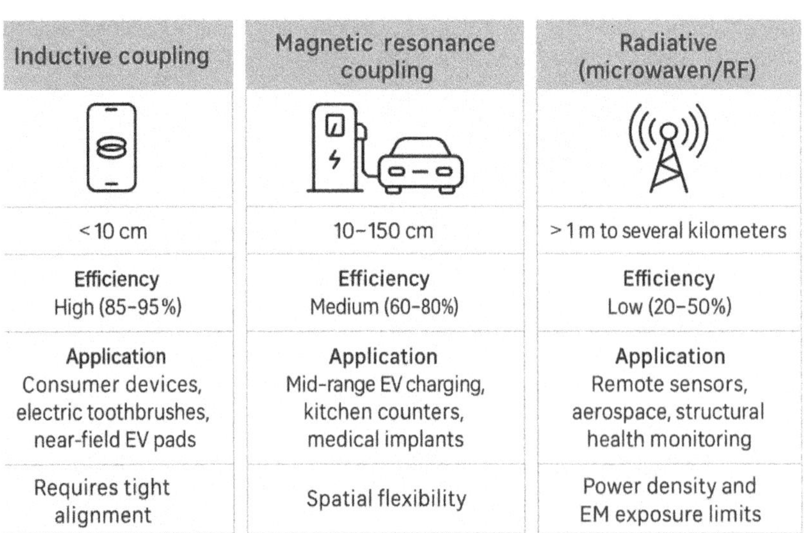

Figure 8.7 Comparison of Wireless Power Transmission Methods by Range and Application

- **Application**: Mid-range EV charging, kitchen counters, medical implants
- **Key Benefit**: More spatial flexibility compared to inductive systems

3) **Radiative (Microwave/RF) Transmission**
    - **Range**: >1 m to several kilometers
    - **Efficiency**: Low (20–50%)
    - **Application**: Remote sensors, aerospace, structural health monitoring
    - **Key Limitation**: Power density and EM exposure limits

8.7.3 Safety, Standards, and Interoperability

WPT and embedded wireless systems must adhere to strict safety and electromagnetic compatibility (EMC) regulations. Key standards include:

- **IEC 61980 Series**: EV wireless charging systems, which covers safety requirements for wireless power transfer systems up to 1000 V AC and 1500 V DC
- **IEC 62311**: Assessment of electronic and electrical equipment related to human exposure restrictions for electromagnetic fields (EMFs) (0 Hz–300 GHz)
- **IEEE 1901 and 802.15.4**: Wireless communication standards for embedded systems.

VE Implications:

- Risk management must include thermal profiles, electromagnetic interference, and fail-safe design protocols.
- Interoperability across manufacturers and future scalability must be assessed during procurement evaluations.
- Additional shielding or spacing requirements may offset perceived gains in some environments.

8.7.4 Industry Adoption and Emerging Use Cases

WPT is increasingly moving from experimental research into targeted real-world applications across a range of domains. Although widespread commercial deployment is still limited, the technology has demonstrated potential value in specific environments where conventional wiring poses physical or operational constraints.

Key Adoption Domains:

- **EV Charging**: WPT is being evaluated in pilot-scale and controlled fleet charging environments. Wireless pads based on magnetic resonance coupling are currently used to automate charging for electric buses and light-duty vehicles in transit depots and high-throughput loading bays (Federal Transit Administration, 2024).
- **Consumer Electronics and Smart Surfaces**: Inductive and resonant WPT is widely implemented in wireless charging pads and embedded furniture solutions. Mobile devices, wearables, and contactless personal electronics increasingly rely on this method for low-voltage charging.
- **Smart Interiors and Modular Buildings**: In modern modular interiors and adaptive architecture, WPT and embedded sensor systems are being trialed to enable wireless power zones for lighting, controls, and occupancy monitoring—particularly in flexible-use spaces where traditional containment is impractical (Gong et al., 2022b).

- **Industrial Robotics and Automated Systems**: WPT systems are under development for powering automated guided vehicles (AGVs), rotary equipment, and sensors in environments subject to vibration, rotation, or restricted maintenance access (Bi et al., 2016).
- **Surface-Embedded Power in Kitchens and Labs**: High-wattage WPT systems are being proposed for transmitting power through benchtops and work surfaces in food service or laboratory settings. These eliminate cord clutter while preserving cleanliness and usability.

8.7.4.1 Market Trends and Geographic Outlook

Regional development of WPT technology is most active in North America and East Asia. National programs on smart cities, public transport electrification, and industry 4.0 have led to early integration efforts. Market forecasts suggest continued double-digit growth through 2032, particularly in sectors involving EV infrastructure and embedded power platforms, although broad adoption depends on standardization and cost convergence

VE Integration Considerations:

- VE assessments should account for energy efficiency losses inherent in wireless power systems (typically 10–30% lower than wired equivalents).
- System-level comparisons must include long-term maintenance benefits, elimination of physical wear interfaces, and layout flexibility.
- Regulatory compliance with EMF exposure limits and power density standards (e.g., IEC 61980, IEC 62311) must be addressed during early-stage VE evaluation.
- WPT and embedded systems should be evaluated in VE processes with consideration for:
  - Infrastructure complexity and retrofit constraints.
  - Maintenance access reduction and safety gains.
  - Energy conversion efficiency and life-cycle cost modeling.
  - System interoperability, especially with wireless communications protocols.
- Stakeholders should be advised that most large-scale WPT systems remain pre-commercial or in pilot-phase deployment, and that final VE assessments must factor in standards maturity, compliance risks, and technology-specific installation conditions.

8.8 Advanced Materials and Nanotechnology in Electrical Systems

Innovations in materials science and nanotechnology are rapidly expanding the performance envelope of electrical systems, offering new pathways for miniaturization, efficiency enhancement, thermal stability, and environmental resilience. These advancements are increasingly relevant to VE as engineers seek solutions that deliver long-term functional value, space savings, and enhanced performance across a system's life cycle.

This section explores the most promising material technologies and their implications for electrical infrastructure, with a focus on emerging conductors, energy conversion materials, and compact integration options.

8.8.1 High-Conductivity and Low-Loss Materials

Traditional conductors such as copper and aluminum dominate electrical applications due to their availability and cost-effectiveness. However, novel materials are being developed with superior electrical, mechanical, and thermal properties:

- **Carbon Nanotubes (CNTs)** and **Graphene**: These nanostructured carbon materials exhibit extremely high conductivity-to-weight ratios, superior mechanical strength, and chemical stability. They are under investigation for use in high-performance cables, electromagnetic shielding, and compact busbars (Novoselov et al., 2004; Saito et al., 1998).
- **Superconductors**: Materials that exhibit zero electrical resistance below a critical temperature. While their widespread application is currently limited to cryogenically controlled environments (e.g., MRI systems, particle accelerators), research is ongoing into high-temperature superconductors for power cables and transformers (Boothroyd, 2015).

VE Implications:

- Enable smaller cable cross-sections and lightweight wiring in space-constrained or mobile installations.
- Offer potential for lossless transmission, reducing long-term operating cost in high-current applications.
- Require careful LCCA due to high production cost and sensitivity to environmental conditions.

8.8.2 Next-Generation Photovoltaic Materials

Beyond conventional silicon solar cells, several emerging materials offer promising improvements in efficiency, flexibility, and form factor:

- **Perovskite Solar Cells**: Synthetic crystalline materials that can be printed onto flexible substrates with high light absorption efficiency. They show potential for integration into building materials, fabrics, or irregular architectural surfaces (Kojima et al., 2009; National Renewable Energy Laboratory, 2020).
- **Organic Photovoltaics (OPVs)**: Carbon-based polymers capable of generating electricity under low-light and diffused sunlight conditions. Suitable for façade applications and portable systems (Brabec et al., 2001).

VE Implications:

- May enable low-profile, building-integrated generation options that reduce facade material costs.
- Expand design flexibility for renewable integration in retrofit projects.
- Require detailed risk assessment of durability, life cycle, and fire safety properties.

8.8.3 Thermally Stable and Flame-Retardant Polymers

Advances in polymer science have produced insulating materials with improved thermal ratings, flame resistance, and lower toxicity in combustion:

- **Cross-linked polyethylene (XLPE)** and **thermoplastic elastomers (TPEs)** are increasingly used in power cables for environments with elevated ambient temperatures or exposure to oils and solvents.

- **Halogen-free flame-retardant (HFFR)** materials reduce smoke and toxic gas release in confined spaces such as tunnels, data centers, or transport systems.

VE Implications:

- Longer service life reduces frequency of replacement.
- Improve system safety in high-density environments.
- May qualify for green certification points or insurance discounts depending on jurisdiction.

8.8.4 Nanostructured Coatings and Additives

Nanotechnology also enhances electrical system performance via protective coatings and functional additives:

- **Hydrophobic coatings** to reduce corrosion and surface leakage in humid or marine environments.
- **Nanoceramic or metal oxide coatings** for improving dielectric strength or thermal emissivity on enclosures and bus ducts.
- **Nano-enhanced thermal interface materials** for efficient heat transfer in power electronics and LED lighting systems (Zhang et al., 2014).

VE Implications:

- Support reliability and maintenance reduction in hostile environments.
- Enable downsizing of cooling systems and enclosures.
- Must be evaluated for cost-effectiveness based on application-criticality and service environment (Figure 8.8).

This figure presents a side-by-side comparison of five conductive materials across three key engineering properties:

1) **Electrical Conductivity**
   - Normalized to copper (100% IACS*) for relative benchmarking
   - Carbon nanotubes and graphene exhibit exceptionally high theoretical conductivity

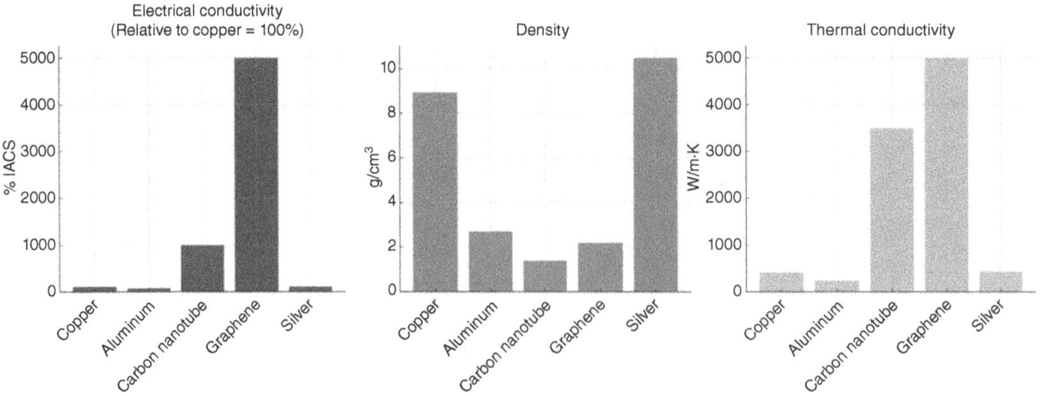

Figure 8.8 Comparative Properties of Advanced Conductive Materials

Table 8.5 Material Selection Matrix for VE Applications in Electrical Systems

Material	Electrical Performance	Thermal Performance	Density (g/cm^3)	Cost Class (Relative)	Availability	Application Suitability
Copper	High	High	8.96	Moderate	Global, high	Standard power cables, busbars
Aluminum	Moderate	Moderate	2.70	Low	Global, high	Overhead lines, weight-sensitive enclosures
Carbon Nanotube	Very high	Exceptional	1.40	Very high (experimental)	R&D phase	Next-gen conductors, aerospace, defense
Graphene	Exceptional	Exceptional	2.20	Very high (experimental)	R&D phase	Flexible electronics, printed circuits
Silver	Very high	High	10.49	Extremely high	Limited industrial use	Specialist electronics, high-performance RF systems

Data compiled from a combination of academic sources (Ashby, 2011; Callister, 2018; Kalantar-Zadeh et al., 2021), industry datasheets from leading material suppliers (e.g., IEEE Material Database, Matmatch, Cambridge Engineering Selector), and relevant manufacturer specifications as of 2023.

2) **Density**
   - Relevant for weight-sensitive applications such as aerospace, modular components, and mobile enclosures
   - Carbon-based materials offer very low mass compared to traditional metals
3) **Thermal Conductivity**
   - Important for thermal management in high-current applications and electronics
   - Graphene significantly outperforms conventional materials

* IACS stands for International Annealed Copper Standard.

It is a benchmark used to express the electrical conductivity of materials relative to pure annealed copper, which is defined as 100% IACS. For example:

Copper = 100% IACS (baseline)
Aluminum ≈ 61% IACS
Silver ≈ 106% IACS

This chart supports material selection in VE assessments where trade-offs between weight, efficiency, and thermal performance must be evaluated (Table 8.5).

8.9 Conclusion

The integration of digital technologies and advanced materials into electrical system design represents a transformative evolution in engineering practice. As shown throughout this chapter, VE must now operate within a broader framework—one that balances technical performance, life-cycle cost, environmental impact, adaptability, and future-readiness.

Tools such as BIM, AI, and IoT platforms provide engineers with unprecedented access to real-time data, predictive analytics, and immersive design environments. These capabilities allow VE to move from traditional cost-cutting exercises toward a proactive, systems-oriented approach to value optimization.

Meanwhile, emerging technologies such as WPT, ESS, and advanced materials—including graphene, carbon nanotubes, and perovskite PVs—offer new opportunities to enhance system resilience, efficiency, and integration flexibility. However, these opportunities also introduce new VE challenges: uncertainty in cost, evolving standards, and the need for rigorous risk-based assessments.

In practical application, this means that:

- **Early-stage VE workshops** should incorporate digital workflows and parametric modelling to evaluate real-time scenarios.
- **Life-cycle analyses** must be broadened to account for maintenance, energy use, and carbon performance alongside initial capital cost.
- **Multidisciplinary collaboration** is essential, as many of the technologies discussed operate across mechanical, architectural, and IT domains.
- **Standards and compliance considerations** must evolve in parallel with emerging technologies, ensuring safety and reliability are maintained.

Ultimately, the application of VE in this evolving context requires both a shift in mindset and a deepening of technical competence. Engineers must understand not only conventional design economics but also the functional and operational implications of integrating advanced, digitally enabled solutions.

References

Agarwal, A., Habeebullah Tunde, A., Yusuf, S. O., Taiwo, A. I., Ocran, G., & Paul-Adeleye, A. H. (2024). AI-driven innovations in energy efficiency: Transforming smart buildings and urban areas through technology and digital transformation. *World Journal of Advanced Research and Reviews*, 24(1), 141–152. https://doi.org/10.30574/wjarr.2024.24.1.2921

Ahmad, A., Alam, M. S., & Chabaan, R. (2021). A comprehensive review of wireless charging technologies for electric vehicles. *Journal of Energy Storage*, 38, 102564.

Aivaliotis, P., Georgoulias, K., & Chryssolouris, G. (2019). The use of Digital Twin for predictive maintenance in manufacturing. *International Journal of Computer Integrated Manufacturing*, 32(11), 1067–1080. https://doi.org/10.1080/0951192X.2019.1686173

Ashby, M. F. (2011). *Materials selection in mechanical design* (4th ed.). Butterworth-Heinemann.

Bi, Z., Kan, T., Mi, C. C., Zhang, Y., Zhao, Z., & Keoleian, G. A. (2016). A review of wireless power transfer for electric vehicles: Prospects to enhance sustainable mobility. *Applied Energy*, 179, 413–425.

Boothroyd, C. (2015). *High-temperature superconductors: Materials, properties, and applications*. Springer.

Brabec, C. J., Sariciftci, N. S., & Hummelen, J. C. (2001). Plastic solar cells. *Advanced Functional Materials*, 11(1), 15–26.

Callister, W. D. (2018). *Materials science and engineering: An introduction* (9th ed.). Wiley.

Chalhoub, J., & Ayer, S. K. (2018). Using mixed reality for electrical construction design communication. *Automation in Construction*, 86, 1–10. https://doi.org/10.1016/j.autcon.2017.10.028

Chen, L., Chen, Z., Zhang, Y., Liu, Y., Osman, A. I., & Farghali, M. (2023). Artificial Intelligence-Based Solutions for Climate Change: A Review. *Environmental Chemistry Letters*, 21, 2525–2557. https://doi.org/10.1007/s10311-023-01617-y

Du, J., Shi, Y., Zou, Z., & Zhao, D. (2018). CoVR: Cloud-based multiuser virtual reality headset system for project communication of remote users. *Journal of Construction Engineering and Management*, 144(2), 04017109. https://doi.org/10.1061/(ASCE)CO.1943-7862.0001414

EN 12464-1. (2011). *Light and lighting – Lighting of work places – Part 1: Indoor work places.* European Committee for Standardization.

Federal Transit Administration. (2024). *FTA report no. 0270: Wireless charging for transit fleets.* https://www.transit.dot.gov/sites/fta.dot.gov/files/2024-12/FTA-Report-No-0270.pdf

Federal Transit Administration. (2024). *Wireless power transfer for transit applications: Technology assessment report.* U.S. Department of Transportation.

Getuli, V., Capone, P., Bruttini, A., & Isaac, S. (2020). BIM-based immersive virtual reality for construction site planning: A case study. *Visualization in Engineering*, 8(1), 1–18. https://doi.org/10.1186/s40327-020-00064-4

Gharaibeh, A., Zorchenko, N. V., Tyupina, T. G., & Parshutin, M. E. (2024). Technologies used by General Electric to create digital twins for energy industry. *Power Technology and Engineering*, 58, 521–526. https://doi.org/10.1007/s10749-024-01839-2. [link.springer.com]

Gong, L., Zhang, R., & Wang, S. (2022b). Review of wireless energy transfer for modular smart building applications. *Sensors*, 22(15), 5573.

Huang, X. Y., Zhao, C. Z., Kong, W. J., Yao, N., Mu, W. Q., Zhang, X. Y., ... Zhang, Q. (2025). Tailoring polymer electrolyte solvation for 600 Wh kg^{-1} lithium batteries. *Nature*, 646, 343–350. https://doi.org/10.1038/s41586-025-09565-z

IEC. (2019). *IEC 62311:2019 Assessment of electronic and electrical equipment related to human exposure restrictions for electromagnetic fields (0 Hz – 300 GHz).* International Electrotechnical Commission.

IEC. (2020). *IEC 61980-1:2020 Electric vehicle wireless power transfer (WPT) systems - Part 1: General requirements.* International Electrotechnical Commission.

IEC 60364. (2020). *Low-voltage electrical installations – Part 1: Fundamental principles, assessment of general characteristics, definitions.* International Electrotechnical Commission.

IEC 61980. (2020). *Electric vehicle wireless power transfer systems – General requirements.* International Electrotechnical Commission.

IEC 62311. (2019). *Assessment of electronic and electrical equipment related to human exposure restrictions for electromagnetic fields.* International Electrotechnical Commission.

IEC 62619. (2022). *Secondary cells and batteries containing alkaline or other non-acid electrolytes – Safety requirements for secondary lithium cells and batteries.* International Electrotechnical Commission.

IEEE 1901. (2010). *IEEE standard for broadband over power line networks: Medium access control and physical layer specifications.* IEEE Standards Association.

IEEE 802.15.4. (2020). *IEEE standard for low-rate wireless networks.* IEEE Standards Association.

International Electrotechnical Commission (IEC). (2011). *IEC 60092-360:2011 - Electrical installations in ships - Part 360: Insulating and sheathing materials for shipboard and offshore units, power, control, instrumentation, telecommunication and data cables.*

Kalantar Zadeh, K., Xie, W., Allioux, F. M., Ou, J. Z., Miyako, E., & Tang, S. Y. (2021). Gallium-based liquid metal particles for therapeutics. *Trends in Biotechnology*, 39(6), 624–640. https://doi.org/10.1016/j.tibtech.2020.10.005. [eprints.soton.ac.uk]

Kojima, A., Teshima, K., Shirai, Y., & Miyasaka, T. (2009). Organometal halide perovskites as visible-light sensitizers for photovoltaic cells. *Journal of the American Chemical Society*, 131(17), 6050–6051.

Kumar, K. V. S., Vennela, V. L., Venkatesh, M., Ashajyothi, K., & Naidu, M. S. (2024). IoT-based smart home energy management systems. *Journal of Digital Economy and Engineering*, 21(3), 1–12.

Kurs, A., Karalis, A., Moffatt, R., Joannopoulos, J. D., Fisher, P., & Soljačić, M. (2007). Wireless power transfer via strongly coupled magnetic resonances. *Science*, 317(5834), 83–86.

Liu, T., Liu, J., Zheng, H., & Bao, J. (2023). Visual analytics for digital twins: a conceptual framework and case study. *Journal of Intelligent Manufacturing*, 35, 1671–1686. https://doi.org/10.1007/s10845-023-02135-y [link.springer.com]

National Renewable Energy Laboratory (NREL). (2020). Best research-cell efficiency chart. https://www.nrel.gov/pv/cell-efficiency.html

National Renewable Energy Laboratory (NREL) and Lawrence Berkeley National Laboratory (LBNL). (2024). Survey of the state of smart charge management for electric vehicles in the United States. U.S. Department of Energy, Vehicle Technologies Office. Golden, CO: National Renewable Energy Laboratory (available at : NREL Researchers Survey the State of Smart Charge Management Nationwide | NREL)

Novoselov, K. S., Geim, A. K., Morozov, S. V., et al. (2004). Electric field effect in atomically thin carbon films. *Science*, 306(5696), 666–669.

Olabi, A. G., Wilberforce, T., Sayed, E. T., Abo-Khalil, A. G., Maghrabie, H. M., Elsaid, K., & Abdelkareem, M. A. (2022). Battery energy storage systems and SWOT (strengths, weakness, opportunities, and threats) analysis of batteries in power transmission. *Energy*, 254(Part A), 123987. https://doi.org/10.1016/j.energy.2022.123987

Sadeghian, O., Oshnoei, A., Mohammadi-ivatloo, B., Vahidinasab, V., & Anvari-Moghaddam, A. (2022). A comprehensive review on electric vehicles smart charging: Solutions, strategies, technologies, and challenges. *Journal of Energy Storage*, 54, 105241. https://doi.org/10.1016/j.est.2022.105241

Saito, R., Dresselhaus, G., & Dresselhaus, M. S. (1998). *Physical properties of carbon nanotubes*. Imperial College Press.

Saniuk, S., Grabowska, S., & Gajdzik, B. (2020). Digital Twin and BIM synergy for predictive maintenance in smart building engineering systems development. *World Journal of Advanced Research and Reviews*, 8(2), 406–421. https://doi.org/10.30574/wjarr.2020.8.2.0409

Saredakis, D., Szpak, A., Birckhead, B., Keage, H. A. D., Rizzo, A., & Loetscher, T. (2020). Factors associated with virtual reality sickness in head-mounted displays: A systematic review and meta-analysis. *Frontiers in Human Neuroscience*, 14, 96. https://doi.org/10.3389/fnhum.2020.00096

Sharma, J., Mittal, M. L., & Soni, G. (2022). Condition-based maintenance using machine learning and role of interpretability: A review. *International Journal of System Assurance Engineering and Management*. https://doi.org/10.1007/s13198-022-01843-7

Shukla, A. K., Sudhakar, K., & Baredar, P. (2017). Recent advancement in BIPV product technologies: A review. *Energy and Buildings*, 140, 188–195. https://doi.org/10.1016/j.enbuild.2017.02.015

UL 9540. (2022). *Standard for energy storage systems and equipment*. UL Standards.

Xu, J., Cai, X., Cai, S., Shao, Y., Hu, C., Lu, S., & Ding, S. (2023). High-energy lithium-ion batteries: Recent progress and a promising future in applications. *Energy & Environmental Materials*, 6(5), e12450. https://doi.org/10.1002/eem2.12450

Yang, R. J., & Zou, P. X. W. (2016). Building integrated photovoltaics (BIPV): Costs, benefits, risks, barriers and improvement strategy. *International Journal of Construction Management*, 16(1), 39–53. https://doi.org/10.1080/15623599.2015.1117709

Yuan, Z., et al. (2019). A review of all-vanadium redox flow battery durability: Degradation mechanisms and mitigation strategies. *International Journal of Energy Research*, 43(22). https://doi.org/10.1002/er.4607

Zhang, X., Wang, L., & Zhang, D. (2014). Nanostructured coatings for corrosion protection: A review. *Surface and Coatings Technology*, 258, 1–8.

Zhang, Z., & Donaldson, N. (2021). Wireless power transfer for implantable medical devices: A comprehensive review. *Biomedical Engineering Online*, 20(1), 1–25.

9

Regulatory and Compliance Challenges in Value Engineering of Electrical Systems

9.1 Introduction and Terminology

9.1.1 Distinguishing Codes, Regulations, and Standards

In electrical engineering and construction, precise understanding of the terms *codes*, *regulations*, and *standards* is essential. These terms are often used interchangeably but carry distinct meanings, legal implications, and operational roles (Table 9.1).

Figure 9.1 illustrates the interaction between these three layers in shaping compliant system design.

Conceptual diagram showing concentric layers: Regulations (outer layer) enforce Codes (middle layer) which reference Standards (inner core) for technical detail.

Understanding this hierarchy is foundational to value engineering (VE) efforts that aim to modify or optimize electrical designs while maintaining full legal and safety compliance.

9.1.2 Why Regulatory Understanding Is Vital to VE

In the context of VE for electrical systems, regulatory knowledge is not merely a passive requirement but an active design parameter. VE proposals—such as equipment substitutions, topology simplifications, or system downsizing—must be assessed not only for cost-effectiveness and performance but also for their alignment with:

- Prescriptive safety codes (e.g., minimum clearance distances, segregation requirements)
- Energy efficiency mandates
- Life safety and emergency power obligations
- Industry-specific compliance (e.g., healthcare, data centers, hazardous areas)

Failure to integrate regulatory considerations early can result in:

- Delays in approval by authorities having jurisdiction (AHJs)
- Rework due to nonconformance
- Legal liabilities and financial penalties
- Loss of certification or insurance validity

Therefore, regulatory fluency is essential for VE teams aiming to achieve optimized, functional, and compliant designs. As subsequent sections will show, regulations are not only constraints—they are also opportunities for innovation, especially in areas such as sustainability, digital compliance integration, and performance-based design methodologies.

Value Engineering for Electrical Systems in Construction, First Edition. Hazem Hussein.

Companion website: www.wiley.com/go/hazem

Table 9.1 Distinctions Between Codes, Regulations, and Standards in Electrical Engineering

Aspect	Codes	Regulations	Standards
Definition	Standards that establish minimum technical requirements for the safe installation of electrical wiring and equipment (ESFI, 2023; NECA, 2024).	Statutory instruments that define legally enforceable obligations under relevant legislation (IET, 2017).	Consensus-based technical documents outlining specifications to ensure safety, quality, and consistency (IEEE, 2024; ISO, 2024).
Issuing Authority	Developed or adopted by recognized professional institutions or government-appointed bodies (e.g., IET, NFPA).	Issued directly by government agencies or statutory authorities (e.g., UK HSE, EU Commission).	Published by formal standards organizations (e.g., IEC, ISO, IEEE, BSI).
Compliance	Legally binding when adopted into legislation or referenced in regulatory frameworks.	Mandatory and enforceable under national or regional law.	Voluntary unless formally adopted by law, regulation, or contractual requirement.
Purpose	To establish minimum technical provisions that ensure public safety, health, and operational integrity.	To implement and enforce public policies through specific technical and procedural rules.	To promote interoperability, efficiency, reliability, and quality across systems and industries.
Examples	• BS 7671 (UK Wiring Regulations) (BSI, 2018) • National Electrical Code (NEC)—Building Code of Australia (BCA)	• UK Electricity Safety, Quality and Continuity Regulations • EU Low Voltage Directive (LVD)	• IEC 60364 (Low-voltage Electrical Installations) (IEC, 2005) • ISO 9001 (Quality Systems) • IEEE 802.11 (WLAN)

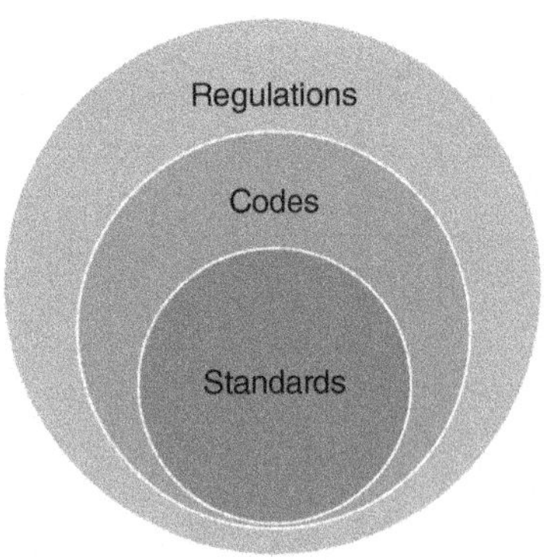

Figure 9.1 Relationship Between Codes, Regulations, and Standards

9.2 Global Regulatory Frameworks in Electrical Design

9.2.1 Overview of Major Frameworks

The design and execution of electrical systems are subject to national and regional codes that incorporate technical standards, often tailored to local legal, environmental, and operational conditions. While the underlying principles—such as safety, efficiency, and reliability—are globally consistent, the implementation varies widely.

This section provides a comparative overview of key electrical regulatory frameworks that influence VE decisions across different jurisdictions (Figure 9.2) (Table 9.2).

This diagram illustrates how international standards bodies—primarily the *International Electrotechnical Commission (IEC)* and the *Institute of Electrical and Electronics Engineers*

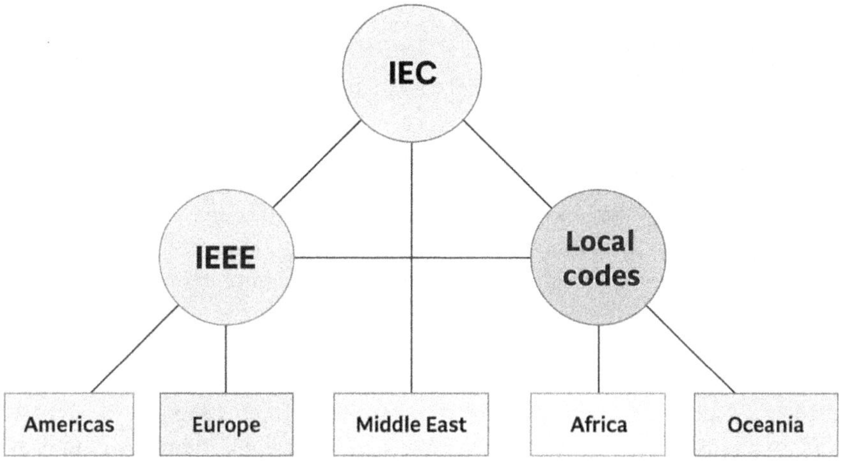

Figure 9.2 Global Influence of Electrical Regulatory Frameworks

Table 9.2 Summary of National Electrical Regulatory Frameworks

Region	Primary Code/ Standard	Issuing Body	Legal Status	Notes
United States	National Electrical Code (NEC)	National Fire Protection Association (NFPA)	Adopted into law by states and municipalities	Updated every 3 years; serves as a model for several international adaptations (NFPA, 2022).
United Kingdom	BS 7671 (IET Wiring Regulations)	Institution of Engineering and Technology (IET)/BSI	Enforced via the Electricity at Work Regulations (EAWR) 1989	Mandated for all electrical installations in the United Kingdom; harmonized with IEC 60364 (HSE, 2015).
France	NF C 15-100	AFNOR/UTE	Mandatory national standard	Covers safety, distribution, and energy efficiency in residential and commercial sectors (AFNOR, 2024).
Middle East	IEC 60364 (adapted)	Local authorities based on IEC/ ISO framework	Varies by country	GCC countries commonly adopt IEC standards with national customization (CIBSE, 2014).
Australia/ NZ	AS/NZS 3000 (Wiring Rules)	Standards Australia/ Standards New Zealand	Enforceable under state/territory legislation	Covers design, construction, verification; includes safety services, earthing, and insulation (Intertek Inform, 2023).

(IEEE)—influence electrical codes and practices across global regions. While most countries adopt or adapt IEC standards into their national codes, IEEE standards are frequently used in sector-specific applications such as telecommunications, power systems, and industrial electronics. Local codes, often derived from or harmonized with international standards, remain the enforceable legal instruments in each jurisdiction.

9.2.2 Comparative Impacts on Project Design

Despite global convergence in safety principles, several key factors influence how VE strategies must adapt to different national frameworks:

- **Technical Prescriptiveness**: Some codes, such as the NEC, are highly prescriptive, limiting the scope for performance-based VE solutions. In contrast, others like BS 7671 and IEC 60364 incorporate risk-based approaches, allowing for greater engineering judgment (HSE, 2015; NFPA, 2022).
- **Certification Requirements**: Many jurisdictions mandate certified equipment—such as CE marking in Europe and UL listing in the United States—restricting the ability to propose lower-cost alternatives without equivalent compliance documentation (NFPA, 2022).
- **Inspection and Enforcement Regimes: Projects in the United States and the United Kingdom often undergo formal** inspections by AHJs, necessitating comprehensive documentation to prove compliance (HSE, 2015; NFPA, 2022). In developing regions, enforcement may be less consistent, placing greater responsibility on the design team
- **Climate and Environmental Adaptation**: Middle Eastern and tropical regions often impose specific regulatory provisions addressing temperature ratings, insulation, and humidity resilience (CIBSE, 2014).
- **Energy Efficiency and Sustainability Mandates**: Australia, the United Kingdom, and parts of the European Union enforce performance benchmarks tied to national energy strategies, such as Part L in the United Kingdom, influencing electrical system sizing, control strategies, and the integration of renewable energy sources (UK Government., 2023).

➢ ***Engineering Insight***

For internationally active VE teams, a strong grasp of these regional nuances is critical. Designs must be adapted not only for technical and commercial optimization but also for code compliance within jurisdiction-specific frameworks. Moreover, early coordination with local experts and regulators is essential to validate equivalency arguments for any proposed alternates or innovative designs.

▪ ***Clarification: Role of IEC and IEEE***

It is important to distinguish that the **IEC** and **IEEE** are *not* regulatory bodies. They are **standards development organizations** that publish consensus-based technical specifications.

- The **IEC** develops globally applicable standards for electrical and electromechanical systems (e.g., *IEC 60364*), often adopted or adapted by national bodies into codes or regulations (IEC, 2022).
- The **IEEE** issues standards predominantly used in power systems, communications, and electronics sectors (e.g., *IEEE 802.11*, *IEEE 519*) (IEEE, 2022).

These standards are voluntary by default but may become legally enforceable when incorporated into national codes or referenced in regulations.

For clarity, neither IEC nor IEEE has enforcement power. Their influence depends on national adoption or sectoral recognition by AHJs.

9.3 The Impact of Regulations on VE Decisions

9.3.1 Compliance as a Constraint and Design Driver

Regulatory frameworks significantly shape the landscape in which VE decisions are made. While VE is intended to identify cost-effective, functional, and efficient solutions, these must be evaluated within the bounds of applicable electrical codes, safety regulations, and performance standards.

In many cases, regulations specify:

- **Minimum safety thresholds**, such as fault current withstand ratings, cable sizing, and fire separation requirements
- **Mandatory system features**, such as dual power paths for life safety systems or surge protection for sensitive electronics
- **Component certifications**, such as CE marking (Europe), UKCA (United Kingdom), or UL listing (United States)
- **Installation conditions**, including clearance zones, earthing configurations, and access requirements

These prescriptive elements place natural constraints on the VE process. However, when well understood, regulations can also be leveraged as a design driver, prompting:

- The selection of modular or pre-certified systems to streamline approval
- The use of performance-based approaches (e.g., functional safety assessments)
- The identification of cost-saving alternatives that remain fully compliant

9.3.2 Risks of Noncompliance in VE Decisions

Where VE decisions are made without adequate regulatory oversight, the risks are substantial (Table 9.3).

Case Example—Generator Redundancy and Fire Safety Compliance

Scenario:

A VE proposal for a commercial development aimed to reduce capital expenditure by removing one of two standby generators originally intended for full-load redundancy.

Regulatory Issue:

Local fire safety codes, aligned with *EN 1838* and *BS 8519*, required **redundant power supply** for life safety systems, such as emergency lighting and fire pumps. The revised design—though sufficient for average load—**did not meet the mandatory redundancy requirements** for critical systems.

Resolution:

Rather than reintroducing full-load redundancy, the VE team implemented a **load prioritization and automatic shedding scheme**, ensuring:

Table 9.3 Risks of Noncompliance in VE Decisions

Risk Area	Description
Code Violation	Substitution of components or changes to topology that breach prescriptive requirements
Certification Gaps	Use of uncertified or improperly documented components leading to nonacceptance by AHJs
Project Delays	Rework due to failed inspections, incomplete documentation, or approval process rejection
Legal and Financial Penalties	Exposure to litigation, loss of insurance validity, or financial penalties for noncompliance
Operational Risk	Failure of critical systems (e.g., UPS, fire alarm) due to nonconforming design or installation

These risks necessitate early and continuous integration of compliance validation within the VE workflow, especially where substitutions or reconfigurations are proposed.

- Continuous power to life safety loads during failure
- Compliance with minimum backup duration and selectivity requirements
- Optimized generator sizing for efficiency

Outcome:

- 22% capital cost saving over the original design
- Full code compliance preserved
- Enhanced system selectivity and control

Key Takeaway:
VE must align with legal and performance-based safety requirements. Intelligent load zoning, sequencing, and protection design can unlock cost savings without breaching compliance.

9.4 Integrating Compliance into the VE Process

9.4.1 Compliance Integration Across Project Phases

To ensure that VE outcomes are not only cost-effective but also fully compliant, regulatory considerations must be integrated into every phase of the design and implementation process. The goal is to transition compliance from a post-design checkpoint into a proactive design parameter (Table 9.4).

Table 9.4 VE–Compliance Integration Workflow

VE Phase	Compliance Integration Activities
Initial Design	Identify applicable codes and standards; consult compliance officers; generate preliminary checklists
Function-Cost Analysis	Evaluate design alternatives for both performance and compliance; eliminate nonconforming options
Design Development	Conduct staged compliance reviews; use BIM-based code checking where applicable
Implementation Planning	Develop detailed compliance plans; align procurement and sequencing with statutory requirements
Post-Implementation	Conduct commissioning audits and regulatory inspections; document compliance for certification

This structured approach ensures that each VE proposal is assessed for technical validity and regulatory acceptability before moving forward.

9.4.2 Role of Compliance Officers and Legal Advisors

A multidisciplinary approach is essential to successful VE. Compliance officers and legal advisors serve a critical function in translating legislative and regulatory frameworks into project-specific technical decisions (ASU Online, 2025).

Key Responsibilities:

- **Regulatory Interpretation:** Identify how applicable codes and directives affect the technical solution.
- **Early Risk Identification:** Flag noncompliant VE proposals before they proceed to design or procurement.
- **Documentation and Traceability:** Ensure that decisions, approvals, and assumptions are fully recorded to support audit trails.
- **Liaison with Authorities:** Serve as the contact point with inspectors, certifiers, and regulatory agencies.

Early integration of compliance stakeholders prevents costly redesigns and ensures that design flexibility is preserved where possible.

➢ **Best Practice Tip:**
Formalize the role of compliance reviewers in design stage gates, ensuring they are signatories in VE decision registers.

9.4.3 Leveraging Technology for Regulatory Validation

Digital platforms and design automation tools are powerful enablers of proactive compliance management within VE processes (Table 9.5).

Recommended Tools and Technologies

Table 9.5 Digital Tools for Compliance Management in VE

Tool/System	Functionality
Building Information Modeling (BIM)	Enables rule-based code-checking, automated clash detection, and compliance tagging
Common Data Environment (CDE)	Centralized document management with version control and traceability of regulatory references
Compliance Management Software	Tracks regulation updates; flags conflicts with project specifications; supports audit trail generation
Scenario Planning Engines	Simulates design variants against multiple regulatory scenarios (e.g., for uncertain or evolving codes)

These tools promote early issue detection, reduce manual error, and ensure that VE proposals do not compromise legal obligations.

9.4.4 Embedding Compliance into Design Culture

Beyond process and tools, regulatory compliance must be embedded into the **project culture** as a core design value.

Strategies to Cultivate a Compliance-Driven VE Culture:

- **Training and Capacity Building:** Offer continuing education on evolving codes and jurisdictional changes.
- **Knowledge Sharing Forums:** Encourage lessons-learned workshops and post-project reviews that include compliance insights.
- **Recognition of Regulatory Innovation:** Reward teams that deliver compliant, innovative solutions within VE constraints.
- **Stakeholder Transparency:** Engage clients, contractors, and authorities in open dialogue regarding compliance trade-offs and decisions.

This cultural integration ensures that compliance is not perceived as a constraint, but rather a strategic dimension of engineering quality and project success.

9.5 Sustainability and Energy Efficiency Regulations

9.5.1 Role of Sustainability Certifications in Value Engineering

The global emphasis on sustainable construction has introduced new regulatory and performance-driven parameters into the electrical design process. Certifications such as Leadership in Energy and Environmental Design (LEED) and the Building Research Establishment Environmental Assessment Method (BREEAM) impose requirements that extend beyond traditional safety or functionality (BRE, 2016; USGBC, 2020). These standards are increasingly integrated into project briefs as mandatory targets.

Table 9.6 Electrical Systems and LEED/BREEAM Criteria

Category	Electrical Systems	LEED/BREEAM Criteria
Energy Efficiency	• High-efficiency lighting • Building automation systems	• Optimize energy performance • Lighting power density • Energy monitoring
Materials and Resources	• Low-embodied carbon cabling • Recyclable electrical components	• Building lifecycle impact reduction • Responsible sourcing of materials
Indoor Environmental Quality	• Daylighting controls • Low-VOC electrical products	• Daylight and views • Low-emitting materials
Innovation in Design	• On-site renewable energy • Advanced energy metering	• Innovative design • Management (energy efficiency)

For VE Processes, This Creates Dual Imperatives:

- Identify **cost-effective solutions** that comply with green building standards.
- Integrate **sustainability principles** early to unlock long-term operational and lifecycle value.

Key Sustainability Areas Influencing Electrical Design:

- **Material Selection**: Preference for low embodied carbon materials and components with environmental product declarations (EPDs) (BRE, 2016; USGBC, 2020).
- **Lighting Systems**: Mandatory compliance with lighting power density limits; integration of daylight harvesting and control systems (USGBC, 2020).
- **Energy Metering**: Inclusion of submetering to support building performance verification credits (USGBC, 2020).
- **Renewables Readiness**: Infrastructure for future Photo Voltaic (PV) integration or compliance with mandated onsite renewable generation (BRE, 2016; USGBC, 2020).

These elements impact equipment choice, spatial allocation, control topology, and electrical infrastructure capacity—all of which are evaluated in VE proposals. Table 9.6 illustrates how key electrical system components contribute directly to credit achievement under LEED and BREEAM frameworks.

9.5.2 Energy Efficiency Mandates and Lifecycle Cost Planning

In parallel with voluntary certifications, mandatory energy efficiency requirements are embedded in building codes and regional legislation. Examples include:

- UK Building Regulations Part L, which defines energy efficiency requirements for building envelopes, HVAC systems, and lighting (UK MHCLG, 2021).
- EU Energy Performance of Buildings Directive (EPBD), which mandates minimum performance requirements, inspection obligations, and energy certification (European Parliament and Council., 2024).
- ASHRAE Standard 90.1, establishing energy efficiency criteria for mechanical, lighting, and envelope systems (ASHRAE, 2022).

VE Design Parameters Affected (Table 9.7)

Table 9.7 Regulatory Design Parameters and VE Implications

Design Element	Regulatory Requirement	VE Implication
HVAC Integration	Use of high-efficiency chillers, heat pumps, and demand control ventilation	May require investment in controls, BMS integration, and harmonized power topology
Lighting Efficiency	Limits on lighting power density (e.g., W/ m^2) and mandatory use of LEDs	LED selection and control interface design become VE focus areas
Building Automation	Controls for zoning, daylight response, occupancy, and energy dashboards	VE must balance cost of controls vs. LEED/ BREEAM credits and user needs
Renewables and Storage	Targets for solar PV contribution, net-zero energy readiness, or battery storage	Impact sizing of main switchboards, inverters, and protection schemes

9.5.3 Lifecycle and Whole-Life Cost Approaches in VE

Sustainability-oriented VE requires a shift from first-cost optimization to whole-life value analysis. Decisions regarding lighting, controls, power distribution efficiency, and renewables integration must consider capital cost vs. operational cost, maintenance cycles, energy use intensity over the building lifecycle, and carbon reduction potential across embodied and operational phases. This approach aligns with internationally recognized lifecycle costing methodologies (ISO, 2017). Decisions regarding lighting, controls, power distribution efficiency, and renewables integration must consider:

- **Capital Cost (CAPEX)** vs. **Operational Cost (OPEX)**
- **Maintenance cycles** and ease of upgrade
- **Energy use intensity (EUI)** over the building lifecycle
- **Carbon reduction potential** across embodied and operational phases

Example Applications

- Selecting a higher-cost luminaire with 10+ year lifespan and low maintenance vs. a cheaper product with 3–4 year replacement cycle.
- Using copper busduct systems to reduce transmission losses in large vertical buildings.
- Designing control systems for future upgrades to net-zero operational targets.

9.6 Regulatory Challenges in Emerging Technologies

9.6.1 Regulatory Uncertainty in Electrical Innovation

The accelerated deployment of renewable energy, energy storage, smart grids, and digital automation in modern electrical infrastructure presents considerable regulatory challenges. While the technologies evolve rapidly, applicable codes and standards often lag, resulting in incomplete or fragmented compliance frameworks for many emerging systems (IEA-PVPS, 2023; NIST, 2014). This regulatory misalignment introduces technical, legal, and financial risks during VE, particularly when proposing innovative design solutions intended to improve system efficiency, resilience, and sustainability (IEC, 2022; ISO, 2022; NFPA, 2023).

Examples of Regulatory Gaps

These uncertainties introduce **technical, legal, and financial risk** during VE, particularly when proposing innovative designs intended to improve efficiency or sustainability (Table 9.8).

Table 9.8 Regulatory Gaps in Emerging Electrical Technologies

Technology	Regulatory Challenge
Solar PV Systems	Inconsistent national requirements for grid interconnection, protection coordination, and islanding prevention mechanisms (IEEE, 2018; IEA-PVPS, 2023).
Battery Energy Storage	Fragmented adoption of safety and performance standards addressing thermal risk, ventilation, lifecycle evaluation, and end-of-life management (IEC, 2022; NFPA, 2023).
Smart Grids and IoT	Limited regulatory enforcement and uneven adoption of protocols for data privacy, cybersecurity, and interoperability (NIST, 2014).
EV Charging Infrastructure	Continuously evolving technical requirements related to fault protection, load coordination, billing integration, and bi-directional energy flow (Federal Register, 2023; ISO, 2022).

9.6.2 Value Engineering Strategies in Unregulated or Evolving Domains

In the face of incomplete or transitional regulatory frameworks, VE must apply adaptive strategies to maintain system value, future-proof design intent, and avoid compliance pitfalls.

Recommended VE Strategies (Table 9.9)

Table 9.9 VE Strategies for Unregulated or Evolving Technologies

Strategy	Description and Application
Proactive Regulatory Engagement	Establish early dialogue with authorities and grid operators to confirm anticipated standards
Scenario-Based Design	Develop alternate configurations based on possible regulatory evolutions (e.g., lithium vs. sodium-ion storage)
Flexible Infrastructure Planning	Design modular systems (e.g., inverter-ready PV bays, battery-ready switchboards) to accommodate upgrades
Exceed Minimum Standards	Use conservative safety factors or design criteria beyond current code to avoid future rework
Cross-Referencing Global Best Practice	Where national codes are silent, refer to IEC, NFPA, UL, or IEEE for validated precedents

9.6.3 Digital Tools to Manage Emerging Tech Compliance

Digital engineering tools increasingly support VE processes by enabling continuous regulatory monitoring, dynamic rule-checking, and multiscenario compliance evaluation. While such tools do not resolve existing regulatory gaps, they enhance risk identification, facilitate transparent documentation, and assist in maintaining regulatory alignment as standards evolve (NIST, 2014).

- **BIM-integrated energy simulations**: Evaluate VE proposals for energy performance under multiple future load profiles.
- **Automated rule-checking engines**: Flag inconsistencies between proposed designs and regulatory updates.
- **Regulation monitoring platforms**: Track international code developments in battery systems, PV, EV infrastructure, etc.

These tools do not resolve gaps in regulation but help teams identify exposure areas and generate design options with traceable assumptions

This matrix Figure 9.3 illustrates the alignment of VE strategies with common regulatory challenges associated with four major emerging electrical technologies: solar photovoltaic systems, battery energy storage, electric vehicle (EV) charging infrastructure, and smart grids with IoT integration. The strategies—proactive regulatory engagement, flexible infrastructure planning, exceeding minimum standards, and digital tool utilization—are positioned against each technology, highlighting how VE can mitigate uncertainty and future-proof system design in the context of evolving regulatory landscapes.

	Proactive regulatory engagement	**Flexible infrastructure planning**	**Exceed minimum standards**
Solar PV	Clarify grid tie-tie requirements early	Inverter-ready PV panel boardos	Oversize to accommodate future regulation
Battery energy storage	Obtain interin authority guidance	Battery-ready switchboards	Simulate different interconnection scenarios
EV charging	Verify evolving OCPP billing requirements	Allow DER integration for V2G	Select GFCIs for TT & N- systems
Smart grids & IoT	Engage cybersecurity agencies	Standard-communication protocols	Monitor code changes for NEV equipment

Figure 9.3 VE Strategy Matrix for Emerging Technologies

9.7 Navigating Global and Local Regulatory Variations

9.7.1 Diversity of Regulatory Frameworks

Electrical infrastructure design operates under a broad spectrum of regulatory frameworks ranging from highly prescriptive codes to flexible, risk-based systems. The degree of technical prescription, enforcement strength, and harmonization varies significantly across jurisdictions, creating both challenges and opportunities for VE teams working internationally (IEC, 2023; RISE, 2022).

Prescriptive regulatory systems such as the US National Electrical Code (NEC) focus on highly detailed, rule-based compliance requirements (NFPA, 2023), whereas performance-based or risk-informed models such as the IEC 60364 (IEC, 2018) and BS 7671 Wiring Regulations in the United Kingdom (IET, 2022) emphasize design outcomes and safety objectives with greater engineering discretion. Standards adoption also varies: some regions directly adopt IEC standards (e.g., Middle East, Africa), while others apply national adaptations or hybrid frameworks (IEC, 2023).

Sustainability codes add further diversity. In the United Kingdom, compliance with Part L of the Building Regulations mandates minimum energy efficiency performance (MHCLG, 2021), while voluntary certifications such as BREEAM (BRE, 2016) and LEED (USGBC, 2020) supplement mandatory codes in multiple jurisdictions. Licensing models similarly differ, with some countries enforcing national licensing and inspection regimes, while others employ less formalized contractor certification (World Bank, 2020).

Key Regulatory Variables Across Jurisdictions (Table 9.10)

Table 9.10 Comparative Regulatory Variables Across Regions

Regulatory Dimension	Variation Factors
Code Structure	Fully prescriptive (e.g., NEC) vs. risk-based or performance-driven (e.g., IEC, BS 7671)
Enforcement Mechanism	Strong AHJ oversight (e.g., United Kingdom, United States) vs. discretionary or underdeveloped enforcement
Standards Adoption	Direct adoption of IEC (e.g., Middle East, Africa) vs. adapted national standards
Sustainability Requirements	Mandatory green codes (e.g., Part L United Kingdom) vs. voluntary certifications (e.g., BREEAM, LEED)
Electrical Licensing Models	National licensing and inspection regimes vs. less formalized contractor certification

This variation affects everything from component selection and protection strategies to inspection protocols and commissioning procedures.

9.7.2 Implications for International Projects

Projects spanning multiple jurisdictions or involving international stakeholders must resolve the tension between standardization for efficiency and customization for local compliance. VE teams must be able to differentiate between aspects of the design that can be harmonized globally and those that require local adaptation.

Common Challenges and Responses (Table 9.11)

Table 9.11 VE Adaptation Responses to Global Compliance Challenges

Challenge	VE Response
Incompatible cable standards or voltage levels	Design dual-rated components; specify to the strictest common denominator
Differing breaker coordination or arc-flash rules	Use software-based protection studies tailored to each national context
Conflicting installation practices	Engage local engineers during early design stages to guide constructability compliance
Green certification credits vary by country	Align baseline design with globally recognized frameworks; adapt credits per region

9.7.3 Role of Local Expertise and Partnerships

Incorporating local technical, legal, and cultural expertise remains essential for effective VE execution in global projects. Local engineering consultants validate design compliance, facilitate negotiations with authorities, and support integration with local supply chains and construction practices. Where regulatory enforcement remains weak, project sponsors often require private adoption of international standards such as IEC and ISO as internal benchmarks to safeguard technical integrity and reduce future legal exposure (ITU-T, 2016).

Recommended Best Practices:

- **Hire Local Compliance Consultants:** Use in-country engineering professionals to validate design compliance and act as intermediaries with authorities.
- **Partner with Local Firms:** Collaborate with local design-build entities to access in-region supply chains and procedural knowledge.
- **Cultural and Legal Sensitivity:** Understand legal liability, documentation formats, and project authority hierarchies that differ by country.

➢ **Engineering Note:**
In regions where enforcement is weak, project sponsors may still require compliance with international standards (e.g., IEC, ISO) as a risk mitigation measure, effectively establishing a private compliance benchmark

This map Figure 9.4 illustrates the dominant influences on electrical regulations worldwide. IEC-based standards are widely adopted across Europe, Asia, Africa, and the Middle East. North America is governed predominantly by the NEC and UL frameworks.

This map illustrates **how different regions approach electrical safety standards**, combining two types of electrical requirements (GT Engineering, 2023):

IEC-Based Regions (Most of the World):

- Follow international standards developed by global experts.
- Use performance-based guidelines that countries adapt locally.
- Philosophy: "Collaborate globally on electrical safety goals."

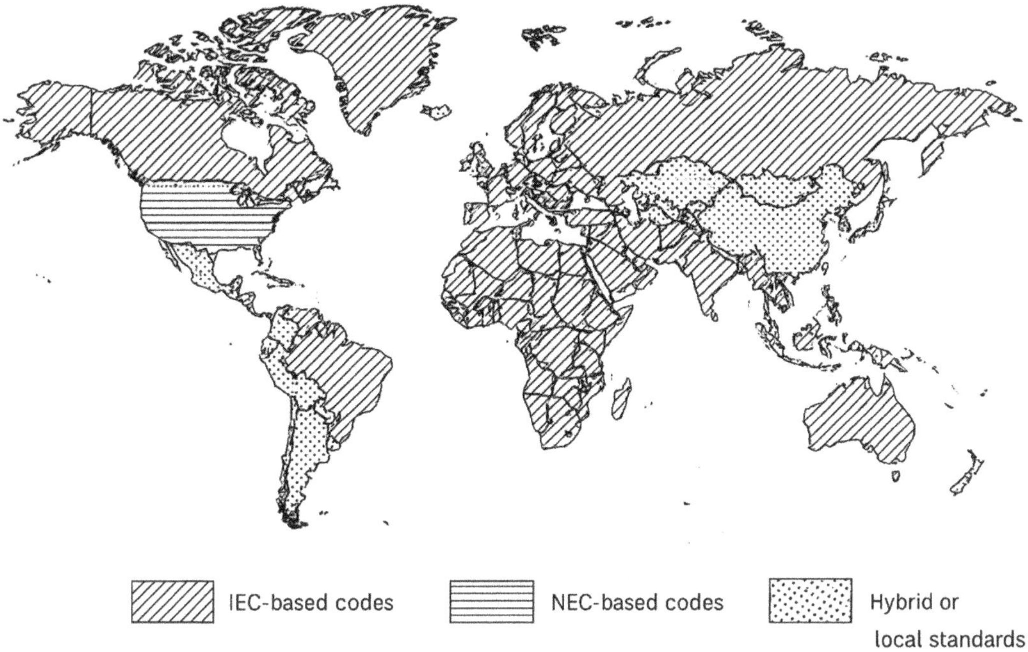

Figure 9.4 Global Compliance Frameworks in Electrical Design

The International Electrotechnical Commission's influence covers more than 97% of the world's population (Eland Cables, 2024).

NEC-Based Regions (Primarily North America):

- Follow detailed rules developed in the United States.
- Use prescriptive requirements with specific implementation steps.
- Philosophy: "Apply proven methods with precise specifications."

The National Electrical Code is used throughout the United States, including territories, with countries in the Caribbean, Central and South America, and some Asia-Pacific countries also adopting it (Earley, 2023).

Key Difference:

- **IEC approach:** "Meet these safety objectives using appropriate methods" (GT Engineering, 2023).
- **NEC approach:** "Follow these specific procedures for safety compliance" (GT Engineering, 2023).

Both approaches effectively ensure electrical safety through different methodologies—international consensus vs. regional expertise.

Note: These distinctions are becoming less defined as North American products increasingly adopt IEC standards, while IEC regions develop their own national codes. This map represents the predominant regional approaches to electrical safety.

9.8 Best Practices for Regulatory Compliance in VE

9.8.1 Establishing a Compliance-First VE Culture

For VE to succeed without compromising regulatory alignment, organizations must embed compliance into their technical culture. This principle extends beyond document control or periodic review; it must influence how project decisions are conceived, evaluated, and approved (OECD, 2020).

■ ***Foundational Cultural Elements***

- **Compliance as a Design Driver**: Treat codes, standards, and regulations as inputs to innovation, not constraints.
- **Early Engagement**: Involve compliance advisors, legal teams, and authority liaisons from concept design onward.
- **Proactive Education**: Conduct continuous training on emerging regulatory requirements across technical disciplines.

Best Practice Insight
Projects that identify critical regulatory risks during VE ideation save time and avoid design reversals at late stages (US DoD, 2025).

9.8.2 Regulatory Impact Assessment in the VE Process

Integrating regulatory assessments into every stage of the VE workflow is essential to ensure systematic compliance. The regulatory impact assessment (RIA) approach enables risk identification and structured decision-making aligned with evolving regulatory frameworks (OECD, 2020)

Recommended RIA Workflow Integration (Table 9.12)

Table 9.12 RIA Across the VE Process

VE Phase	Regulatory Integration Activity
Functional Analysis	Map each system function to mandatory compliance domains (e.g., fire safety, energy codes).
Idea Generation	Eliminate non-compliant options early; prioritize proposals with known regulatory headroom.
Cost-Function Analysis	Evaluate technical options using total compliance cost (direct + indirect).
Design Optimization	Use digital rule-checkers and authority pre-review where available.
Implementation Planning	Include regulatory milestones in project delivery plans.

9.8.3 Leveraging Digital Tools for Compliance Oversight

Modern projects increasingly employ digital platforms to automate portions of the compliance process. When properly implemented, such tools improve efficiency, reduce manual errors, and enhance compliance traceability, particularly in complex multi-standard environments (Zhang et al., 2022; Zhao et al., 2023).

Key Tool Categories:

- **BIM-Based Rule Checking**: Automates code compliance using geometry-aware logic engines (e.g., spatial clearances, system zoning) (Zhang et al., 2022).
- **Regulatory Change Trackers**: Monitors and alerts teams to updated codes or regional amendments (e.g., IEC or NEC revisions) (Zhao et al., 2023).
- **Audit Trail Platforms**: Records VE changes, code interpretations, and justification paths for approval authority transparency (Zhao et al., 2023).

➢ **Engineering Tip**

Digital tools don't replace judgment. Use them to augment compliance validation, not to offload responsibility.

9.8.4 Driving Continuous Improvement

A mature VE-compliance strategy includes feedback loops that enhance future performance.

Strategies for Continuous Regulatory Excellence:

- **Post-Project Compliance Reviews**: Capture lessons from inspections, deviations, or delayed approvals.
- **Regulatory Knowledge Base**: Build and maintain a shared VE + compliance repository, accessible across teams.
- **Recognition Systems**: Reward teams who achieve design innovation within challenging regulatory contexts

This flowchart Figure 9.5 illustrates how regulatory compliance checkpoints are embedded within the VE lifecycle. At each phase—ranging from functional analysis to implementation—compliance roles and tools are identified to ensure that VE proposals remain legally valid, technically sound, and fully auditable.

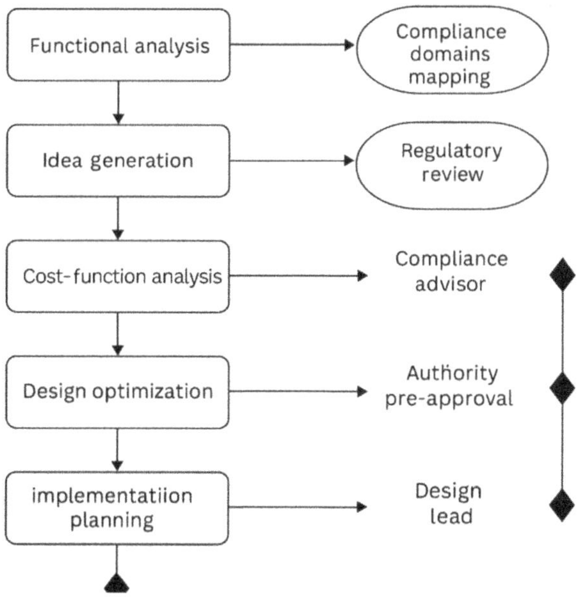

Figure 9.5 Integrated VE + Compliance Workflow

References

AFNOR. (2024). *NF C 15-100: Low-voltage electrical installations.* AFNOR Editions.

ASHRAE. (2022). ANSI/ASHRAE/IES Standard 90.1-2022: Energy Standard for Buildings Except Low-Rise Residential Buildings. American Society of Heating, Refrigerating and Air-Conditioning Engineers.

ASU Online. (2025) *What is a compliance officer? Job duties and more.* https://asuonline.asu.edu/newsroom/online-learning-tips/what-is-compliance-officer

BRE. (2016). *BREEAM international new construction manual 2016.* BRE Group.

BSI. (2018). *BS 7671:2018 requirements for electrical installations. IET wiring regulations* (18th ed.). BSI Standards Limited.

CIBSE. (2014). *Designing MEP systems and code compliance in the Middle East GCC region.*

Earley, M. (2023). Going global: international standards and the NEC. *Electrical Contractor Magazine.* https://www.ecmag.com/magazine/articles/article-detail/codes-standards-going-global

Eland Cables. (2024). FAQ: Countries using IEC cables. *The Cable Lab.* https://www.elandcables.com/the-cable-lab/faqs/faq-which-countries-routinely-use-iec-cables

ESFI. (2023). The national electrical code (NEC), *Electrical Safety Foundation International.* https://www.esfi.org/workplace-safety/industry-codes-regulations/the-national-electrical-code-nec/

European Parliament and Council. (2024). Directive (EU) 2024/1275 on the energy performance of buildings (recast). Official Journal of the European Union, L 2024/1275.

Federal Register. (2023). *National electric vehicle infrastructure standards and requirements.* U.S. Federal Government.

GT Engineering. (2023). IEC 60364 vs NEC - NEC and IEC countries. *GT Engineering Insights.* https://www.gt-engineering.it/en/insights/ul-and-csa-conformity/iec-60364-vs-nec/

HSE. (2015). *The electricity at work regulations 1989.* Health and Safety Executive.

IEA-PVPS. (2023). *Assessment of grid codes and regulations for PV grid integration: International survey report IEA-PVPS T14-12:2023.* International Energy Agency – Photovoltaic Power Systems Programme.

IEC. (2005). *IEC 60364-1:2005 low-voltage electrical installations - Part 1: Fundamental principles, assessment of general characteristics, definitions.* International Electrotechnical Commission.

IEC. (2018). *IEC 60364: Low-voltage electrical installations.* International Electrotechnical Commission.

IEC. (2022). *IEC 60364 series: Low-voltage electrical installations.* International Electrotechnical Commission.

IEC. (2022). *IEC 62933-5-2:2022 electrical energy storage (EES) systems — Safety considerations.* International Electrotechnical Commission.

IEC. (2023). *Global IEC adoption data 2023.* International Electrotechnical Commission.

IEEE. (2018). *IEEE Std 1547-2018: Standard for interconnection and interoperability of distributed energy resources with associated electric power systems interfaces.* Institute of Electrical and Electronics Engineers.

IEEE. (2022). *IEEE standards.* Institute of Electrical and Electronics Engineers. Intertek Inform (2023). *Is AS 3000 a Legal Requirement?*

IEEE. (2024). *IEEE standards association.* https://standards.ieee.org/

IET. (2017). Statutory and non-statutory documents applicable to the electrical industry, *Wiring Matters,* September 2017. https://electrical.theiet.org/wiring-matters/years/2017/67-september-2017/statutory-and-non-statutory-documents-applicable-to-the-electrical-industry/

IET. (2022). *BS 7671: IET wiring regulations* (18th ed.). Institution of Engineering and Technology.

ISO. (2017). ISO 15686-5:2017 Buildings and constructed assets — Service life planning — Part 5: Lifecycle costing. International Organization for Standardization.
ISO. (2022). *ISO 15118-20:2022 road vehicles — Vehicle to grid communication interface — Part 20: 2nd generation network layer and application layer requirements.* International Organization for Standardization. International Organization for Standardization.
ISO. (2024). *Standards.* International Organization for Standardization. https://www.iso.org/standards.html
ITU-T. (2016). *technical report on private standards adoption in national contexts.* International Telecommunication Union.
MHCLG. (2021). *Approved document L: Conservation of fuel and power.* Ministry of Housing, Communities and Local Government.
NECA. (2024). The national electric code (NEC), *National Electrical Contractors Association.* https://www.necanet.org/topics/codesandstandards/the-nec
NFPA. (2022). *NFPA 70: National electrical code (NEC).* National Fire Protection Association.
NFPA. (2023). *NFPA 855:2023 standard for the installation of stationary energy storage systems.* National Fire Protection Association.
NFPA. (2023). *NFPA 70: National electrical code 2023.* National Fire Protection Association.
NIST. (2014). *NIST framework and roadmap for smart grid interoperability standards, release 3.0.* National Institute of Standards and Technology.
OECD. (2020). *Regulatory impact assessment: Best practices in OECD countries.* Organisation for Economic Co-operation and Development.
RISE. (2022). *Regulatory indicators for sustainable energy global report.* World Bank/IEA.
UK Government. (2023). *Approved Document L: Conservation of fuel and power, Volume 1: Dwellings, 2021 edition incorporating 2023 amendments.* Ministry of Housing, Communities & Local Government. https://www.gov.uk/government/publications/conservation-of-fuel-and-power-approved-document-l
US DoD. (2025). *Value engineering: A guidebook of best practices and tools.* United States Department of Defense.
USGBC. (2020). *LEED v4.1 building design and construction guide.* U.S. Green Building Council.
World Bank. (2020). *Doing business 2020: Getting electricity indicators.* World Bank. https://subnational.doingbusiness.org/en/data/exploretopics/getting-electricity/good-practices#Regulating
Zhang, J., Wang, Q., & Li, B. (2022). Automated rule checking for MEP systems based on BIM and knowledge-based management system (KBMS). *Buildings*, **12**(7), 934. https://doi.org/10.3390/buildings12070934
Zhao, H., Wu, X., Li, Q., & Lu, W. (2023). Automated code compliance checking research based on BIM and knowledge graph technology. *Scientific Reports*, **13**, 34342. https://doi.org/10.1038/s41598-023-34342-1

10

Cross-Disciplinary Interfaces with Electrical Systems in Design and Value Engineering

10.1 Overview

Value engineering (VE) has long served as a methodological framework for optimizing cost, function, and performance within individual engineering disciplines. However, as the complexity of construction projects continues to evolve—particularly in the domain of electrical systems—cross-disciplinary applications of VE have become not only advantageous but essential. The electrical infrastructure of a modern facility does not exist in isolation; it interacts with architectural vision, structural design, mechanical services, plumbing systems, information and communication technologies (ICTs), and sustainability strategies. The convergence of these disciplines presents both challenges and unprecedented opportunities for value creation.

This section introduces the foundational principles of cross-disciplinary VE, outlining its scope, relevance, and strategic role in the delivery of high-performance, cost-effective, and sustainable electrical systems. By integrating perspectives from civil, architectural, mechanical, environmental, and digital engineering domains, VE becomes a powerful tool for identifying hidden synergies, resolving interface conflicts, and future-proofing infrastructure.

10.1.1 Strategic Importance of Cross-Disciplinary VE

Modern buildings, infrastructure networks, and mission-critical facilities demand integrated solutions that align across systems. Whether evaluating the structural implications of cable routing, the architectural constraints of luminaire placement, or the thermal loads imposed by electrical switchgear, a cross-disciplinary VE approach ensures that decisions are made with comprehensive technical foresight.

Cross-Disciplinary VE Fosters:

- **Whole-of-system optimization**, replacing discipline-siloed decision-making.
- **Life-cycle performance evaluations**, accounting for long-term operability, resilience, and adaptability.
- **Multi-stakeholder alignment**, aligning the objectives of architects, engineers, contractors, and clients.
- **Interface risk mitigation**, especially at junctions between electrical systems and other disciplines.
- **Informed trade-off analysis**, facilitating design choices that maximize value across technical, functional, and aesthetic dimensions.

Value Engineering for Electrical Systems in Construction, First Edition. Hazem Hussein.

Companion website: www.wiley.com/go/hazem

10.1.2 Applications in Project Life Cycle

The integration of VE across disciplines enhances every stage of the project life cycle (Table 10.1).

Table 10.1 Project Life-Cycle Phases and Cross-Disciplinary VE Opportunities

Phase	Cross-Disciplinary VE Applications
Concept and Feasibility	Aligning architectural massing with electrical distribution zoning; feasibility of renewable integration
Design Development	Collaborative decision-making on service zones, coordination of structural and electrical pathways
Construction Planning	Prefabrication strategies for mechanical, electrical and plumbing (MEP) racks, conduit routing in precast elements
Commissioning	Coordinated testing of power systems, Heating, Ventilation, and Air Conditioning (HVAC), and ICT control interfaces
Operation and Maintenance	Life-cycle costing and adaptability reviews across building systems

10.1.3 Driving Factors for Integration

Several evolving factors underscore the necessity for interdisciplinary VE in electrical design:

- **Electrification of infrastructure**, including HVAC, mobility systems, and water systems
- **Digital convergence**, with building systems integrating through Internet of Things (IoT), building management system (BMS), and cloud-managed services
- **Carbon neutrality targets**, necessitating integration of renewable generation, storage, and efficient load management
- **Modular and offsite construction**, requiring pre-coordinated services embedded in structural components
- **Code and compliance harmonization**, where safety, accessibility, and energy standards affect multiple systems simultaneously

10.1.4 Cross-Disciplinary Collaboration Framework

Effective VE requires structured interdisciplinary collaboration. Best practices include:

- **Integrated project delivery (IPD)** frameworks promoting early contractor and engineer involvement
- **Multi-disciplinary VE workshops** at key design milestones
- **Use of shared BIM models** for coordination, clash detection, and visualization
- **Standardized documentation** of interface assumptions, decisions, and dependencies
- **Feedback loops** from operations and maintenance teams to inform VE analyses

10.1.5 Key Outcomes of Cross-Disciplinary VE

- Reduction of rework and construction delays through proactive coordination.
- Enhanced spatial efficiency via co-optimized routing and access planning.
- Better system performance through harmonized operational parameters.

- Sustainable building outcomes by integrating environmental, mechanical, and electrical VE goals.
- Improved client satisfaction from cohesive, resilient, and adaptable infrastructure.

10.1.6 Conclusion and Transition

The overarching aim of this chapter is to demonstrate how VE—when approached from a cross-disciplinary lens—transforms isolated design efforts into coordinated, high-value systems. The subsequent sections explore this integration in detail, discipline by discipline, to provide practical, technical, and strategic guidance for engineering professionals engaged in complex projects.

10.2 Architectural Coordination with Electrical Systems

10.2.1 Balancing Aesthetic Intent and Functional Utility

The architectural concept of a building fundamentally shapes the parameters within which electrical systems must operate. From spatial layouts and ceiling heights to façade transparency and zoning, architectural decisions directly influence lighting schemes, power distribution pathways, and control strategies. VE facilitates a structured methodology to align aesthetic goals with electrical performance metrics—maximizing system utility without compromising design integrity (Miles, 2015).

Key VE Applications:

- Evaluation of ceiling heights and space provisions for cable trays, luminaires, and access zones.
- Coordination of lighting fixture placement with interior design and user experience.
- Avoidance of architectural features that obstruct service access or impede natural airflow around equipment.

10.2.2 Integrated Lighting and Power Planning

Lighting design is a primary area where architecture and electrical systems intersect. The joint application of VE allows for the harmonization of **natural and artificial lighting**, resulting in enhanced occupant well-being, lower energy consumption, and architectural coherence.

VE Opportunities:

- **Daylighting Strategies**: Incorporation of skylights, light shelves, and glazing aligned with lighting control zones.
- **Smart Lighting Controls**: Occupancy sensors and daylight harvesting systems integrated into lighting layouts.
- **Façade Coordination**: Influence of window-to-wall ratio and surface reflectance on internal lighting requirements.
- **Power Outlet Planning**: Distribution aligned with furniture layouts, workstation zoning, and future adaptability.

10.2.3 Renewable Systems and Electrical Readiness

Architectural planning increasingly includes provisions for on-site renewable energy generation, such as photovoltaic arrays, solar thermal systems, or small-scale wind turbines. These design elements demand early collaboration between architects and electrical engineers to ensure structural, spatial, and electrical compatibility.

Value Engineering Actions:

- Roof orientation and load-bearing assessments for PV system integration.
- Routing of DC cabling and inverter placement in line with electrical room accessibility.
- Coordinated design of green roofs or terraces to coexist with energy systems.
- Life-cycle cost analysis comparing conventional vs. renewable-ready building envelopes.

10.2.4 Fire Safety, Egress, and Regulatory Compliance

Electrical and architectural systems must jointly satisfy fire codes, emergency egress standards, and safety regulations (International Code Council, 2021; National Fire Protection Association, 2020). VE ensures that such compliance is embedded efficiently into the design, without redundancy or design conflicts.

Key Considerations:

- Coordinated placement of **emergency lighting, illuminated signage, and fire alarms** within architectural zones.
- Minimization of conflict between **escape route design and electrical access/containment**.
- Ensuring electrical rooms are sited and dimensioned in line with local code requirements (e.g., door swing, clearance).
- Compliance with **accessibility codes** for control panels and outlets in public buildings.

10.2.5 Adaptive Reuse and Architectural Upgrades

In **adaptive reuse projects**, legacy architecture must be reconciled with modern electrical demands. VE facilitates strategic planning to integrate new electrical systems while respecting architectural constraints or heritage requirements.

Common Challenges and VE Strategies:

- Limited ceiling voids or inaccessible vertical shafts: use of **surface-mounted raceways** or low-profile conduit.
- Preservation of historical features: concealment of wiring and fixtures in a reversible manner.
- Upgrading lighting and data systems in occupied buildings with minimal disruption.
- Integration of building automation systems (BAS) without altering existing finishes.

10.2.6 Elevator and Escalator Systems Integration

Vertical transportation systems interface heavily with both architectural flow and electrical supply. VE allows a joint review of system selection, energy use, and operational logic, enhancing both form and function.

VE Techniques:

- Evaluation of **regenerative drive systems** for elevators and escalators to capture braking energy.
- Incorporation of **smart dispatching and scheduling algorithms** to minimize idle energy.
- Power diversity and redundancy planning within **lift switchboards**.
- Sizing of lift lobbies and shafts to accommodate future equipment upgrades.

10.2.7 Smart Technologies and Flexibility in Design

Smart building systems—including IoT-enabled lighting, HVAC, and security—necessitate **design flexibility** and architectural accommodation for sensors, hubs, and control panels. VE supports proactive identification of future-ready design interventions.

Design-Focused VE Approaches:

- Recessed sensor placements integrated with architectural finishes.
- Zoning strategies that align architectural spaces with control groupings.
- Infrastructure for **PoE (Power over Ethernet)** and centralized building intelligence platforms.
- Modular design of service voids and ceiling systems to allow phased upgrades.

10.2.8 Summary of Architectural–Electrical Integration Principles

VE Opportunity Area	Architectural Consideration	Electrical Integration Outcome
Lighting Design	Façade design, glazing, and ceiling heights	Reduced energy use, improved user comfort
Equipment Room Siting	Core placement, access corridors	Efficient maintenance, reduced cable lengths
Renewable System Readiness	Roof orientation, parapet height	Improved solar yield, reduced retrofit cost
Emergency Planning	Escape route layout, fire compartmentalization	Code compliance, integrated alarm systems
Retrofit Scenarios	Heritage features, limited void space	Minimal disruption, surface-mounting strategy
Smart System Zoning	Architectural partitioning, control zones	Reduced cabling, flexible future expansions

Key Takeaways:

- **Cross-disciplinary coordination between architects and electrical engineers is critical** to delivering both visual and functional value in built environments.
- VE ensures that **aesthetic design is not compromised** by electrical requirements—and vice versa.
- Integrated early planning results in **energy-efficient, code-compliant, and adaptable systems**, adding long-term operational and commercial value.

- The adoption of **smart technologies and renewables requires joint spatial and technical foresight**, supported by VE principles.

Lighting design alternatives and their energy impact are summarized in Table 10.2.

Table 10.2 Lighting Design Alternatives—Façade Coordination and Energy Impact

Design Scenario	Façade Type	Lighting Control Strategy	Estimated Energy Use (kWh/m²/year)	VE Outcome
Scenario A—Full Daylight	High-glazing	Daylight dimming + occupancy	38	High CAPEX, Low OPEX
Scenario B—Mixed Façade	Mid-glazing	Manual override + sensors	45	Moderate CAPEX, Balanced
Scenario C—Minimal Glazing	Low-glazing	Time-based scheduling	52	Low CAPEX, High OPEX

10.3 Structural Engineering Interfaces

10.3.1 Electrical Routing and Structural Penetrations

Electrical systems impose spatial and physical requirements on the structural framework of a building. These include cable risers, trenching, containment supports, and room-specific load zones. Without early coordination, clashes between containment routes and beams, slabs, or structural walls often emerge—leading to costly rework or compromised performance. VE enables proactive resolution of these interface conditions.

VE Applications:

- Early mapping of containment paths (e.g., busducts, cable trays) to coordinate slab penetrations.
- Rationalization of vertical shaft locations aligned with core structural zones.
- Elimination of unnecessary slab cut-outs through prefabricated penetration planning.
- Avoidance of high-cost structural bridging by aligning heavy electrical equipment with beam spans.

10.3.2 Grounding Systems and Foundation Types

The foundation system selected—pile or raft—has direct implications on the electrical grounding design. Electrical engineers depend on stable, low-impedance earthing systems to protect personnel and equipment. VE helps reconcile the structural and electrical constraints.

Pile Foundations:

- Deep foundations may present **challenges for conventional earth pit installations**
- VE options include chemically enhanced ground rods, ring-type earthing systems, or earthing electrodes integrated with steel piles

Raft Foundations:

- Allow **greater flexibility** in locating shallow earth pits or integrating ground grids within slab structures.
- VE supports coordination of foundation rebar design to serve dual purposes (structural and electrical bonding).

VE Supports Assessment of:

- Resistance targets for safety and surge protection.
- Civil–electrical trade-offs for trenching vs. embedded methods.
- Cost and reliability across foundation types.

10.3.3 Generator Rooms and Acoustic Isolation

Generator rooms require careful structural coordination to accommodate heavy loads, vibration isolation, ventilation, and acoustic treatment. Structural VE considerations ensure that civil works support electrical functionality without overdesign.

Key Elements:

- Load-bearing slab design to support generator weight and inertia blocks.
- Floating floor systems and resilient mountings for vibration control.
- Integration of acoustic panels and absorptive wall assemblies for noise control.
- Coordination of exhaust flues, fuel piping, and fresh air intakes in slab and wall cores.

VE evaluates trade-offs between in situ construction and **modular acoustic enclosures**, balancing cost, performance, and space.

10.3.4 Embedded Services in Precast Construction

Precast concrete construction offers speed and quality benefits but demands exacting coordination of embedded electrical components. VE plays a critical role in:

- Optimizing the placement of embedded conduits, junction boxes, and anchor points
- Ensuring tolerances and conduit bending radii match prefabrication capabilities
- Mitigating the cost and inflexibility of misaligned or inaccessible in-slab conduits
- Evaluating hybrid strategies (e.g., embedded vertical conduits + surface-mounted horizontal systems)

Where feasible, VE may recommend **plug-and-play wiring systems** and **modular raceways** to improve adaptability in precast settings.

10.3.5 Equipment Weight, Access, and Floor Loading

Heavy electrical infrastructure, such as switchgear, UPS systems, or transformer units, imposes **static and dynamic loads** on the building structure. Improper coordination may lead to floor overstress or restricted access for maintenance and replacement.

VE Focus Areas:

- Load mapping of all electrical rooms, with reinforcement of slabs where required.
- Use of independent concrete plinths to isolate vibrations and moisture.

- Ensuring unobstructed access paths, clearances, and floor opening sizes for equipment delivery.
- Planning for future replacement cycles by including removable floor hatches or access hoods.

10.3.6 Electrical Trenches and Civil Interface Optimization

Electrical trenches are a fundamental civil feature used to facilitate underfloor cable routing in **substations, main switchrooms, transformer bays, control centers**, and large electrical plant rooms. These cast-in or precast enclosures provide segregated containment for power, control, and communication cables while preserving surface-level space for equipment placement and movement. However, trenches also introduce civil, operational, and safety challenges that must be resolved through coordinated VE processes (Table 10.3).

Table 10.3 Trench Construction Typologies and Applications in Electrical Infrastructure

Trench Type	Typical Application	Construction Notes
Cast-in-situ Concrete	Custom-designed main electrical rooms	Fully integrated with slab; requires formwork coordination
Precast Trench Sections	Outdoor substations, modular plant rooms	Factory-made; rapid install, limited flexibility
Buried Duct Banks	Outdoor to indoor transition, long-distance feeds	Below-grade with manholes or handholes for access

10.3.6.1 VE Considerations in Trench Design

Structural Coordination:

- **Load-bearing capacity of trench covers** must be coordinated with expected foot traffic or equipment rolling loads.
- **Reinforcement detailing** to avoid cracking under point loads (e.g., trolley wheels, transformers).
- **Waterproofing membranes** or admixtures for trenches in wet environments or basement levels.

Cable Management and Segregation:

- Use of **pre-divided cable ducts** to maintain clear segregation (power, extra-low voltage [ELV], control).
- VE analysis to determine optimal trench width and depth based on cable volume and bending radius.
- Optimization of access spacing: fewer but larger trenches vs. many smaller ones.

Drainage and Ventilation:

- **Sump pits and drainage points** to prevent water accumulation.
- Forced or passive ventilation in long trenches to manage heat buildup and gas accumulation (especially where cables run hot).
- VE trade-off: cost of continuous drainage piping vs. isolated local sumps.

Access and Maintainability:

- **Removable heavy-duty trench covers** for key access points (galvanized steel, concrete with lifting points).
- **Anti-slip surface finishes** for safety in operational areas.
- Avoiding cable congestion through phased installation planning and coordinated cable tray elevations.

10.3.6.2 Safety and Operational Efficiency

Trenches can pose **trip hazards, water ingress risks, and operational obstacles** if not designed with end-users in mind. VE teams should assess:

- Compliance with **electrical clearance requirements** inside trenches
- Implementation of **safety barriers or warning lines** near open trench covers during maintenance
- Use of **nonconductive covers or protective mats** in high-voltage areas
- Layout planning to minimize crossovers or overlap of access routes and open trenches

10.3.6.3 Trench Optimization in Specific Use Cases

Substations:

- Integration of **HV and LV segregation** within same trench via concrete partitions.
- Optimized layout to minimize trench crossings under HV switchgear.
- VE focus on reduced excavation and faster trench pour sequence.

Data Centers/Switchrooms:

- Use of raised floor alternatives vs. trenching in critical white space.
- Where trenches are used, evaluate **under-rack cable access points** and hot-aisle alignment.
- Coordination of trench covers with equipment footprints to maintain clearances.

Prefabricated Plant Rooms:

- Use of **precast trench sections with pre-installed conduits.**
- Alignment with skid-mounted equipment baseplates.
- Fast-track deployment with minimal wet works on site.

Key Takeaways:

- Electrical trenches are a **cost-effective and spatially efficient** solution for dense cable routing in large plant areas.
- VE analysis ensures that trench systems are **fit-for-purpose, safe, and maintainable**, without overengineering.
- **Early structural coordination** and **access planning** are vital to reduce operational disruptions and long-term risks.

- Optimized trench design must address **cable segregation, drainage, ventilation**, and **personnel safety** simultaneously.

10.3.7 Use of Structural Reinforcement in Lightning Protection and Earthing Systems

The use of steel reinforcement within building foundations as part of the lightning protection system (LPS) and earthing arrangement is a well-established strategy under IEC 62305-3 and EN 50522 (European Committee for Electrotechnical Standardisation, 2022; International Electrotechnical Commission, 2010). This method—commonly known as a Type B earthing arrangement or foundation earthing—offers a low-resistance, mechanically protected, and cost-effective grounding solution. However, successful implementation requires close coordination between structural and electrical design teams, particularly during early rebar layout, concrete casting, and inspection stages.

10.3.7.1 Foundation Earthing Principles and Standards

According to **IEC 62305-3 (Protection Against Lightning—Chapter 3)** and **EN 50522 (Earthing of Power Installations Exceeding 1 kV AC)**:

- A **foundation earth electrode** consists of **reinforcing bars or a dedicated conductor** embedded in concrete footings or slabs.
- The **electrical continuity** of rebar cages must be **mechanically secured and verified**, typically through welding or clamping.
- Minimum conductor cross-section, concrete cover, and connection requirements must be met to ensure durability and performance.

10.3.7.2 VE Opportunities

VE supports the optimization of foundation earthing by:

- **Eliminating separate earth electrodes** where compliant rebar-based systems are sufficient
- **Integrating equipotential bonding** directly into slab and footing reinforcement
- **Avoiding post-cast excavation**, saving time, labor, and material
- Ensuring rebar arrangements account for both **structural load paths** and **current dissipation needs** (Table 10.4)

Table 10.4 Rebar Coordination Requirements for Integrated Foundation Earthing

Coordination Area	Key Actions
Rebar Layout and Continuity	Ensure closed loops, good electrical contact, and connectivity across footings/slabs
Connection Points	Designate locations for LPS down conductor terminations and earth bar tie-ins
Structural Detailing	Confirm that rebar size, spacing, and location meet both structural and electrical roles
Inspection and Testing	Include continuity checks before concrete pour; ensure test points for resistance checks

10.3.7.3 Typical Implementation Approaches

- **Use of dedicated rebar cages** in foundation edge strips as earthing conductors.
- **Galvanized or copper-clad steel conductor** bonded to rebar grid for enhanced continuity.
- **Embedded test points** routed through slab to allow for post-construction inspection.
- **Combined LPS and power system earthing** where common earth termination is permitted by code.

10.3.7.4 VE in Challenging Soil Conditions

In areas with high soil resistivity or environmental constraints (e.g., rock, brownfield sites):

- Rebar-based foundation earthing may **outperform rod-based systems** due to extensive surface contact.
- VE may recommend **chemically enhanced concrete** or additional external ring conductors to supplement the rebar network.

Structural–electrical coordination principles and their VE outcomes are summarized in Table 10.5.

Table 10.5 Summary Table of Structural–Electrical Coordination Principles

Integration Area	Structural Consideration	VE Outcome
Cable Containment Penetrations	Beam layout, slab thickness, fire zoning	Reduced clashes, optimized core hole layout, lower rework costs
Generator Acoustic Isolation	Vibration isolation, wall thickness, load-bearing slab	Integrated structural-acoustic performance, smaller footprint
Grounding with Foundation System	Soil condition, rebar continuity, concrete type	Durable earthing, cost savings via natural electrode reuse
Embedded Conduits in Precast	Casting tolerances, conduit routing, rebar clash detection	Enhanced prefab coordination, reduced on-site rectifications
Equipment Weight on Slabs	Floor loading limits, plinth design, access openings	Reliable support, life-cycle service access preserved
Electrical Trenches	Load-bearing cover design, waterproofing, access spacing	Safe, maintainable underfloor routing with optimized structural integration
Foundation Earthing & LPS Bonding	Rebar grid continuity, connection locations, concrete coverage	Unified earthing system, reduced material cost, simplified compliance

10.4 Mechanical and HVAC Systems Integration

10.4.1 Overview

The integration of mechanical and electrical systems (M&E)—particularly HVAC services—with power infrastructure is essential in modern building environments. From chiller energy demands to BMS-controlled ventilation, the interdependence of these disciplines necessitates rigorous coordination. VE provides a structured method to identify redundancies, optimize system interaction, and maximize energy efficiency while controlling capital and operating expenditure.

This section addresses cross-disciplinary VE opportunities where HVAC and electrical systems converge—covering equipment selection, control logic integration, spatial coordination, redundancy planning, and energy strategies.

10.4.2 Chiller Selection and Power Distribution Planning

Chillers are among the most energy-intensive components in any facility. Their selection and positioning significantly impact electrical design, infrastructure sizing, and operational efficiency.

VE Considerations:

- **Air-Cooled vs. Water-Cooled Chillers**: Evaluate power draw, start-up current, auxiliary load (e.g., cooling towers, pumps).
- **Location Impact**: Rooftop vs. plantroom placement affects cable length, voltage drop, and containment cost.
- **Power Factor and Harmonics**: Evaluate chiller variable speed drives (VSDs) and harmonics impact on system quality.
- **Part-Load Performance**: Ensure VE reviews life-cycle energy costs, not just peak efficiency.

10.4.3 Electrical Load from Mechanical Systems

Mechanical systems contribute significant electrical loads through:

- HVAC motors (air handling units [AHUs], pumps, fans)
- Compressors, variable frequency drives (VFDs), and drives
- Control panels and actuators
- Heating elements (where not gas-based)

VE Integration:

- Grouping HVAC loads into logical distribution board (DB) zones for balancing and monitoring.
- Use of load diversity factors in calculating panel sizes and generator backup loads.
- Evaluation of shared uninterruptible power supply (UPS) or generator support for BMS, pressurization, and smoke control fans.

10.4.4 Motor Control Centers and Distribution Strategy

HVAC loads are often managed via motor control centers (MCCs). Their location, configuration, and control interface play a critical role in both electrical and mechanical efficiency.

VE Approaches:

- Centralized vs. decentralized MCCs: Balance between cable length and redundancy.
- Evaluate smart MCCs with integrated communication and condition monitoring.
- Standardize motor starter ratings to reduce spares and panel footprint.
- Plan physical access and clearance around MCCs for maintenance and thermal dissipation.

10.4.5 System Redundancy and Operational Resilience

Mechanical systems in critical environments (e.g., data centers, hospitals, laboratories) require redundancy. However, full N + 1 or 2N can incur significant electrical infrastructure cost.

VE Scenarios:

- Use of modular chiller banks with staged loading and partial redundancy.
- Dedicated circuits for life-safety fans with automatic load-shedding schemes.
- Assessment of power availability vs. HVAC criticality to prioritize load shedding or UPS inclusion.
- Use of quick-connect panels for temporary chiller or AHU deployment in failure scenarios.

10.4.6 BMS Integration and Control Strategy Optimization

The BMS is the central nervous system connecting HVAC operation to the electrical infrastructure. Poorly coordinated BMS logic can lead to energy wastage, short cycling, or equipment degradation.

VE Tactics:

- Early co-development of BMS sequence of operation with electrical logic diagrams.
- Use of load-driven control loops to reduce electrical peaks.
- Evaluate placement of control panels for power and network access.
- Incorporate emergency overrides, time-of-use logic, and maintenance modes in BMS programming.

10.4.7 Sustainable Design Synergies

Mechanical and electrical coordination is crucial for meeting sustainability goals such as Leadership in Energy and Environmental Design (LEED) (US Green Building Council, 2024), Building Research Establishment Environmental Assessment Method (BREEAM) (Building Research Establishment, 2023), and local energy codes.

Integrated VE Focus:

- Use of heat recovery ventilators (HRVs) to reduce electrical heating load.
- Passive cooling strategies (e.g., thermal mass) to reduce mechanical load.
- Integration of rooftop solar with HVAC auxiliary systems.
- Electrical infrastructure support for geothermal or solar-assisted chillers.

Table 10.6 highlights key coordination requirements between mechanical and electrical systems, along with value engineering opportunities.

Table 10.6 Mechanical–Electrical Integration VE Summary

Focus Area	MEP Coordination Requirement	VE Opportunity
Chiller Sizing and Placement	Power draw, cable length, location	Optimize energy profile, reduce distribution cost
HVAC Electrical Loads	Motor grouping, panel sizing, load shedding	Efficient DB zoning, resilience planning
MCC and Control Distribution	Space, access, smart control integration	Reduced footprint, improved operations and maintenance (O&M)
Redundancy Strategy	Equipment criticality, backup load hierarchy	Balanced CAPEX/OPEX across life-safety and comfort systems
BMS Control Logic	Wiring routes, override strategy, integration	Minimize peak demand, enable efficient part-load operation
Sustainability Measures	Passive systems, renewables integration	Reduced electrical load, enhanced certification outcomes

10.5 Plumbing–Electrical System Integration

10.5.1 Overview

While often treated as separate design domains, **plumbing and electrical systems are increasingly interdependent**, particularly in high-performance buildings. Water-based systems—including domestic water supply, drainage, wastewater treatment, and fire suppression—rely heavily on electrically powered components for functionality, control, and resilience.

VE in this interface focuses on **integrating water–energy infrastructure** to reduce duplication, optimize equipment selection, and enhance service continuity. This section highlights coordination strategies across pumping, heating, water treatment, rainwater reuse, and emergency systems, with an emphasis on life-cycle energy consumption and system interconnectivity.

10.5.2 Electrically Powered Plumbing Systems

The following systems within plumbing rely directly on electrical infrastructure:

- Domestic water booster pumps
- Hot water circulation pumps
- Greywater treatment units
- Water-cooled heat exchangers and storage tanks
- Electric geysers and heaters
- Sewage ejector pumps
- Rainwater harvesting pumps
- Chemical dosing and sterilization systems
- Fire suppression jockey pumps and deluge valve actuation

VE Considerations:

- Accurate power sizing of pumps and motors to avoid oversizing.
- Shared **MCCs** or standalone control panels.
- Standby power integration (via **UPS** or generator) for critical systems.
- Consolidation of local power and control circuits for ease of operation.

10.5.3 Water Heating and Electrical Demand

Electrical water heating is prevalent in small to mid-sized facilities and in all-electric buildings. It can also serve as a backup or pre-heating mechanism alongside solar or heat pump systems.

VE Review Areas:

- Evaluation of **resistive electric heaters** vs. **heat pump water heaters** for life-cycle energy use (Table **10.7**).
- Coordination of heating element sizing and switching with panel board capacity.
- BMS integration for **time-of-use tariff management** and demand control.
- Peak load reduction through storage tank sizing and staggered heating schedules.

Table 10.7 Comparative Analysis—Electric vs. Heat Pump Water Heating

System Type	CAPEX	OPEX	Efficiency %	Grid Impact	Suitability
Electric Heater	Low	High	~90	High	Small facilities, low duty
Heat Pump Water Heater	Medium	Low	250–300	Low	Sustainable buildings

10.5.4 Fire Protection Systems and Emergency Power Coordination

Fire protection systems contain both passive and active elements, with many requiring **electrical power to function**. These include:

- Electric fire pumps (main and jockey)
- Control panels and alarm interfaces
- Deluge valve solenoids
- Diesel engine batteries and pre-heaters
- Fire suppression system actuation (e.g., gas-based agents)

VE Planning Requirements:

- Dual-feed or ATS (automatic transfer switch) design for fire pumps.
- Proper discrimination and backup prioritization in switchgear.
- Integration of alarm signals to BMS and fire alarm panel.
- Generator fuel sizing considering fire pump runtime per NFPA 20or local standards.

10.5.5 Greywater, Rainwater, and Sustainable Systems

Increasing water reuse through greywater or rainwater harvesting introduces **new electrical demands**:

- UV sterilization, filtration, and chlorination equipment
- Control valves and solenoid actuators
- Pressure boosting and pumping systems
- Level sensors and monitoring instrumentation

VE Strategies:

- Shared panels for control and power of rainwater and greywater systems
- Low-power instrumentation selection
- Load-shedding classification for noncritical recycling equipment
- Renewable power source integration for remote or small-scale systems

10.5.6 Electrical Coordination in Wet Service Areas

Wet areas (e.g., pump rooms, riser shafts, under-tank vaults) pose specific **electrical safety challenges** that must be addressed during VE reviews.

Risk Mitigation Measures:

- Ingress protection (IP) rating compliance for panels and enclosures per IEC 60529.
- Residual current device (RCD) use in wet-accessible circuits in accordance with IEC 61008.

- Clearance zoning around water lines and floor drains.
- Bonding of metallic piping systems to equipotential grounding grid.
- Flood detection and automatic equipment shutdown circuits.

Plumbing–electrical integration opportunities are summarized in Table 10.8.

Table 10.8 Summary of Plumbing–Electrical System Integration Opportunities

Focus Area	Electrical Coordination Requirement	VE Opportunity
Booster and Circulation Pumps	MCC/control panel grouping, backup supply	Reduced cabling, consolidated control, better redundancy
Water Heating Systems	Load calculation, panel impact, tariff scheduling	Peak shaving, energy efficiency, lower OPEX
Fire Protection Powering	Critical load routing, ATS/generator coordination	Code compliance, resilience, reduced installation risk
Greywater and Rainwater Systems	Monitoring and control logic, power supply zoning	Sustainability, modular controls, low-power operations
Wet Area Power Infrastructure	IP rating, bonding, RCD protection	Improved safety, maintainability, compliance

10.6 Environmental Engineering and Sustainability Coordination

10.6.1 Overview

Environmental engineering plays an increasingly pivotal role in the planning and performance of electrical systems. As global mandates for decarbonization, net-zero design, and high-efficiency buildings accelerate (Energy Transitions Commission, 2025; IEA, 2024), electrical engineers must coordinate closely with environmental and sustainability specialists. This section focuses on VE strategies that align electrical design with sustainability objectives, covering energy optimization, renewable integration, material selection, life-cycle costing, and compliance with green certification frameworks.

10.6.2 Energy Efficiency and Load Reduction

Reducing energy consumption at the system level is a cornerstone of sustainable electrical engineering (IEA, 2024). Environmental strategies directly affect electrical load profiles and must be accounted for in VE analysis.

VE Opportunities:

- **High-efficiency LED lighting and daylight harvesting** to reduce lighting loads.
- **Demand-controlled ventilation** and variable-speed fans to cut HVAC electrical demand.
- **Envelope improvements** (e.g., high-performance glazing, insulation) that reduce cooling/heating loads, impacting chiller and pump sizing.
- **Energy-efficient plug loads**: smart appliances, low-loss transformers, and occupancy-controlled devices.

VE reviews quantify the **total energy impact**, enabling right-sizing of electrical infrastructure (e.g., distribution boards, UPS systems, generator capacity).

10.6.3 Renewable Energy System Integration

Environmental engineers and energy consultants often lead the design of on-site renewable generation systems. VE ensures these are **technically integrated into the electrical backbone**.

VE Focus Areas:

- **Photovoltaic (PV) System Placement**: orientation, structural load, inverter placement, shading
- **Battery Energy Storage Systems (BESS)**: autonomy targets, fire separation, charge/discharge logic
- **Hybrid AC–DC distribution considerations**
- Grid-tied vs. off-grid architecture analysis for cost, performance, and code compliance

VE supports **realistic feasibility studies** that account for capital cost, payback period, electrical coordination, and long-term maintainability.

10.6.4 Sustainable Materials in Electrical Design

Environmental engineering also influences material selection, favoring **low-carbon, and recyclable components** across the electrical system.

VE Review Points:

- Copper vs. aluminum cables: cost, embodied carbon, end-of-life recycling
- Halogen-free and low-smoke cables for indoor air quality and fire performance
- Recycled or recyclable enclosures for switchboards, luminaires, and wiring devices
- Avoidance of PVC and other high-impact polymers where feasible

VE tools such as **environmental product declarations (EPDs)** and **embodied carbon calculators** support material decisions that align with sustainability targets.

10.6.5 Life-Cycle Costing and Total Environmental Impact

VE analysis extends beyond initial capital cost to include **operational, maintenance, and disposal impacts** (Table 10.9).

Table 10.9 Life-Cycle Phases and Electrical Considerations in Sustainability

Life-Cycle Stage	Electrical Considerations
Design & Installation	Upfront cost, embedded energy, environmental certification
Operation	Energy usage, fault response, consumables
Maintenance	Spare parts, access, inspection frequency
End-of-life	Recycling potential, dismantling, e-waste handling

Environmental life-cycle costing models help compare conventional and sustainable system options in VE workshops.

10.6.6 Compliance with Green Certification Systems

Major building certifications such as LEED, BREEAM, and WELL Building Standard (BRE Group, 2024; International WELL Building Institute, 2024; US Green Building Council, 2024) assign credits for electrical efficiency, renewable integration, and smart control systems.

VE Contribution:

- Maximizing credit scores related to:
  - Lighting power density (LPD)
  - Renewable energy capacity
  - Energy monitoring and submetering
  - Demand-response readiness
- Ensuring required documentation (e.g., metering plans, controls logic, commissioning records)

VE teams help strategically align design decisions with certification criteria to achieve target ratings without unnecessary cost escalation (Table 10.10).

Table 10.10 Electrical Contributions to Sustainability Certifications

Certification System	Category/Credit Area	Electrical Scope/System Contribution	VE Integration Focus
LEED v4/ v4.1	Energy and Atmosphere (EA) Optimize Energy Performance	Design of energy-efficient lighting, HVAC motors, transformers; submetering infrastructure	Selection of high-efficiency equipment; load calculation accuracy; right-sizing systems
	EA—Renewable Energy Production	On-site PV, wind, or geothermal integration with electrical system	PV placement feasibility, inverter sizing, battery system coordination
	EA—Advanced Energy Metering	Whole-building and system-level metering (lighting, plug loads, HVAC)	BMS coordination, cost-optimized meter layout and gateway connectivity
	EA—Demand Response (DR)	Load shedding logic, DR-enabling equipment, smart controls	VE evaluates participation readiness, relay selection, and automation compatibility
	Indoor Environmental Quality (IEQ)—Lighting Control	Occupant-controlled lighting zones, tunable luminaires	Layout optimization of sensors, switches, DALI bus integration
	Innovation	Smart grid integration, off-grid resilience, intelligent system performance	Novelty of system design, dual-function solutions (e.g., solar + shading + daylighting)
BREEAM (UK/New EU)	Energy—Energy Monitoring	Submetering of major electrical loads and circuits	Cable containment coordination, accessibility for maintenance
	Management—Commissioning and Handover	Electrical system test plans, commissioning records	VE in test planning to minimize redundant test setups and support verification
	Materials—Environmental		

Table 10.10 (Continued)

Certification System	Category/Credit Area	Electrical Scope/System Contribution	VE Integration Focus
	Product Declarations (EPDs)	Switchboards, cables, luminaires with declared environmental impact	Sourcing low-carbon materials with EPDs; support circular economy
	Resilience	Standby generation, climate risk protection for switchgear, solar autonomy	Site-specific VE studies for climate-related electrical design decisions
WELL v2	Light—Glare Control, Circadian Lighting	Lighting systems designed to support occupant well-being	Sensor-integrated tunable lighting, spectral quality optimization
	Energy—Thermal and Electrical Resilience	UPS-backed critical circuits, load prioritization, microgrid readiness	VE supports system topology planning, backup coverage validation

10.6.7 Resilience and Climate Adaptation

Sustainability also includes **resilience planning**: the ability of electrical systems to perform during extreme weather, supply interruptions, or climate stress.

VE Tactics:

- Evaluation of system autonomy (solar + BESS for critical loads)
- Robust protection of outdoor switchgear and cable paths against flooding and heat
- Selection of **corrosion-resistant materials** for coastal or high-humidity environments
- Modular backup systems that enable post-event recovery or upgrade.

Environmental–electrical integration opportunities are summarized in Table 10.11.

Table 10.11 Summary of Environmental–Electrical Integration Opportunities

Focus Area	Environmental Coordination Requirement	VE Outcome
Energy Efficiency	Passive design, envelope upgrades	Reduced electrical loads, smaller infrastructure
Renewables Integration	PV layout, inverter spec, storage planning	Optimized energy mix, decarbonization compliance
Sustainable Material Selection	Embodied carbon, circular economy principles	Lower life-cycle impact, certification alignment
Life-Cycle Cost Optimization	Full system costing beyond CAPEX	Clearer value justification, long-term savings
Resilience and Adaptation	Extreme event tolerance, modularity	Robust infrastructure, climate readiness

10.7 ICT and ELV Systems Synergy

10.7.1 Overview

ICT and ELV systems have become integral to modern building functionality, supporting operations, automation, safety, and occupant experience. These systems include structured cabling, networking, security, voice/data, access control, CCTV, audio-visual (AV), public address (PA), and BAS.

The VE process applied to ICT and ELV focuses on **functional consolidation**, **power and containment coordination**, **interface efficiency**, and **technology adaptability**. The growing convergence of digital systems onto shared networks and infrastructure further increases the opportunity—and need—for early cross-disciplinary collaboration between electrical and ICT/ELV designers. VE also plays a critical role in aligning these systems with electrical power design, earthing strategies, spatial layout planning, and compliance with life-safety standards.

10.7.2 Structured Cabling and Containment Integration

Structured cabling systems (category 6/6A/7 and fiber optics) form the backbone of digital connectivity (ANSI/TIA, 2024; ISO/IEC, 2024). These systems require close integration with architectural and electrical pathways to ensure maintainability, isolation, and future-proofing.

VE Opportunities:

- **Shared cable tray or ladder pathways**, with appropriate separation from power circuits
- Use of **modular containment systems** for fiber and copper bundles
- **Pre-terminated cabling** or plug-and-play panels in high-density environments
- Centralized **telecom rooms** per floor for load balancing and containment efficiency

10.7.3 Power and Redundancy Strategy for ELV Systems

ICT and ELV systems often have sensitive power requirements and varying levels of redundancy. VE supports correct power classification, distribution topology, and backup strategy.

Key Systems:

- Network switches and PoE devices (IEEE, 2018)
- Security systems: access control, CCTV, intrusion detection
- Fire alarm panels and detectors
- Voice evacuation and PA systems
- Servers, storage arrays, and core IT infrastructure

VE Actions:

- Consolidated **centralized UPS** systems vs. distributed UPS units
- **Low-voltage DC busbars** for telecommunications or control equipment
- Identification of **noncritical devices** for load-shedding during power loss
- Use of **ATS** or generator-tied circuits for life-safety ELV
- Classification of systems into **life-safety**, **critical operational**, and **noncritical** loads to optimize protection, cable routing, and backup power strategy (Table 10.12).

Table 10.12 Redundancy Strategy and Backup Planning for ELV Systems

ELV System	Backup Requirement	Typical Power Source	VE Considerations
Fire Alarm System	Mandatory	Generator + UPS	Prioritize life safety loads with automatic transfer switching
CCTV	Recommended	UPS/PoE switch	Combine with access control for shared infrastructure
Access Control	Required	UPS/DC supply	Ensure fail-safe unlocking and egress power continuity
PA/Voice Evacuation	Required	Generator + battery backup	Integrate into fire alarm logic; maintain EN54 compliance
BMS Server	Recommended	Centralized UPS	Modular UPS recommended with isolated output to reduce interference
Telecom Rack (Core)	Required	Central UPS + secondary battery	Evaluate runtime autonomy and fire-rated containment access

10.7.4 ICT–Electrical Room Zoning and Spatial Planning

Proper spatial planning of telecommunications and ICT rooms ensures:

- **EMI (electromagnetic interference) protection** from power equipment (IEC, 2024)
- Easy access to floor-level structured cabling and backbone fiber
- Adequate cooling and heat dissipation infrastructure
- Routing segregation (power, comms, fire)

VE Considerations:

- **Avoid colocating ICT rooms** with high-voltage or noisy equipment
- Include **separate earthing points** for sensitive systems
- Allow for **rack scalability and future switchgear upgrades**
- Implement **equipotential bonding grids** coordinated with electrical engineers to meet both telecom and electrical code requirements

10.7.5 ELV System Convergence and Digital Integration

The shift toward **building convergence**—integrating ELV systems onto a common IP backbone—offers functional and cost efficiencies.

Converged ELV Systems Include:

- IP-based CCTV and access control
- Unified communications (telephony + video conferencing)
- BMS/BAS
- Lighting control (DALI, KNX)
- AV, PA, and digital signage

VE Benefits:

- Reduction in total cabling and conduit runs
- Shared switchgear and PoE power design
- Reduced equipment rooms, patch panels, and intermediate points
- Easier integration with BMS and centralized dashboards

10.7.6 Technology Readiness and Future Adaptability

VE supports future-resilient ICT/ELV design by incorporating:

- **Minimum 20% spare capacity** in rack units, conduits, and fiber pairs
- Flexible room layouts and raised floors
- Modular patch panels and switch configurations
- Design for hybrid networks: wired + wireless + cloud interface

Technology adaptability is essential for:

- Tenant turnover in commercial buildings
- Adoption of 5G, edge computing, or IoT systems
- Upgrades to smart building or digital twin infrastructure.

A summary of VE opportunities for ICT and ELV systems is provided in Table 10.13.

Table 10.13 Summary of ICT and ELV VE Opportunities

VE Focus Area	Coordination Objective	Expected Outcome
Cabling and Containment	Pathway optimization, separation, modularity	Easier installation, flexibility, code compliance
Power Supply and Redundancy	UPS topology, PoE design, load prioritization	Reliable operation, right-sized backup
Room Zoning and Access	ICT–HV segregation, EMI protection	Reduced signal interference, safer installations
Converged Network Architecture	Functional integration of ELV systems	Lower CAPEX/OPEX, enhanced automation
Technology Flexibility	Future-proofing for upgrades and reconfiguration	Reduced disruption, maximized long-term system value
Life-Safety Systems Integration	Power supply coordination, redundancy classification	Enhanced compliance, resilience, and risk prioritization

10.8 Construction Techniques and Modular Integration (Cross-Disciplinary VE in Prefabricated Electrical Deployment)

10.8.1 Overview

As construction increasingly adopts modular and prefabricated methods, electrical systems are now frequently designed, assembled, and partially tested off-site. The value of these systems depends not only on the quality of the electrical assemblies themselves but critically on their integration across disciplines—particularly civil, structural, mechanical, and architectural packages. VE must therefore address interface risks, sequencing, spatial constraints, and delivery readiness holistically to ensure successful modular deployment (Gibb and Isack, 2003; Pan and Sidwell, 2011).

10.8.2 Modular Riser Systems

Prefabricated riser modules—housing containment, cable trays, and, in some cases, even pre-pulled wiring—enable vertical power, lighting, and ICT infrastructure to be delivered with high repeatability and lower on-site labor.

Cross-Disciplinary VE Coordination:

- **Structural**: Coordination of riser shaft size, insert plates, and tolerance stacking during concrete pour stages.
- **Fire Safety**: Factory-integrated fire collars and sleeves to meet FRR (fire resistance rating) and sealability.
- **Architectural**: Finish interfaces, access panels, and floor transitions.
- **Mechanical**: Space coordination with HVAC duct risers or hydronic piping in mixed riser zones.

VE Benefits:

- Predictable installation sequence across typical floors.
- Reduced slab penetration works and rework risk.
- Firestopping and access requirements designed in from the start.

When effectively coordinated, these elements reduce slab penetrations, improve installation predictability, and simplify compliance with firestopping and inspection requirements (BCA, 2021).

10.8.3 Construction Sequencing and Assembly Logic

One of the most overlooked challenges in modular electrical deployment is how to sequence trade activities to accommodate large, preassembled units.

VE Tasks Include:

- Determining handover stages between civil shell works, electrical module install, and final cable terminations.
- Developing lifting and craning strategies for switchgear rooms, risers, and containment frames.
- Designing modules for early power-up, allowing temporary or phased commissioning.
- Identifying hold points for inspection and quality assurance before final enclosure of services.

Interdisciplinary Dependencies:

- Civil works must deliver cast-in supports, slab embeds, and shaft openings accurately.
- Mechanical installations (e.g., chillers, piping) may block access routes to electrical modules if not properly sequenced.
- Architect must account for floor loading, hatch clearances, and future maintenance zones.

10.8.4 Transport and Site Logistics for Electrical Assemblies

Prefabricated electrical systems—particularly skids, risers, and fully enclosed switchgear units—require careful transport planning to avoid costly delays or on-site improvisation.

VE Factors:

- Delivery dimensions must align with road clearance, hoisting plans, and crane envelope.
- Ensure access paths, laydown areas, and module landing points are free from obstruction.
- Modular interfaces (e.g., busduct flanges, pre-installed terminations) must be coordinated with adjacent mechanical and structural systems.

- Electrical engineers must coordinate temporary power supply and protection settings for early testing and energization.

Inadequate logistics coordination can erode the productivity gains of prefabrication and delay critical commissioning milestones (Pan & Sidwell, 2011).

10.8.5 Plant Equipment Access, Delivery, and Replacement Planning

Large and immovable electrical assets—such as transformers, main LV switchboards, UPS systems, and battery banks—must be planned not only for initial delivery but also for long-term maintenance, repair, and replacement.

VE Considerations:

- Coordinate floor loading capacity with structural engineers based on transformer mass or UPS weight.
- Provide **removable architectural panels, louvers, or access hatches** to allow for future equipment extraction.
- Maintain minimum corridor widths and turning radii for equipment handling equipment (e.g., dollies, pallet jacks).
- Incorporating embedded rails or track systems in large electrical rooms to facilitate safe rigging and future replacement.
- Avoiding obstructed equipment positioning that would necessitate major structural demolition for eventual removal.

Poor planning of future access requirements frequently leads to costly structural alterations many years after handover, especially when heavy or immovable equipment fails or requires upgrade (Shah et al., 2010). A VE checklist for modular electrical deployment interfaces is provided in Table 10.14.

Table 10.14 VE Checklist: Modular Electrical Deployment Interfaces

Item	Discipline Lead	Coordination Required With	VE Risk Focus
Shaft Size and Riser Fit	Structural	Electrical, fire consultant	Misalignment, inadequate clearances, rework during installation
Lifting Strategy and Weights	Electrical	Civil, logistics	Delayed delivery, crane oversizing, unsafe rigging
Early Testing and Energization	Electrical	BMS, safety, main contractor	Incomplete power sequencing, access conflicts during live testing
Slab Opening and Fixing Insertions	Civil	Electrical	Clash with reinforcement, core drilling post-pour
Fire Collar Pre-Installation	Fire engineer	Electrical, structural	Failed inspections, noncompliance with fire code
Equipment Replacement Access	Electrical	Architecture, structural	Costly demolition, inaccessible routes for future upgrades

10.9 Common Coordination Challenges and VE Resolutions

10.9.1 Overview

Despite advances in design tools and delivery models, large infrastructure and building projects often encounter coordination breakdowns at the interfaces between disciplines. These challenges frequently arise between electrical, mechanical, structural, architectural, and ICT domains—especially where spatial, operational, or compliance requirements overlap.

This section identifies common coordination failure points in electrical system design and implementation, and demonstrates how VE methods can resolve or pre-empt these issues through structured collaboration, phased reviews, and performance-based decision-making.

10.9.2 Misalignment of Equipment and Architectural Constraints

Common Issues:

- Electrical rooms sized too small or incorrectly proportioned
- Panel clearances, door swings, or ceiling heights in conflict with code
- Switchgear not aligned with architectural grid or columns

VE Resolutions:

- Early spatial coordination workshops with detailed BIM overlays
- 3D clash detection and walk-throughs with discipline leads
- Application of modular layout standards with scalable footprints
- Flexible walling systems for transformer/substation rooms

10.9.3 Structural Interface Conflicts

Common Issues:

- Cable trays clashing with beams or rebar
- Slab penetrations too small, misplaced, or omitted
- No allowance for equipment plinths or recessed trenching

VE Resolutions:

- Structural–electrical coordination checklists during RIBA Stage 3/4
- Predefined cable zone corridors in structural design packages
- Cast-in-sleeve strategies agreed during early VE studies
- Integration of rebar grounding strategies to reduce duplication

10.9.4 Mechanical Load and Space Overlaps

Common Issues:

- AHU ductwork obstructing switchgear access
- No space for thermal dissipation or air movement in UPS rooms
- Pumps and pipework blocking riser access for cabling

VE Resolutions:

- Joint power-load room zoning (electrical + mechanical)
- Minimum clearance guides and heat dispersion overlays
- Spatial VE review of critical plantrooms, risers, and ceiling voids
- Shared access corridors or ceiling height adjustments where feasible

10.9.5 Fire Strategy Conflicts

Common Issues:

- Incorrect fire rating of cable paths or penetrations
- Electrical risers not compliant with compartmentation
- ELV cabling routes exposed in non-fire-rated ceiling voids

VE Resolutions:

- Integrated fire stopping schedules included in VE workshop documentation
- Specification of tested firestop systems compatible with electrical containment
- Use of fire-resistant cable and sleeving materials pre-approved by consultants
- Modular riser collars and penetration sleeves with certification

10.9.6 Grounding and Earthing Overlaps

Common Issues:

- Grounding bar locations clashing with floor finishes or columns
- Lack of continuity across prefabricated structural steel
- Separate grounding systems for ICT, LPS, and power creating loops

VE Resolutions:

- Unified grounding strategy led by electrical team but reviewed by structural and ICT disciplines
- Earthing grids modeled in BIM with conductor routing and bonding plans
- Coordination with BMS and telecom teams for surge protection and equipotential bonding

10.9.7 Documentation and Design Version Discrepancies

Common Issues:

- Mismatched drawing revisions between trades
- Construction sequencing not reflected in service access plans
- Missing cable tray tags or containment references in architectural sets

VE Resolutions:

- Common data environment (CDE) and naming standards agreed during VE phase
- Discipline-specific drawing checklists with shared release protocols
- Tagging conventions across electrical and architectural sheets
- BIM federated model reviews at milestone design gates.

The following matrix highlights typical coordination risk levels between design disciplines based on common interface challenges encountered during electrical system integration (Table 10.15).

Each cell in the matrix indicates the severity of potential design conflicts between the two intersecting trades. The table supports early-stage VE by highlighting high-priority coordination zones where proactive planning and interdisciplinary workshops can yield the greatest benefit.

Table 10.15 Pairwise Interface Clash Risk Heatmap (Disciplines Overview)

From/To	Architectural	Structural	Mechanical	Fire	ICT	Electrical
Architectural	–	Medium	Medium	Low	Low	High
Structural	Medium	–	High	Low	Low	High
Mechanical	Medium	High	–	Low	High	High
Fire	Low	Low	Low	–	Low	⚠ Medium
ICT	Low	Low	High	Low	–	⚠ Medium
Electrical	High	High	High	Medium	Medium	–

Legend: Low Risk Medium Risk High Risk.

Notes:

- Risk levels are **based on historical clash trends and VE workshop outcomes** from complex projects.
- "High Risk" indicates interfaces where spatial, functional, or compliance misalignments are frequently observed and require structured resolution.
- The **diagonal entries are blank** (–), as disciplines do not clash with themselves.
- VE teams should use this table during **coordination kick-off sessions**, BIM modeling alignment, and construction staging reviews.

Table 10.16 summarizes common cross-disciplinary interface challenges and recommended VE-based coordination actions.

Table 10.16 Summary of Common Cross-Disciplinary Interface Challenges and VE Actions

Interface Type	Conflict Risk	VE-Based Coordination Action
Electrical–Architectural	Space, access, orientation	Room sizing templates, BIM validation, door swing simulation
Electrical–Structural	Slab openings, beam clashes, weight loading	Load coordination sheets, riser casting guides, trench integration
Electrical–Mechanical	Spatial overlap, thermal ventilation	Shared zoning plans, heat maps, void allocation plans
Electrical–Fire Consultant	Fire rating compliance	Firestop matrix, certified modular collars, riser seal schedules
Electrical–ICT	Bonding, segregation, UPS integration	Grounding topology VE, PoE vs. UPS separation logic
Cross-Discipline Docs	Design mismatches and tag misalignment	Centralized coordination log, shared mark-ups, discipline interface sheets

References

ANSI/TIA. (2024). *ANSI/TIA-568-D.1: Commercial building telecommunications cabling standard.* Telecommunications Industry Association.

BCA. (2021). *Prefabricated mechanical, electrical and plumbing (MEP) systems guidebook.* Building and Construction Authority.

BRE Group. (2024). *BREEAM International new construction technical manual.* Building Research Establishment.

Building Research Establishment. (2023). *BREEAM International new construction technical manual.* BRE Global.

Energy Transitions Commission. (2025). *Achieving zero-carbon buildings: Electric, efficient and flexible.* Energy Transitions Commission.

European Committee for Electrotechnical Standardisation. (2022). *EN 50522:2022 Earthing of power installations exceeding 1 kV a.c.* CENELEC.

Gibb, A. G. F., & Isack, F. (2003). Re-engineering through pre-assembly: Client expectations and drivers. *Building Research and Information*, **31**(2), 146–160.

IEC. (2024). *IEC 61000-4-6: Electromagnetic compatibility (EMC) - Testing and measurement techniques - Immunity to conducted disturbances.* International Electrotechnical Commission.

IEEE. (2018). *IEEE 802.3bt: DTE power via MDI over 4-pair infrastructure.* Institute of Electrical and Electronics Engineers.

International Code Council. (2021). *2021 International building code (IBC).* ICC Digital Codes.

International Electrotechnical Commission. (2010). *IEC 62305-3:2010 Protection against lightning – Part 3: Physical damage to structures and life hazard.* IEC.

International Electrotechnical Commission. (2013). *IEC 60529:2013 Degrees of protection provided by enclosures (IP code).* IEC.

International Electrotechnical Commission. (2018). *IEC 61008-1:2018 Residual current operated circuit-breakers without integral overcurrent protection for household and similar uses.* IEC.

International Energy Agency. (2024). *Technology and innovation pathways for zero-carbon-ready buildings by 2030.* IEA Publications.

International WELL Building Institute. (2024). *WELL building standard v2.* International WELL Building Institute.

ISO/IEC. (2024). *ISO/IEC 11801-1: Information technology - Generic cabling for customer premises.* International Organization for Standardization.

Miles, L. D. (2015). *Techniques of value analysis and engineering* (3rd ed.). McGraw-Hill Professional.

National Fire Protection Association. (2020). *NFPA 70: National electrical code.* NFPA.

Pan, W., & Sidwell, R. (2011). Exploring the potential of offsite construction for improving construction logistics. *Construction Management and Economics*, **29**(8), 748–765.

Shah, S., Kumar, S., & Lavy, S. (2010). Maintainability factors in facility design for electrical and mechanical systems: Literature review and future research directions. *Journal of Facilities Management*, **8**(2), 91–103.

U.S. Green Building Council. (2024). *LEED v4.1 Building design and construction reference guide.* U.S. Green Building Council.

11

Value Engineering for Existing Electrical Systems (Renovation, Brownfield Integration, and Retrofit Optimization)

11.1 Introduction

In the fast-evolving landscape of construction and electrical infrastructure, the conclusion of a project does not mark the end of engineering responsibility. Existing buildings—especially those several decades old—continue to serve critical functions within urban, industrial, and institutional environments. However, they often rely on outdated technology, undocumented modifications, and aging components that were never designed for today's operational demands.

These conditions introduce a complex mix of constraints and opportunities. Legacy systems must be adapted to meet modern energy efficiency goals, safety standards, operational continuity expectations, and digital infrastructure requirements. This is where **value engineering (VE)** becomes an essential tool—not only during initial design and construction, but as a strategic methodology for assessing, upgrading, and optimizing existing electrical systems.

This chapter addresses how VE principles can be systematically applied to **renovation**, **brownfield**, and **retrofit** projects. It provides a framework for evaluating legacy infrastructure, identifying upgrade priorities, analyzing cost-benefit options, and implementing technically sound and economically viable improvements—all while maintaining safety, compliance, and operational integrity in active environments.

11.1.1 Purpose and Scope

The application of VE to existing electrical systems presents a distinct set of challenges and opportunities compared to new-build scenarios. In operational environments, VE must address not only cost-effectiveness and technical performance but also life-cycle extension, regulatory compliance, safety enhancements, and the preservation of operational continuity. Legacy constraints, physical limitations, evolving codes, and accumulated obsolescence often require tailored VE methodologies specific to existing systems.

A structured framework is established to support the identification, analysis, and execution of VE interventions within operational facilities and existing infrastructures. The principles discussed apply across diverse building types and operational contexts, covering both targeted system upgrades and comprehensive electrical refurbishments. Emphasis is placed on technical optimization, capital efficiency, minimal service disruption, and long-term regulatory compliance.

Value Engineering for Electrical Systems in Construction, First Edition. Hazem Hussein.

Companion website: www.wiley.com/go/hazem

11.1.2 Key Characteristics of Existing System VE

VE applied to existing systems is shaped by the following distinguishing characteristics:

- **Operational Constraints**: Unlike greenfield projects, existing facilities often remain operational during interventions, requiring phased works, isolation protocols, and risk mitigation plans.
- **Space and Access Limitations**: Electrical rooms, risers, and cable routes may lack spare capacity or clearances needed for conventional upgrades, necessitating compact or modular retrofit solutions.
- **Legacy Compliance Gaps**: Older installations may not meet current safety and performance standards (e.g., earthing, discrimination, fault levels), requiring prioritized upgrades to mitigate nonconformance risks.
- **Unknown Asset Conditions**: Aging systems frequently lack reliable documentation or asset health records, increasing the importance of comprehensive surveys and diagnostic testing prior to design.

11.1.3 Objectives of VE in Existing Electrical Systems

The primary objectives guiding VE in existing systems are:

- **Maximizing System Performance and Reliability:** Improve power quality, resilience, and maintainability without full-scale system replacement.
- **Reducing Life-Cycle Operating Costs:** Identify retrofit options that reduce energy consumption, downtime risk, and maintenance burden.
- **Enhancing Safety and Code Compliance:** Address noncompliant installations and improve personnel and asset safety through targeted upgrades.
- **Enabling Adaptability for Future Demands:** Introduce modularity, smart metering, and infrastructure provisions to support evolving building and tenant requirements.
- **Minimizing Disruption and Downtime:** Apply engineering strategies that allow works to proceed under live or partially live conditions where possible.

11.2 Assessment of Existing Electrical Infrastructure

A rigorous and systematic assessment is the cornerstone of VE in existing electrical systems. Unlike new designs where parameters are established by the design team, retrofit scenarios must begin with a precise understanding of the current system's performance, limitations, and compliance status.

This section outlines the methodologies, tools, and evaluation criteria required to establish a reliable baseline for further VE analysis.

11.2.1 Condition Assessment Methodologies

Visual and Physical Inspections:

- Inspection of switchgear, panelboards, cables, and terminations for signs of overheating, corrosion, mechanical wear, and contamination.

- Verification of labeling, access clearances, and physical integrity of enclosures and raceways.
- Documentation review (e.g., single-line diagrams, test certificates, maintenance logs).

Thermal Imaging and Electrical Diagnostics:

- Use of infrared thermography to detect thermal anomalies at terminations, breakers, and busbars.
- Earth loop impedance and insulation resistance testing to validate integrity of protective devices and earthing systems.
- Power quality analyzers to measure voltage imbalance, harmonics, flicker, and transients.

Load Measurement and Logging:

- Deployment of portable or fixed meters to capture peak, average, and base loads over a representative time period (typically 1–4 weeks).
- Phase-wise load balancing assessment.
- Identification of transient loads and nonlinear load patterns.

Asset Health and Maintenance History:

- Review of maintenance history, fault logs, and prior interventions.
- Evaluation of known failure patterns, e.g., breaker nuisance tripping or voltage drop in distant circuits.
- Risk-based prioritization of aged or unsupported components (e.g., discontinued switchgear).

11.2.2 Baseline Performance and Benchmarking

System Efficiency and Losses:

- Estimation of system losses through comparative metering (utility vs. sub-distribution).
- Benchmarking against modern efficiency metrics (e.g., transformer and cable losses per IEC 60076-20, IEC 60287) (IEC, 2017).

Power Quality and Reliability Metrics:

- Identification of harmonics, voltage sags/swells, and flicker affecting sensitive loads.
- Quantification of system reliability using standard indices (IEEE, 2022):
  - SAIFI (System Average Interruption Frequency Index)
  - SAIDI (System Average Interruption Duration Index)
- Comparison with IEEE 1366 or IEC 61000-2-2 benchmarks for similar facilities.

Compliance Gaps and Performance Limitations:

- Mapping of actual installation against current standards (e.g., IEC 60364, NEC, BS 7671).
- Assessment of fault level compliance, discrimination, and device coordination using software simulation tools (e.g., ETAP, Amtech, etc.).
- Identification of capacity bottlenecks, such as overloaded circuits or undersized cable trays.

11.2.3 Documentation and Digital Modeling

As-Built Drawing Validation:

- Redlining of discrepancies between as-built drawings and field conditions.
- Update of single-line diagrams (SLDs) and panel schedules to reflect actual configurations.

Digital Twin Creation (If Applicable):

- Generation of a building information modeling (BIM)-ready model from validated field data to support scenario planning.
- Integration of asset condition tags and smart monitoring data, where available.

11.2.4 Component Life-Cycle Reference and Replacement Planning

A critical aspect of VE for existing systems involves understanding the typical operational lifespan of electrical components. While actual service life may vary due to environmental conditions, loading factors, and maintenance regimes, standard lifespan ranges provide a reference for prioritizing inspections, refurbishments, and phased replacements (Table 11.1).

Table 11.1 Typical Lifespan of Electrical Components

Component	Typical Lifespan	Remarks
Transformers	20–40 years	Subject to insulation degradation and thermal cycling. IEC/IEEE models estimate 20.55 years at full load, extendable with maintenance (CHK Power Quality, 2022).
MV/LV Switchgear	30–40+ years	Lifespan depends on mechanical operation count and enclosure conditions. IEEE guidelines and Schneider Electric recommend proactive maintenance for >40 years.
Circuit Breakers	20–40 years	Mechanical wear and contact degradation are key failure modes. Manufacturer data and NFPA 70E indicate 20–40 years with annual exercising (IEEE PES, 2024).
Cables	25–50+ years	Accelerated aging due to thermal loading or moisture ingress. ICEA/IEEE/IEC tests show modern cables can last 50+ years (Bajbor, 1987; IEEE, 2022; IEC 60502).
Busbars	30–40 years	Often durable; inspection for joint integrity is critical. Typical lifespan cited in industry practice; maintenance-dependent (ABB, 2014).
UPS Systems	10–15 years (electronics), battery replacement required sooner	Limited by battery aging and inverter component degradation. IEEE 944 outlines application/testing; typical electronics lifespan 10–15 years (IEEE, 1986).
Batteries (Lead-acid)	3–7 years (UPS use)	Performance declines with cycle count and temperature exposure. Real-world service life 3–7 years, often below manufacturer's 5–10 year design life.

Table 11.1 (Continued)

Component	Typical Lifespan	Remarks
Batteries (Li-ion)	8–15 years (UPS use)	Longer cycle life but more sensitive to voltage and thermal management. Typically last 2–3 times longer than lead-acid, with 8–15 years common in UPS.
Generators	20–30 years	Dependent on runtime hours and maintenance quality. Industry consensus; IEEE 446 for specifics (IEEE, 1995).
Lighting Fixtures (LED)	25,000–50,000 hours (∼10–20 years at 8 hours/day)	Gradual lumen depreciation; drivers often fail earlier. IEC 62722-2-1 standard and manufacturer data (IEC, 2014; Philips, 2024).
Solar Panels	25–30 years	Output efficiency declines ∼0.5% per year on average. IEC 61215 standard for performance warranty (IEC, 2021).
Sockets and Receptacles	15–25 years	Contact fatigue or insulation breakdown are failure modes. Industry practice, based on wear and environmental exposure.
Switches	15–20 years	Affected by mechanical wear and arcing. Manufacturer data and industry practice.
Timers and Sensors	10–15 years	Varies by application and exposure to voltage surges or environmental factors. Manufacturer data; varies by type and environment.
Smoke Detectors	8–10 years	NFPA 72 and manufacturer recommendation (NFPA, 2022).

Note: Lifespan figures are compiled from a mix of manufacturer datasheets and application notes, standards-body publications, peer-reviewed or government-funded studies, and reputable industry white papers. These values represent typical design, warranty, or field-observed expectations—not guarantees. Actual service life can vary greatly with installation environment (temperature, humidity, and vibration), loading and duty cycle, maintenance quality, manufacturing tolerances, and compliance with applicable codes. Validate against site-specific data and the original equipment manufacturer's guidance before making end-of-life or replacement decisions.

Implications for VE:

These lifespan benchmarks support decision-making in retrofit projects by identifying:

- **Obsolescence risks** that affect safety, reliability, or maintainability.
- **Components beyond economic repair**, where replacement offers better long-term value.
- **Phasing strategies** for staged upgrades based on criticality and condition.

VE principles recommend that replacements or upgrades should:

- Enhance functional performance and reliability.
- Reduce maintenance burdens or operational risks.
- Provide cost-effective life-cycle value through durable materials or advanced technology.
- Facilitate integration with future systems (e.g., digital controls, energy monitoring).

By aligning component renewal strategies with these goals, VE enables stakeholders to make informed decisions that balance capital expenditure with operational excellence. A visual representation of typical component lifespans is shown in Figure 11.1.

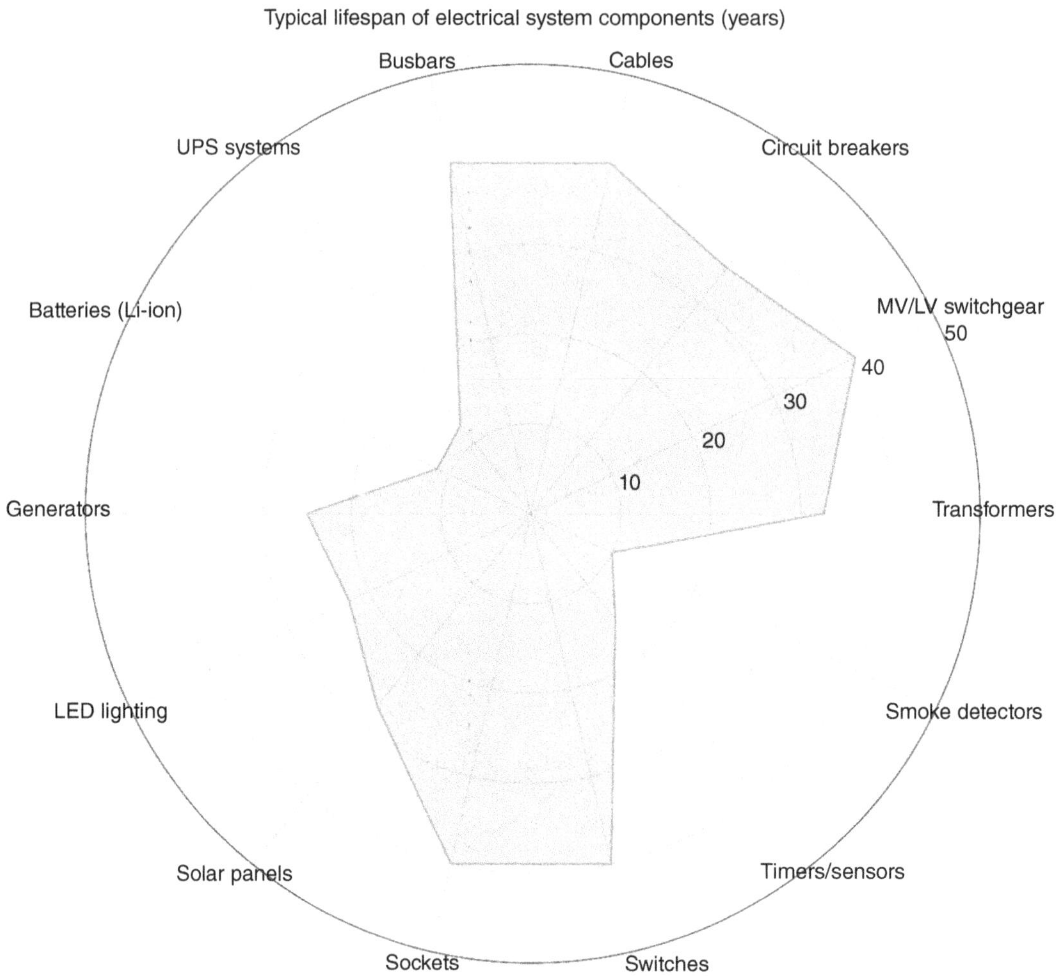

Figure 11.1 Typical Lifespan of Key Electrical System Components. This radar chart illustrates the average expected service life of major and minor electrical components under standard operating conditions. The chart aids in visualizing comparative longevity and supports life cycle-based planning for retrofit and VE interventions.

Key Takeaways:

- A comprehensive site and system assessment is essential to ground any VE proposal in factual performance data.
- Assessments should use a blend of noninvasive diagnostics, metering, and documentation reviews.
- Benchmarking against modern performance and safety standards reveals hidden upgrade opportunities with high value return.

11.3 Strategic VE Opportunities

Following the comprehensive assessment phase, strategic VE interventions can be identified across multiple layers of the electrical system. Unlike new-build contexts where design freedom is higher, retrofitting demands a more surgical approach—maximizing existing asset value while upgrading deficient areas within space, operational, and budget constraints.

11.3.1 System-Wide Optimization Areas

Panelboard Rationalization:

- **Objective**: Reduce the number of underloaded or fragmented boards through load reallocation.
- **Approach**: Consolidate circuits to improve load density, free up space, and minimize energy losses.
- **VE Benefit**: Reduces distribution losses, simplifies maintenance, and enables better monitoring integration.

Load Redistribution and Phase Balancing:

- **Objective**: Eliminate phase imbalance that leads to neutral overloading and inefficiency.
- **Approach**: Analyze historical load profiles and redistribute circuits across phases.
- **VE Benefit**: Enhances power quality, extends transformer life, and optimizes generator and UPS performance.

Enhanced Metering and Monitoring Integration:

- **Objective**: Enable real-time visibility of energy consumption and power quality.
- **Approach**: Install digital meters, transducers, and gateway interfaces on key panels.
- **VE Benefit**: Enables energy optimization, anomaly detection, and supports facility-wide smart building integration.

Redundancy Rationalization:

- **Objective**: Optimize redundant paths to align with actual criticality needs.
- **Approach**: Perform a load criticality audit and revise dual-feeding, ATS, or STS schemes accordingly.
- **VE Benefit**: Avoids overspecification and unnecessary CAPEX/OPEX in noncritical areas.

11.3.2 Specific Upgrade Opportunities

Transformer Refurbishment or Replacement:

- **Approach**: Assess winding condition, oil dielectric strength, and core losses. Replace with higher-efficiency (Eco Design Tier 2) units if below performance benchmarks.
- **VE Benefit**: Reduces core losses, improves voltage stability, and extends life cycle.

Switchgear and Panel Upgrades:

- **Approach**: Retrofit obsolete breakers, relays, and busbar sections. Where noncompliant, replace entire panels using modular, front-accessible designs.
- **VE Benefit**: Improves protection selectivity, enhances operator safety (e.g., arc flash containment), and allows space-efficient replacements.

UPS and Generator System Retrofits:

- **Approach**: Replace monolithic UPS with scalable modular units; reassess generator sizing based on actual load profile and duty cycle.
- **VE Benefit**: Reduces energy waste (especially at low loads), improves maintainability, and aligns with phased capacity growth.

Cable Life Extension and Routing Strategies:

- **Approach**: Use partial discharge testing and insulation resistance assessments. Where routing constraints exist, explore fire-rated trunking or underfloor containment solutions.
- **VE Benefit**: Extends usable life of legacy cables and improves maintainability without full cable replacement.

Controls and Automation Enhancement:

- **Approach**: Replace local switches/timers with programmable logic controllers (PLCs) or building management system (BMS) interfaces.
- **VE Benefit**: Enables centralized control, reduces manual interventions, and increases adaptability.

11.3.3 Upgrade Strategies for Future Readiness

In retrofit scenarios, futureproofing must be implemented pragmatically. This subsection presents VE-aligned strategies to introduce adaptability within legacy system constraints:

- **Smart Metering and Communication Provisions**: Install devices with Modbus/BACnet output and gateway compatibility to support energy analytics and digital integration.
- **Modular UPS and Generator Design**: Select scalable systems with hot-swappable modules and modular paralleling capabilities for future capacity alignment.
- **Space-Proofing and Provisioning**: Reserve space for additional panels, bus couplers, and cable entries to avoid costly future reworks.
- **Conduits for Renewables and EV Infrastructure**: Lay spare conduits and future stub-ups to rooftop and parking areas to support solar PV or EV charger rollouts.
- **IoT and BMS Interface Readiness**: Pre-install data points, local I/O space, and auxiliary power for future sensors, gateways, and edge computing devices.

Key Takeaways:

- VE strategies in existing systems must balance technical merit with installation feasibility and operational continuity.
- Opportunities range from optimization of existing assets to targeted component upgrades with long-term performance benefits.
- Forward-thinking interventions should be integrated wherever possible to minimize rework and support future adaptability.

> To support the structured evaluation of existing electrical systems, a full Electrical Retrofit Master Checklist Framework has been developed. This checklist, presented in **Appendix 11.A**, consolidates best practice inspection criteria across distribution systems, emergency power, lighting, ICT, medium-voltage systems, earthing, fire safety, and cable management domains.

11.4 Regulatory and Safety Compliance

Retrofitting existing electrical systems necessitates careful consideration of current safety standards and regulatory requirements. Unlike new builds, legacy installations often predate the latest editions of national and international codes, leading to compliance gaps that can pose legal, operational, and safety risks.

This section outlines the key regulatory considerations and safety interventions that must be addressed through VE during renovation or upgrade works.

11.4.1 Relevant Standards and Codes

International and Regional Frameworks:

- **IEC 60364:** Low-voltage electrical installations (EU, international) (International Electrotechnical Commission).
- **NEC NFPA 70:** National Electrical Code (United States).
- **BS 7671 (IET Wiring Regulations):** UK compliance standard for electrical installations.
- **IEEE 1584:** Guide for arc flash hazard calculations (Institute of Electrical and Electronics Engineers).
- **ISO 45001:** Occupational health and safety (OHS) management systems (for construction works).
- **EN 61439:** Low-voltage switchgear and controlgear assemblies (International Electrotechnical Commission).

11.4.1.1 Local Codes and Project-Specific Requirements

In addition to these international standards, all retrofit projects must comply with:

- **Local Statutory Regulations** issued by national and municipal authorities having jurisdiction (AHJ), including but not limited to Civil Defence regulations, ministry-level safety codes, or national construction standards.
- **Client-Specific Specifications** or operational standards are developed by facility operators, asset owners, or industrial sectors with stricter internal compliance frameworks.
- **Utility Company Regulations** governing metering, interconnection, protection coordination, and service entrance configurations.
- **Fire and Life Safety Codes** issued by local fire authorities, which may include firestopping systems, emergency egress lighting, smoke control systems, and emergency power classification requirements beyond international norms.

Local codes may impose stricter or additional requirements compared to international standards, particularly regarding:

- Fire-rated cable specifications
- Emergency generator redundancy
- Earthing system configurations
- Lightning protection testing intervals
- Arc flash boundary labeling and Personal Protective Equipment (PPE)classifications

Compliance with project-specific requirements must be validated at the design stage and verified during implementation through coordinated review with the local AHJ.

Compliance Strategy for Mixed Jurisdictions: For multi-site retrofit projects spanning different jurisdictions, establish a compliance matrix identifying the most restrictive requirements across all locations to standardize design approaches while ensuring local compliance.

11.4.1.2 Application to Existing Systems

While existing installations may remain compliant under legacy code provisions, any material upgrade typically triggers partial or full compliance with the current edition of the standard. For instance:

- Replacing a main LV panel generally requires compliance with EN 61439-1/2.
- Introducing new protective devices requires coordination and discrimination compliance per IEC 60947-2.
- Fire protection retrofits must meet NFPA 70 or BS 5839 requirements, depending on jurisdiction.

Common compliance issues in existing electrical systems and VE-oriented mitigation strategies are summarized in Table 11.2.

Table 11.2 Common Compliance Issues in Existing Electrical Systems and VE Responses

Issue	Cause	VE-Oriented Mitigation
Inadequate Earthing or Bonding	Legacy TT or IT system, corroded terminals	Reinforce earth electrodes, inspect bonding continuity
Outdated Protection Coordination	Breakers from different generations/manufacturers	Conduct full coordination study, replace critical devices
Noncompliant Cable Installations	Unsupported or exposed wiring, derated sizes	Re-route via fire-rated containment, re-calculate ampacity
Arc Flash Risk Not Evaluated	Lack of energy labelling and PPE specification	Perform IEEE 1584-based study, label panels, train staff
Nonisolated Maintenance Provisions	Live work on boards without safe isolation paths	Install test sockets, isolators, maintenance bypass schemes
Overloaded Circuits	Organic growth in loads over decades	Recalculate design load, reinforce circuits, update SLDs

11.4.2 VE Integration with Safety Upgrades

Arc Flash Risk Mitigation:

- **VE Strategy**: Replace manually resettable MCCBs with electronic trip units and add remote racking systems in high-risk switchgear.
- **VE Benefit**: Enhances safety, supports predictive maintenance, and reduces need for de-energizing panels.

Selective Protection and Cascading:

- **VE Strategy**: Optimize upstream/downstream breaker trip curves and introduce cascading configurations to reduce fault stress.
- **VE Benefit**: Prevents unnecessary upstream tripping, reduces equipment stress, and meets selectivity requirements.

Fire Protection Enhancements:

- **VE Strategy**: Retrofit smoke detectors and temperature sensors in distribution rooms and cable voids; use fire-rated enclosures.
- **VE Benefit**: Complies with fire safety regulations (e.g., BS 5839), mitigates business continuity risks.

Earthing System Reinforcement:

- **VE Strategy**: Conduct soil resistivity tests and improve earthing via additional rods, meshes, or chemically enhanced systems.
- **VE Benefit**: Reduces impedance, improves surge protection, and mitigates shock risk.

Key Takeaways:

- Compliance upgrades are often mandatory when retrofitting legacy systems, especially where critical functions are involved.
- VE principles can guide the integration of safety enhancements with minimal disruption and maximum long-term benefit.
- Coordination with local AHJ and certified testing bodies is essential for final sign-off.

11.5 Cost-Benefit Analysis for Retrofits

11.5.1 Overview

Cost-benefit analysis (CBA) constitutes a core decision-making tool within the VE process when evaluating electrical system retrofits. Following comprehensive condition assessment and deficiency identification, proposed retrofit measures must be subjected to a structured financial appraisal that quantifies not only direct costs, but also downstream operational, risk, compliance, and performance impacts over the projected service life of the installation (Dell'Isola, 1997).

The objective of VE-driven CBA extends beyond the selection of the lowest initial-cost solution. Instead, it aims to optimize total value by balancing capital investment against quantifiable operational savings, reduction of failure risks, deferred maintenance liabilities, regulatory exposures, and nonfinancial performance improvements over a clearly defined analysis horizon.

11.5.2 Key Components of Electrical Retrofit CBA

A comprehensive CBA for electrical retrofits typically includes the major elements listed in Table 11.3.

11.5.3 Financial Evaluation Methods

Selection of an appropriate financial evaluation model depends on project scope, data availability, and intended decision framework. Table 11.4 summarizes the primary methods applicable to VE-based retrofit

Table 11.3 Comprehensive CBA Components for Electrical Retrofits

Element	Description
Capital Cost (CapEx)	Total cost of implementation, including material procurement, installation labor, engineering design, project management, testing, and commissioning.
Operational Cost Savings (OpEx Reduction)	Reduction in energy consumption, maintenance efforts, downtime risk, and penalties (e.g., poor power factor charges) (Kelly & Male, 2001).
Life-Cycle Cost (LCC)	Total cost of ownership over a defined period (typically 10–20 years), including replacements and recurring maintenance (Woodward, 1997).
Risk Avoidance Value	Financial quantification of risk reductions (e.g., avoiding arc flash incidents, preventing business interruptions).
Regulatory and Compliance Value	Costs avoided through compliance with updated standards, fire codes, insurance conditions (Kelly & Male, 2001).
Residual Asset Value	Extended useful life and improved reliability of existing assets due to retrofit interventions.
Nonquantifiable Benefits	Improvements to safety culture, brand reputation, or staff productivity, even if not directly monetized.

Table 11.4 Financial Evaluation Methods for Retrofits

Method	Application
Simple Payback Period (SPP)	Indicates elapsed time for cumulative savings to equal initial investment; useful for rapid screening of alternatives (Woodward, 1997).
Net Present Value (NPV)	Calculates present value of all future cash flows (savings and costs), discounted to reflect time value of money; regarded as the most robust VE financial method (Woodward, 1997).
Internal Rate of Return (IRR)	Derives effective annual return rate on investment; facilitates comparative evaluation of competing alternatives (Woodward, 1997).
Life-Cycle Costing (LCC)	Captures total financial burden over full service life, inclusive of capital cost, operating expenditures, periodic replacements, and maintenance liabilities (Woodward, 1997).

Note: NPV and LCC methods are recommended for professional electrical retrofits, as they account for time value of money and future cost trends.

11.5.4 Example Applications of CBA in Electrical Retrofits

Example 1: UPS System Replacement:

- **CapEx:** €75,000 for new modular UPS system.
- **OpEx Saving:** €8,000/year energy savings + €2,000/year maintenance reduction.
- **SPP:** ~7.5 years.
- **Additional Benefits:** Improved reliability, future scalability.

Example 2: Lighting Retrofit (Fluorescent to LED):

- **CapEx:** €50,000.
- **OpEx Saving:** €12,000/year energy savings + €2,500/year reduced relamping costs.
- **SPP:** ∼3.5 years.
- **Additional Benefits:** Better lighting quality, reduced cooling load.

Example 3: Installation of Active Harmonic Filters:

- **CapEx:** €30,000.
- **OpEx Saving:** €5000/year improved equipment life expectancy + improved power factor penalties avoided.
- **SPP:** ∼6 years.

11.5.5 Key Factors Influencing Retrofit Financial Outcomes

When performing a CBA for electrical retrofits, the following critical factors should be accounted for:

- **Energy Tariff Escalations:** Future increases in energy costs amplify OpEx savings.
- **Maintenance Cost Trajectories:** Older systems often have escalating maintenance costs.
- **Risk Exposure Costs:** Potential losses due to system failure, fire incidents, or downtime must be realistically included.
- **Regulatory Pressure Trends:** Upcoming regulations may force action—early retrofits can pre-empt costly penalties.
- **Technology Advancements:** Emerging technologies (e.g., IoT-enabled monitoring, modular switchgear) offer longer-term flexibility and cost containment.

11.5.6 Practical Recommendations for CBA in Electrical VE Projects

- Always evaluate at least two or three alternatives to establish comparative value.
- Include sensitivity analysis to account for uncertainty (e.g., varying energy price trends).
- Present results to stakeholders using both financial and nonfinancial metrics.
- Ensure that cash flow impacts and operational risks are transparently communicated.
- Update the CBA if project scope changes during detailed design or procurement phases.

Key Takeaways:

- CBA transforms subjective upgrade ideas into financially justified retrofit strategies.
- Net present value (NPV) and life-cycle costing (LCC) offer the most complete financial evaluation methods.
- A proper CBA strengthens client confidence and supports informed decision-making.
- Integrating CBA into the VE process ensures optimized investment for both performance and cost efficiency.

11.6 Implementation Planning and Risk Management

11.6.1 Overview

Effective implementation planning is crucial to the success of electrical retrofit projects. Unlike new installations, retrofits must be executed within live operational environments where electrical safety, continuity of service, and stakeholder coordination are paramount.

Risk management is an integral part of implementation planning, ensuring that the retrofit works do not compromise personnel safety, building functionality, or legal compliance at any stage.

This section outlines the structured approach to planning and risk mitigation necessary for successful execution of value-engineered retrofit strategies.

11.6.2 Objectives of Implementation Planning

The main objectives of retrofit implementation planning are:

- Minimize disruption to building operations during electrical works.
- Ensure safety of occupants, maintenance personnel, and contractors.
- Sequence works logically to manage temporary supplies, shutdowns, and critical load transfers.
- Comply with all OHS regulations and electrical standards.
- Deliver project milestones on time and within the approved risk tolerance.

11.6.3 Risk Assessment During Implementation

Before any retrofit activities commence, a detailed project-specific risk assessment must be conducted (Table 11.5).

Reference Standards for Risk Assessment:

- ISO 45001 (Occupational Health and Safety Management Systems)
- IEC 60364-1 Clause 131.3 (Protection against electric shock and thermal effects)
- NFPA 70E (Electrical Safety in the Workplace)
- Local OHS and fire safety regulations

Table 11.5 Risk Assessment During Renovation Implementation

Risk Area	Assessment Focus
Personnel Safety	Risks of electric shock, arc flash, falls, hot works, and confined spaces.
Operational Continuity	Risks of unscheduled power loss, network interruptions, or system failures.
Fire Safety	Risks from cutting, drilling, or cable works penetrating fire barriers.
System Compatibility	Risks of integrating new equipment into legacy infrastructure without proper interface verification.
Material Availability	Risks due to long lead times for critical equipment (e.g., switchgear, UPS).

Each identified risk must be rated for likelihood and severity, and corresponding control measures must be documented.

11.6.3.1 Risk Prioritization Matrix

To systematically prioritize identified risks during electrical retrofit planning, a standard 5 × 5 risk assessment matrix is applied in accordance with ISO 31010. This matrix evaluates risks based on two dimensions:

- **Likelihood**: the probability of occurrence.
- **Severity**: the consequence of the event should it materialize.

Each identified risk is assigned a category based on the intersection of these two parameters. Risks rated as **high** or **very high** require immediate corrective action before any retrofit works commence, while **medium** risks must be actively mitigated and monitored. **Low** risks are considered tolerable but should still be periodically reviewed.

The typical risk assessment matrix applicable to electrical retrofit projects is shown in Table 11.6.

Table 11.6 Typical Electrical Retrofit Risk Matrix

Severity ↓ / Likelihood →	Very Low (Rare)	Low (Unlikely)	Medium (Possible)	High (Likely)	Very High (Almost Certain)
Minor (e.g., trivial injury, no equipment damage)	Low	Low	Low	Medium	Medium
Moderate (e.g., minor injury, isolated equipment damage)	Low	Low	Medium	Medium	High
Major (e.g., serious injury, major equipment damage)	Low	Medium	Medium	High	High
Critical (e.g., life-threatening injury, major system shutdown)	Medium	Medium	High	High	Very high
Catastrophic (e.g., multiple fatalities, total system loss, fire)	Medium	High	High	Very high	Very high

11.6.4 Phased Execution Strategies

Retrofit works must often be sequenced in phases to ensure minimum disruption. Typical phasing approaches are listed in Table 11.7.

Table 11.7 Phasing Strategies for Implementing Electrical Upgrades

Phasing Strategy	Description
Zone-Based Execution	Dividing the facility into logical zones (e.g., by floor or building wing) and working sequentially.
Time-Based Execution	Performing disruptive works during nights, weekends, or scheduled downtime windows.
Load-Based Execution	Prioritizing critical loads for temporary support or early commissioning before secondary systems.
Temporary Systems Deployment	Installing temporary generators, UPS, or distribution panels to bridge critical loads during retrofits.

Planning documents should include detailed method statements, switchover procedures, fallback strategies, and emergency recovery plans.

11.6.5 Stakeholder Coordination

Successful retrofits require coordinated planning with all relevant stakeholders, including:

- Facility management
- Building occupants and tenants
- IT and communications teams
- Security, fire safety, and emergency response teams
- Utility providers (where primary supply works are involved)

Stakeholder communication must include early notification of scheduled shutdowns, risk summaries, contingency plans, and clear points of contact during execution.

11.6.6 Safety Controls During Execution

The following electrical safety controls must be rigorously implemented during retrofit works, consistent with established standards: ISO 45001, NFPA 70E, and BS 9999 (Fire Safety Code of Practice):

- **Isolation Verification**: Lockout-Tagout (LOTO) systems applied before starting work.
- **Live Working Avoidance**: No live work permitted except under formally risk-assessed and approved conditions.
- **Temporary Earthing**: Application of portable earths where required.
- **Arc Flash PPE**: Personnel equipped with PPE appropriate to incident energy analysis results.
- **Fire Safety Measures**: Hot work permits, fire extinguishers, and fire watches where drilling or cutting activities are involved.

11.6.7 Documentation and Handover Planning

Implementation plans must include:

- Updated as-built drawings post-retrofit.
- Revised SLDs showing changes.
- Commissioning records, testing certificates, and EICRs.
- Updated maintenance manuals and emergency response procedures.

These documents must be reviewed and approved before the final project closeout and formal handover.

Key Takeaways:

- Retrofit implementation must be phased, risk-mitigated, and precisely sequenced to ensure safety and continuity.
- Risk assessment and stakeholder coordination are non-negotiable components of retrofit success.
- LOTO, temporary supplies, phased shutdowns, and fallback plans must be integral to project planning.
- Thorough documentation and commissioning sign-offs are essential for full life-cycle value delivery.

11.7 Case Studies

The following case studies[1] illustrate practical applications of VE principles in the retrofit of existing electrical systems. Each case study demonstrates how structured assessment, strategic VE intervention, and systematic risk management can achieve tangible improvements in safety, operational performance, and long-term value.

11.7.1 Case Study 1—Retrofit of Main Electrical Infrastructure in an Operational Commercial Building

11.7.1.1 Background

The facility was a **15-storey commercial office building** constructed in the early 1990s, located in a major metropolitan city in the Gulf Cooperation Council (GCC) region. The building served multiple tenants across financial services, law firms, and technology companies.

The original electrical infrastructure comprised:

- Main distribution boards (MDBs) nearing 30 years of operation.
- A network of sub-distribution boards (SDBs) and outdated cabling.
- No centralized surge protection.
- Emergency systems powered by a single generator with a limited runtime capacity, sized without de-rating for high ambient temperatures common in the region.

Planned retrofits needed to be completed while the building remained fully operational, with minimal disruption to tenants, and all upgrades required compliance with **local Civil Defence fire and life safety regulations** in addition to international standards.

11.7.1.2 Challenges Identified

The initial site audit revealed several critical issues:

- **Aging Switchgear**: Breakers with no available spare parts; mechanical wear is evident.
- **Overloaded Sub-Boards**: Load growth from IT expansion had not been properly managed.
- **Poor Power Quality**: Elevated levels of third and fifth harmonics were detected, causing transformer overheating.
- **Noncompliant Emergency Systems**: No battery autonomy compliance with updated standards (BS EN 50171).
- **Fire Barrier Breaches**: Cable penetrations through rated walls without firestopping.
- **High Risk of Arc Flash**: Lack of proper protection coordination and arc flash energy mitigation.

Without intervention, the building faced increased risk of service failure, noncompliance penalties, and insurance challenges.

11.7.1.3 VE Opportunities Implemented

Following a structured VE analysis, the strategic retrofit measures given in Table 11.8 were adopted.

In Addition:

- Annual maintenance costs decreased by approximately **25%**.
- Building insurance premiums were reduced following the compliance upgrade confirmation.
- Tenant satisfaction surveys showed improved ratings regarding power reliability post-retrofit.

---

1 While based on real-world engineering practice, specific project names, client identities, and sensitive details have been anonymized to preserve confidentiality.

Table 11.8 VE Summaries for Case Study 1

VE Strategy	Action
Switchgear Modernization	Replacement of MDBs and major SDBs with modular, withdrawable-type switchgear, improving serviceability.
Load Rebalancing and Segregation	Reallocation of IT and HVAC loads onto separate new SDBs to prevent circuit overloading.
Harmonic Mitigation	Installation of active harmonic filters at the main LV panels to control total harmonic distortion (THD).
Emergency Power Upgrade	Replacement of the generator with a dual-fuel model and installation of compliant central battery systems for lighting.
Fire Safety Enhancements	Comprehensive fire stopping installed at all cable wall and floor penetrations.
Arc Flash Risk Reduction	New protection coordination study performed; arc flash protection relays installed on main panels.

A phased execution strategy was developed, executing works outside peak operational hours, and temporary bypass panels were installed to maintain continuous supply during switchgear changeover (Table 11.9).

Table 11.9 Results Achieved for Case Study 1

Metric	Before Retrofit	After Retrofit
Switchgear Reliability Rating	Moderate (MTTR >48 hours)	High (MTTR <8 hours with modularity)
Power Quality (THD Voltage)	8.5%	3.2%
Compliance with Fire Safety Standards	Partial	Full (BS 9999 and EN 1366-3 compliant)
Emergency Power Runtime	45 minutes	3 hours minimum, compliant with critical load requirements
Arc Flash Incident Energy	>12 cal/cm^2	<8 cal/cm^2 with faster clearing times

Lessons Learned:

- Early involvement of tenants and building management was critical to scheduling shutdowns and minimizing complaints.
- Temporary bypass strategies greatly reduced perceived downtime risks.
- Modular switchgear significantly enhanced maintenance flexibility and future scalability.
- Comprehensive risk assessments before and during execution were essential to preventing live circuit accidents.

11.7.2 Case Study 2—Upgrade of Emergency Power Systems and Fire Safety Circuits in a Critical Facility (GCC Region)

11.7.2.1 Background

The project involved a **specialized healthcare facility** located in a major city within the GCC region. The facility provided 24/7 emergency services, intensive care units (ICUs), surgical suites, and critical pharmaceutical storage.

Originally built in the early 2000s, the electrical infrastructure had become increasingly strained:

- A primary diesel generator installed during initial construction.
- UPS systems with VRLA batteries exceeding 10 years of operation.
- Fire alarm power supplies based on obsolete and decentralized battery systems.
- Lack of compliant fire-rated cabling for emergency systems.

Given the essential nature of medical operations, **zero unplanned downtime** was permitted. All works had to comply fully with **local Civil Defence fire and life safety codes**, in addition to international standards such as IEC 60364 and ISO 8528.

11.7.2.2 Challenges Identified

The detailed electrical audit revealed several urgent risks:

- **Aging UPS Systems**: Battery degradation leading to reduced backup times (<15 minutes vs. 1-hour requirement).
- **Single-Point Generator Failure**: No redundancy on emergency generation.
- **Noncompliant Emergency Lighting Circuits**: Standard cables used instead of fire-rated wiring.
- **Outdated Automatic Transfer Switch (ATS)**: Transfer times exceeding permitted thresholds (>15 seconds).
- **Fire Alarm System Vulnerability**: Critical panels powered from non-fire-rated circuits with inadequate backup.

Environmental conditions (ambient temperatures exceeding 45 °C) exacerbated system stress, particularly on batteries and generator performance.

11.7.2.3 VE Opportunities Implemented

Following VE analysis, key strategies given in Table Table 11.10 were deployed.

All retrofits were performed in live operational conditions with detailed phased planning, temporary backup systems, and 24/7 site supervision (Table 11.11).

Table 11.10 Deployed VE Strategies for Case Study 2

VE Strategy	Action
UPS System Modernization	Replacement with modular, high-temperature-rated UPS systems (designed for ambient up to 50 °C) using lithium-ion batteries for longer life cycle and resilience.
Generator System Redundancy	Installation of N+1 diesel generators with synchronized operation capability, de-rated appropriately for peak summer conditions.
Fire-Rated Cabling for Emergency Circuits	Complete rewiring of fire alarm, evacuation lighting, and essential systems using two-hour fire-resistant cables (BS 7629-1, IEC 60331 compliant).
Upgraded ATS Panels	Installation of high-speed ATS panels with test bypass capability and integrated monitoring (UL 1008 certified).
Fire Alarm Power Supply Centralization	Consolidation of fire alarm power supplies onto a monitored central battery system compliant with BS EN 54-4 and local Civil Defence standards.

Table 11.11 Results Achieved for Case Study 2

Metric	Before Retrofit	After Retrofit
UPS Backup Time	12–15 minutes (unstable)	90 minutes minimum (stable across critical systems)
Generator Redundancy	Single generator (no backup)	N+1 configuration (full redundancy)
Emergency Lighting Compliance	Noncompliant wiring	Full compliance with Civil Defence fire rating standards
ATS Transfer Time	>15 seconds	<6 seconds (with test bypass ready)
Fire Alarm System Resilience	High risk of failure	Fully redundant, centralized, monitored supply

Additional Qualitative Benefits Included:

- Enhanced building fire safety certification from local Civil Defence authorities.
- Improved operational readiness audits for healthcare accreditation purposes.
- Reduced battery replacement costs due to lithium-ion upgrade (projected 10-year savings).
- Elevated resilience against regional power instability during peak load periods.

Lessons Learned:

- Early and close coordination with local Civil Defence authorities significantly accelerated approval processes.
- Modular UPS and N+1 generator designs proved highly effective for facilities requiring uninterrupted operations during staged retrofits.
- Fire-rated cabling upgrades had to account not only for resistance to fire but also for mechanical protection in congested retrofit routes.
- Temporary UPS and generator support systems were crucial to managing zero-downtime during transition phases.

References

ABB. (2014). *Busbar systems: design and application guide.*

BS. (2017). *BS 9999:2017 – Code of practice for fire safety in the design, management and use of buildings*. BSI.

CHK Power Quality. (2022). *Transformer loss of life.* https://www.chkpowerquality.com.au/

Dell'Isola, A. J. (1997). *Value engineering: practical applications for design, construction, maintenance & operations*. Romans.

Bajbor, Z. Z. (1987). Cable life expectancy calculation—a practical approach. *IEEE Transactions on Electrical Insulation*, 22(4), 485–487. Available at:. https://doi.org/10.1109/TEI.1987.298909

IEC. (2021). *IEC 60502-1:2021 Power cables with extruded insulation and their accessories for rated voltages from 1 kV (Um = 1.2 kV) up to 30 kV (Um = 36 kV) – Part 1: Cables for rated voltages of 1 kV (Um = 1.2 kV) and 3 kV (Um = 3.6 kV)*. Geneva: International Electrotechnical Commission.

IEEE. (2022) *Issue of aging equipment (end of life, diagnostics, etc.).* IEEE Power & Energy Society. Available at: https://ieee-pes.org/trending-tech/issue-of-aging-equipment/ (Accessed: 25 October 2025).

IEC. (2014). *IEC 62722-2-1: Luminaire performance – LED luminaires.*

IEC. (2017). *IEC TS 60076-20:2017 power transformers - Part 20: Energy efficiency.* International Electrotechnical Commission.

IEC. (2020). *IEC 60364-1:2020 – Low-voltage electrical installations – Part 1: Fundamental principles, assessment of general characteristics, definitions.* International Electrotechnical Commission.

IEC. (2021). *IEC 61215: Terrestrial photovoltaic (PV) modules – Design qualification and type approval.*

IEEE. (1986). *IEEE Std 944-1986: Application and testing of uninterruptible power supplies for power generating stations.*

IEEE. (1995). *IEEE Std 446-1995: Recommended practice for emergency and standby power systems for industrial and commercial applications.*

IEEE. (2022). *IEEE 1366-2022 - IEEE guide for electric power distribution reliability indices.* Institute of Electrical and Electronics Engineers.

IEEE PES. (2024). *Issue of aging equipment* (end of life, diagnostics, etc.)', *IEEE Power & Energy Society.* Available at: https://ieee-pes.org/trending-tech/issue-of-aging-equipment/

Institute of Electrical and Electronics Engineers. (2018). *IEEE 1584-2018: Guide for performing arc-flash hazard calculations.* IEEE Standards Association.

International Electrotechnical Commission. (2005). *IEC 60364-1:2005 Low-voltage electrical installations - Part 1: Fundamental principles, assessment of general characteristics, definitions.* IEC.

International Electrotechnical Commission. (2020). *IEC 61439-1:2020 Low-voltage switchgear and controlgear assemblies - Part 1: General rules.* IEC.

ISO. (2018). *ISO 45001:2018 – Occupational health and safety management systems – Requirements with guidance for use.* International Organization for Standardization.

ISO. (2019). *ISO 31010:2019 – Risk management – Risk assessment techniques.* International Organization for Standardization.

Kelly, J., & Male, S. (2001). *Value management in design and construction: The economic management of projects.* Routledge.

NFPA. (2022). *NFPA 72: National fire alarm and signaling code.*

NFPA. (2024). *NFPA 70E:2024 – Standard for Electrical Safety in the Workplace.* National Fire Protection Association.

Philips. (2024). *LED lighting lifespan.* https://www.lighting.philips.com/

Woodward, D. G. (1997). Life cycle costing—theory, information acquisition and application. *International Journal of Project Management*, 15(6), 335–344.

12

Value Engineering for External Electrical Infrastructure Works

12.1 Introduction to Electrical Infrastructure Value Engineering

12.1.1 Purpose of Electrical Infrastructure Value Engineering

Electrical infrastructure forms the critical backbone of any development, underpinning operational reliability, safety, future adaptability, and compliance with regulatory frameworks. In large-scale construction and infrastructure projects, electrical works external to the building envelope represent a significant portion of project cost, complexity, and risk.

Applying value engineering (VE) methodologies during the planning, design, and implementation phases of electrical infrastructure enables systematic identification of opportunities to optimize function, reduce costs, enhance reliability, and accommodate future needs without compromising project requirements. The purpose of VE for electrical infrastructure is to deliver the required performance standards in the most cost-effective manner while embedding flexibility, maintainability, and sustainability into the system from the outset.

When applied proactively, VE contributes to:

- Early identification of cost and performance risks.
- Rationalization of service entry points and pathways.
- Optimization of underground and overhead systems.
- Improvement of long-term operation and maintenance (O&M) efficiency.
- Support for future expansions, technology upgrades, and resilience needs.

12.1.2 Role of Electrical Infrastructure in Project Value

Electrical infrastructure holds a unique and pivotal role in shaping the overall value of a project. While often less visible than architectural or interior elements, the infrastructure fundamentally governs. Infrastructure contributions to project value by aspect are summarized in Table 12.1.

Decisions made at the infrastructure stage often have irreversible consequences for a project's life cycle. Inadequate initial investment or poor coordination can lead to operational inefficiencies, high maintenance costs, retrofit challenges, and regulatory non-conformities.

Thus, infrastructure VE is a strategic lever to optimize both immediate construction expenditure and long-term operational value.

Value Engineering for Electrical Systems in Construction, First Edition. Hazem Hussein.

Companion website: www.wiley.com/go/hazem

Table 12.1 Infrastructure Contribution to Project Value by Aspect

Aspect	Infrastructure Impact
Operational Reliability	Ensures continuous power supply and critical system functionality
Safety	Provides earthing, lightning protection, and compliance with life safety standards
Life-Cycle Cost	Influences energy efficiency, maintenance requirements, and system durability
Expandability	Determines ease of future capacity upgrades and technology integration
Environmental Sustainability	Supports smart controls, renewable integration, and reduced carbon footprint

12.1.3 Common Challenges in Electrical Infrastructure Design

Without structured VE application, electrical infrastructure works are frequently prone to the following challenges:

- **Overdesign and Cost Escalation**: Excessive system capacities or redundant pathways not justified by actual risk or usage.
- **Fragmented Coordination**: Poor alignment between electrical, civil, utility, and authority stakeholders causing conflicts and rework.
- **Inefficient Routing**: Unoptimized trenching, duct banks, and cabling increasing material and labor costs.
- **Inflexible Systems**: Infrastructure that does not accommodate future technology adoption or load growth.
- **Permitting and Regulatory Risks**: Delays and redesigns caused by misalignment with utility company or Authority Having Jurisdiction (AHJ) requirements.
- **High O&M Costs**: Due to inefficient layouts, difficult access, or substandard materials.

Addressing these risks proactively through VE workshops, functional analyses, and integrated planning frameworks is critical to successful infrastructure delivery.

12.1.4 Opportunities for Value Optimization

VE in electrical infrastructure focuses on extracting greater value across the project life cycle. Major optimization opportunities include:

- **Utility Coordination and Service Entry Rationalization**: Engaging utilities and AHJs early to optimize intake points, reduce connection fees, and expedite approvals.
- **Underground Infrastructure Efficiency**: Minimizing trench runs, duct bank sizes, and unnecessary manholes or pull boxes.
- **System Topology Optimization**: Designing medium voltage (MV) and low voltage (LV) networks for maximum reliability with minimum cost through strategic redundancy.
- **Earthing System Optimization**: Adapting designs to actual soil conditions rather than defaulting to conservative, costly assumptions.

- **Lighting System Efficiency**: Implementing energy-saving exterior lighting designs and intelligent control systems.
- **Space Provisioning for Future Needs**: Including spare ducts, oversized conduits, and upgradeable switchgear to reduce future disruption.
- **Standardization of Components**: Specifying common panel types, fittings, and ductbank modules to lower procurement and maintenance costs.

Systematic exploration of these areas during early project stages enables targeted interventions that preserve or enhance function while optimizing cost, constructability, and future readiness.

12.2 Scope of Electrical Infrastructure Works

12.2.1 Definition of Electrical Infrastructure Works

Electrical infrastructure refers to the collective systems, components, and assemblies associated with the external electrical distribution network of a development, located outside the primary building footprint. These systems form the essential backbone that connects buildings or facilities to utility grids, standby power systems, and external service networks.

When discussing the construction, installation, or maintenance of these external electrical systems within the context of a broader construction or development project, they are commonly categorized under the umbrella terms **infrastructure works** or **infrastructure development**.

Electrical infrastructure works typically include, but are not limited to:

- MV primary distribution systems
- LV sitewide distribution systems
- External grounding and bonding networks
- Utility intake facilities and metering stations
- External lighting systems (street, area, parking, architectural lighting)
- Associated underground infrastructure (duct banks, manholes, pull boxes)

These works interface directly with both utility providers and building internal systems, requiring meticulous coordination to ensure compliance, operational reliability, and long-term system resilience.

12.2.2 Typical Elements of Electrical Infrastructure

Electrical infrastructure works encompass a wide range of systems and assemblies that are essential to provide reliable and safe electrical services to the built environment. The typical elements forming electrical infrastructure can be categorized as given in Table 12.2.

12.2.3 Boundaries and Interfaces

Clear definition of boundaries and interfaces is essential for the effective planning, coordination, and implementation of electrical infrastructure works. Misalignment or ambiguity at the demarcation points between infrastructure systems and building internal systems can lead to design conflicts, operational inefficiencies, and costly rework.

Table 12.2 Typical Components of Electrical Infrastructure Categories

Category	Typical Components
Medium Voltage (MV) Systems	MV cables, MV switchgear, compact substations, ring main units (RMUs), primary transformers
Low Voltage (LV) Systems	LV feeder cables, main distribution boards (MDBs), outdoor LV panels, feeder pillars
Utility Interface Equipment	Service intake panels, metering cabinets, current transformer (CT) chambers
Underground Infrastructure	Duct banks, handholes, manholes, pull boxes, junction structures
Earthing and Bonding Systems	Earth rods, ground grids, earthing pits, bonding conductors
External Lighting Systems	Street lighting poles, area lighting masts, bollard lights, floodlights, architectural luminaires
Exterior Control Systems	Photocell controllers, timer relays, programmable lighting control panels
Communications Ductwork	Fiber optic ducts, telecom service ducts, manholes for data cabling
Emergency and Standby Systems	Outdoor generator sets, automatic transfer switches (ATS), associated fuel systems

Each of these categories requires careful technical evaluation during VE exercises to ensure optimized function, minimum cost, ease of maintenance, and future scalability.

In the context of VE for electrical infrastructure, boundary management focuses on ensuring that responsibilities, design ownership, and functional integration across systems are precisely understood and documented.

12.2.3.1 Boundary Definition

The primary boundary between **electrical infrastructure works** and **building internal electrical systems** is typically established at the point where external power services physically enter the building envelope. This point may vary based on project-specific configurations but generally falls into one of the categories given in Table 12.3.

Table 12.3 Boundary Types, Locations, and Associated Components

Boundary Type	Typical Location	Key Components at Boundary
Service Entry Point	External wall or basement entry	MV cables, LV feeder cables, service panels
Transformer or RMU Connection	Outdoor transformer yard or substation	Transformer secondary terminals, LV bus risers
Main Switchboard Entry	Indoor main distribution board (MDB) room	MDB incomers, utility metering, main earthing bar

The precise definition of these boundaries must be established early during the design phase and reflected consistently across civil, structural, mechanical, electrical, and utility coordination drawings (Figure 12.1).

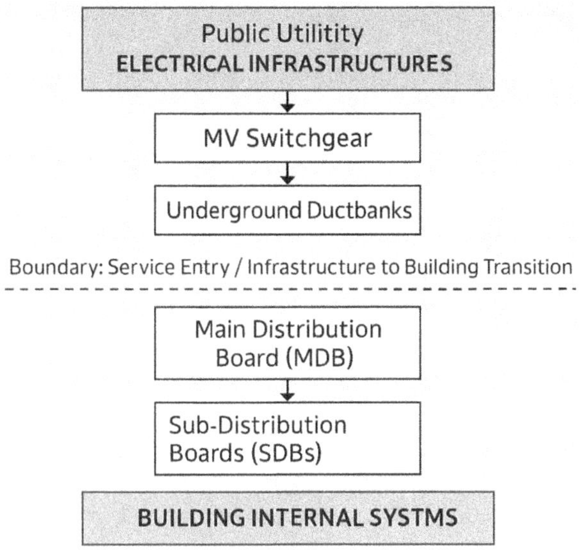

Figure 12.1 Logical Separation of Electrical Infrastructure Works and Building Internal Systems

12.2.3.2 Interface Considerations

The functional and physical interfaces between infrastructure works and building internal systems typically involve the following critical considerations:

- **Service Responsibility Split:** Clearly identify which entity (utility provider, main contractor, specialist subcontractor, end user) is responsible for design, supply, installation, testing, and commissioning of components at the boundary.
- **Load Assumptions and Capacity Planning:** Ensure that infrastructure works are sized based on realistic building load assessments and future growth projections to avoid overdesign or early obsolescence.
- **Earthing System Coordination:** Integrate external grounding networks with building internal earthing systems to provide a unified, compliant protective earth.
- **Protection and Discrimination Alignment:** Coordinate settings and fault levels between external MV/LV protection devices and internal distribution boards to ensure system selectivity and fault isolation.
- **Access and Maintenance Provisions:** Provide sufficient access for future maintenance, upgrades, or testing of boundary components (e.g., service panels, pull boxes, transformers).
- **Regulatory and Utility Compliance:** Ensure that all boundary elements meet applicable utility company regulations, electrical standards (e.g., IEC, IEEE), and local authority requirements.

Best Practice Recommendations:

- Conduct early joint workshops with utility companies, civil engineers, architects, and electrical consultants to define and document interfaces.
- Develop and maintain an interface responsibility matrix (IRM) throughout design and construction stages.

- Include clear boundary demarcation details in tender documents, construction drawings, and commissioning protocols.
- Verify boundary component readiness during progressive inspections and milestone reviews to prevent later commissioning delays.

12.3 Key Principles of VE in Electrical Infrastructure

Effective application of VE principles in electrical infrastructure works demands a systematic, disciplined approach that balances functionality, reliability, cost efficiency, regulatory compliance, and future adaptability. The VE methodology, originally developed for manufacturing applications, has been adapted for construction and infrastructure projects (Miles, 1972). Infrastructure systems, given their scale and criticality, require special attention to long-term performance and risk management beyond the immediate construction phase.

This section outlines fundamental VE principles that should be applied during the planning, design, and procurement stages of electrical infrastructure projects, based on industry best practices and professional experience.

12.3.1 Functional Analysis

At the core of VE lies functional analysis—the process of rigorously defining and understanding what each infrastructure element must achieve. This approach follows established VE methodologies that emphasize function over form (Dell'Isola, 1997) The functional analysis process steps are summarized in (Table 12.4).

Application Example:

- Assessing whether redundant MV switchgear bays are truly needed based on load criticality and utility redundancy policies.

12.3.2 Cost-Function Optimization

VE seeks to optimize cost relative to function over the entire system life cycle, not just during construction. This life-cycle approach is increasingly recognized in infrastructure planning (Flanagan et al., 1989).

Table 12.4 Functional Analysis Process Steps

Step	Action
Identify Primary Functions	Determine the essential operational purpose of each system (e.g., power delivery, fault protection, lighting provision).
Identify Secondary Functions	Recognize supplementary functions (e.g., aesthetics for lighting poles, future capacity for ducts).
Evaluate Necessity	Challenge each function's necessity: Is it mandatory by code? Critical to operation? Optional but desirable?
Establish Function-Cost-Worth Relationships	Analyze whether the current design approach delivers the function at optimum cost relative to its criticality.

Key Strategies Include:

- **Life-Cycle Costing (LCC):** Consider initial construction cost, maintenance, energy consumption, replacement cycles, and disposal costs. International standards provide frameworks for life-cycle costing analysis (ISO 15686-5, 2017).
- **Modularization:** Design systems (e.g., ductbanks, lighting control panels) to allow incremental expansion without major reconstruction.
- **Avoid Over-Specification:** Resist unnecessarily conservative design margins (e.g., oversizing cables or transformer capacities) unless justified by risk or future growth analysis.
- **Selective Redundancy:** Apply redundancy where it delivers quantifiable risk reduction; avoid redundant systems that only marginally enhance resilience at high cost. *[Professional recommendation based on industry practice.]*

12.3.3 Risk-Based Design Adjustments

Infrastructure systems are inherently exposed to external risks (e.g., flooding, utility failures, public access damage). Risk-based design approaches are supported by established risk management frameworks (ISO 31000, 2018) Risk-based design adjustments for infrastructure projects are summarized in (Table 12.5).

Table 12.5 Risk-Based Design Adjustments in Infrastructure Projects

Risk Area	Risk Mitigation VE Measures
Environmental Hazards	Elevate transformers and critical panels above flood levels; use corrosion-resistant materials.
Utility Outage Risks	Design dual supply feeds or integrate standby generators where outage impact is critical.
Future Capacity Risks	Provide spare conduits and oversized cable trays to accommodate future expansions cost-effectively.
Civil Coordination Risks	Engage civil engineers early to align underground routes and avoid costly rework.

VE workshops should incorporate structured risk assessments to identify opportunities for risk-weighted design optimizations. *[Recommended practice based on professional experience.]*

12.3.4 Futureproofing Strategies

Futureproofing is critical for infrastructure systems with long lifespans and high disruption costs when retrofitting. The concept of designing for adaptability has been explored in architectural and infrastructure literature (Brand, 1994).

Best Practice Strategies Include:

- **Spare Capacity Provision:** Install spare ducts, allocate additional space in underground handholes and pull boxes. *[Industry best practice]*
- **Scalable Architecture:** Specify modular switchgear and scalable lighting control systems that can grow without major replacement. *[Professional recommendation]*
- **Technology Readiness:** Design with future smart grid, renewable energy integration, and electric vehicle (EV) charging readiness in mind. *[Current industry trend]*

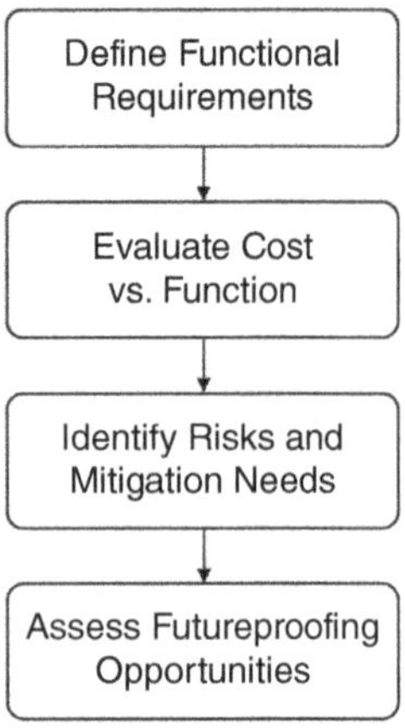

Figure 12.2 Key Principles of VE in Electrical Infrastructure

- **Standards Evolution Awareness:** Anticipate evolving energy efficiency, fault protection, and monitoring standards to avoid premature obsolescence. *[Professional recommendation]*

Futureproofing interventions, when planned correctly, are generally recognized to deliver higher value compared to reactive retrofits, though quantitative analysis is often project-specific (Figure 12.2).

12.4 Technical Considerations

Designing and executing electrical infrastructure works requires careful integration of technical, regulatory, spatial, and coordination aspects. Early-stage decision-making significantly influences project value, reliability, life-cycle costs, and the ability to expand or modify systems in the future.

Each technical consideration must be evaluated not only for its immediate performance but also for its impact on construction phasing, utility coordination, long-term operations, and regulatory approvals.

The following subsections provide detailed guidance on the core technical areas requiring focused VE assessment for electrical infrastructure works.

12.4.1 Utility Coordination and Design Integration

12.4.1.1 Importance of Utility Coordination

Successful electrical infrastructure delivery depends heavily on early and continuous coordination with external utility providers and authorities having jurisdiction (AHJs). According to the National Electrical Code (NEC) an AHJ is defined as "an organization, office, or individual responsible for enforcing the requirements of a code or standard, or for approving equipment, materials, an installation, or a procedure" (NEC, 2023). Misalignment with utility technical requirements or approval processes often results in costly redesigns, delayed energization, and compromised system performance.

VE in this domain focuses on optimizing service intake configurations, expediting approvals, and ensuring that utility interface works are efficient, compliant, and future-proof.

Key Areas of Coordination (Table 12.6)

Table 12.6 Key Utility Coordination Areas and Focus Points

Area	Typical Focus Points
Service Intake Location	Optimal placement relative to load centers, site access, and aesthetic considerations
Supply Capacity and Redundancy	Defining demand load, diversity factors, required backup or dual feed arrangements
Metering Requirements	CT/VT metering specifications, placement of metering cabinets, and space allowances
Protection Requirements	Utility-side protection (e.g., fault levels, breaker types) and interface with client-side protection
Physical Access and Safety	Accessibility for utility personnel and segregation of public and utility access routes
Compliance and Testing	Utility inspections, testing protocols, and energization processes

12.4.1.2 Service Intake Point Coordination

Service intake points must be coordinated both functionally and physically to ensure minimal disruption to building operation and external siteworks. Considerations include:

- Determining whether MV or LV intake is required, based on load size and utility regulations.
- Sizing and configuring intake facilities (e.g., transformer yards, service panels) to meet current and projected future loads.
- Aligning intake point locations with civil and architectural layouts to minimize trenching distances and civil works impacts.
- Allowing sufficient clearance zones around transformers or switchgear as per utility guidelines.

12.4.1.3 Coordination with AHJ

AHJ involvement is critical in obtaining the necessary construction and operational permits for electrical infrastructure works. Typical areas of AHJ interaction include:

- Building code compliance (electrical clearances, fire safety integration).
- Environmental protection regulations (e.g., trenching impact, flood mitigation).
- Public right-of-way management for external ducting or cabling.
- Safety inspections and final system acceptance.

Failure to adequately coordinate with AHJs can lead to refusal of occupancy permits, commissioning delays, and legal liabilities.

12.4.1.4 Civil Works and Utility Interface Coordination

Electrical infrastructure must be harmonized with broader civil works, including roadways, pathways, landscaping, and stormwater systems. Key considerations include:

- Aligning ductbank and manhole routes with approved civil alignments.
- Avoiding clashes with other buried utilities (water, gas, and sewer) through composite utility plans.
- Minimizing road crossings and reinstatement works through intelligent routing.
- Planning for installation phasing (e.g., install ducts before paving finalization).

Proactive civil-utility-electrical coordination maximizes construction efficiency, reduces disruption risk, and optimizes trenching and excavation costs.

12.4.2 Underground Utility Infrastructure

Underground utility infrastructure forms the critical hidden backbone of sitewide electrical distribution systems. It facilitates the safe, protected, and coordinated routing of electrical power cables, communication lines, and grounding networks across a project site.

Effective VE in this domain focuses on optimizing duct bank designs, reducing unnecessary manholes and pull boxes, managing spatial coordination, and future-proofing underground capacity (Table 12.7).

Table 12.7 Key Components of Underground Utility Infrastructure

Component	Purpose
Duct Banks	Bundled conduit systems for routing MV, LV, and communication cables underground
Manholes	Large underground chambers enabling cable pulling, splicing, and maintenance access
Handholes	Smaller underground enclosures for branching or cable pulling in low-traffic areas
Pull Boxes	Intermediate pulling points in long duct runs to ease cable installation
Junction Structures	Specialized chambers for connecting different duct routes or transitioning between services

Each of these components must be designed, routed, and constructed in coordination with civil, structural, and landscaping works to avoid conflicts and ensure long-term accessibility.

12.4.2.1 Duct Bank Design Principles

Duct banks must be designed to balance functionality, constructability, cost, and future expansion flexibility. Key design considerations include:

- **Duct Sizing and Quantity:**
  - Determine based on initial load requirements plus 20–30% spare ducts for future use. In some installations, an entire spare duct may be included for each service type (e.g., one spare MV duct, one spare LV duct). This represents established engineering practice for futureproofing underground infrastructure, similar to spare capacity recommendations for electrical distribution systems (Institution of Engineering and Technology, 2024).
  - Size conduits appropriately for cable diameters plus pulling clearance.
- **Separation and Arrangement:**
  - Maintain physical separation between MV and LV ducts per local codes.
  - Group ducts logically (e.g., MV cables together, fiber optics separately).
- **Materials Selection:**
  - Common duct materials include PVC, HDPE, or fiber-reinforced concrete conduits.
  - Select based on soil conditions, mechanical loading, and chemical exposure risks.
- **Depth and Cover Requirements:**
  - Follow utility and code requirements for minimum burial depths.
  - Provide additional cover under roadways or in high-traffic zones.
- **Thermal Considerations:**
  - Assess cable derating factors due to mutual heating in duct banks.
  - Consider using concrete encasement or thermally stable backfill materials where necessary.

12.4.2.2 Manholes, Handholes, and Pull Boxes

Manholes, handholes, and pull boxes serve vital operational and construction purposes but can significantly impact costs and construction timelines.

Best Practice Guidelines:

- **Spacing Optimization:**
  - Space manholes based on maximum cable pulling distances, typically 150–200 m for MV cables (IEEE, 2016).
  - Avoid unnecessary handholes unless dictated by topography or routing complexity.
- **Standardization:**
  - Standardize manhole sizes and designs across the site to simplify construction and maintenance.
- **Structural Design:**
  - Design for anticipated traffic loads (e.g., vehicular-rated covers for roadway locations).
- **Waterproofing and Drainage:**
  - Provide drainage solutions or waterproof designs where groundwater risks exist.
- **Access and Safety:**
  - Ensure manholes have safe entry designs (steps, ladders) and appropriate confined space entry markings (Table 12.8).

Table 12.8 Cable Pulling Factors and Recommended Manhole Spacing

Factor	Typical Value / Consideration	Impact on Manhole Spacing
Maximum Pulling Tension (MV Cables)	6000–9000 N (depending on cable type and size) (IEEE, 2019)	Limits total cable length between pulling points
Maximum Sidewall Bearing Pressure (SWBP)	~500 N/m (cable-dependent) (IEEE, 2019)	Bends and curves increase sidewall pressure; closer spacing may be needed
Duct Material and Surface	PVC, HDPE, concrete—with or without lubricant	Lower friction (HDPE, lubricated) allows slightly longer pulls
Cable Route Geometry	Straight vs. multiple bends	More bends = reduced allowable pulling distance
Installation Method	Manual pull or motorized winch with tension control	Winch pulling can slightly extend maximum pull distance safely
Standard Practice	Based on individual cable pulling calculations considering tension limits, SWBP, and route complexity	Varies significantly based on specific installation conditions

Notes:

- **Straight Duct Runs:** Manholes spaced approximately 150–200 m.
- **Ducts with One or More Significant Bends:** Reduce spacing to approximately 100–150 m depending on bend severity.
- **Special Installations:** Complex routes may require detailed pulling tension calculations based on duct friction, cable weight, and tension limits.

12.4.2.3 Futureproofing Underground Infrastructure

To maximize value over the life cycle:

- **Install Spare Ducts:**
  - Install additional empty conduits for future services at minimal incremental cost.

- **Coordinate with Other Utilities:**
  - Where feasible, combine trench works with other utilities (water, gas, telecom) to reduce site disruption and trenching costs (composite ductbanks).
- **Provide Mapping and As-Built Documentation:**
  - Maintain precise underground utility maps (including GPS coordinates where possible) to support future expansion and maintenance activities.

12.4.3 Medium Voltage Systems

MV systems form the critical backbone of electrical infrastructure distribution networks. They facilitate the efficient transmission of electrical energy across larger site areas at voltages typically ranging from 1 to 36 kV, depending on local regulations and project requirements (Table 12.9).

VE in MV systems targets optimizing system configuration, equipment specification, protection coordination, redundancy planning, and future expansion capabilities.

Table 12.9 Key Components of Medium Voltage Systems

Component	Function
MV Incoming Feeder	Connection point from public utility or on-site generation source
MV Switchgear	Controls, protects, and isolates MV circuits
MV Distribution Cables	Transmit electrical energy between substations and distribution points
Transformers	Step-down devices converting MV to LV for downstream use
Protection Devices	Relays, circuit breakers, fuses ensuring system protection and fault isolation

Note

This section provides detailed design and VE guidance for MV systems, building upon the high-level evaluation principles introduced in Chapter 3, Section 3.4.1 (Power Distribution MV/LV).

While Chapter 3 focused on system selection, topology evaluation, and VE assessment criteria, this section addresses practical implementation considerations, detailed technical parameters, and infrastructure-specific design strategies essential for successful MV system delivery.

12.4.3.1 Design Parameters for MV Systems

Several key parameters must be defined at the outset of MV system design:

- **Voltage Class Selection:**
  - Common classes include 11, 22, or 33 kV depending on regional utility standards and site load requirements.
  - Select based on distance, load magnitude, and voltage drop considerations.
- **Load Estimation and Diversity Factors:**
  - Base MV system sizing on realistic load forecasts, applying appropriate diversity factors to avoid overdesign.

- **Short-Circuit Level Calculations:**
  - Determine maximum prospective fault currents to specify switchgear, cables, and protection ratings appropriately.
- **System Topology:**
  - **Radial Systems:** Simpler, lower cost, but single-point-of-failure risk.
  - **Ring Main Systems:** Improved resilience; faults can be isolated without full outages.
  - **Meshed Systems:** Highest reliability; typically used only for very large or critical facilities.

12.4.3.2 MV Switchgear Selection

MV switchgear plays a pivotal role in system protection, operational flexibility, and maintenance safety.

Key Selection Criteria:

- **Type:**
  - Metal-clad switchgear (air-insulated or gas-insulated) for high-reliability applications.
  - Compact RMUs for space-constrained external sites.
- **Insulation Medium:**
  - SF_6 gas-insulated switchgear for compactness and maintenance reduction (subject to environmental regulations).
  - Air-insulated switchgear for lower cost and simpler systems.
- **Internal Arc Classification:**
  - Ensure switchgear complies with IEC 62271-200 (IEC, 2021) internal arc standards for personnel safety.
- **Accessibility Requirements:**
  - Front access only vs. full rear access considerations based on available space and maintainability.
- **Future Expansion:**
  - Specify switchgear with provision for additional feeder bays where expansion is anticipated.

12.4.3.3 MV Cable Design

MV cables must be selected and routed considering electrical performance, mechanical robustness, installation feasibility, and environmental resilience (Table 12.10).

Table 12.10 MV Cable Design Considerations

Parameter	Design Consideration
Cable Size	Based on load current, voltage drop limits, and future capacity margins
Conductor Type	Copper for higher conductivity and mechanical strength; aluminum for cost efficiency
Insulation Type	XLPE insulation widely used for MV applications (thermally robust, water resistant)
Installation Method	Direct buried, ducted, or in underground cable trays depending on site conditions
Pulling and Termination	Pulling tension limits to be respected; stress cones or sealing ends for terminations

12.4.3.4 Protection and Fault Management

MV protection design must ensure rapid fault detection and isolation to minimize damage and maintain operational continuity.

Typical Protection Elements:

- Overcurrent and earth fault protection relays.
- Differential protection for critical transformers.
- Directional protection for interconnected or ring systems.
- Remote monitoring and control interfaces via Supervisory Control and Data Acquisition (SCADA) or building management systems (BMS).

Protection schemes must be coordinated across the entire infrastructure from MV intake to final LV distribution to ensure selectivity and prevent nuisance trips.

12.4.4 Substation Design and Location Considerations

Substations serve as the critical nodes within electrical infrastructure, enabling the transition from MV to LV distribution. They house essential components such as MV switchgear, transformers, protection equipment, and, in some cases, auxiliary systems such as battery banks or communication panels.

Effective VE in substation design focuses on optimizing substation type selection, location, configuration, future expandability, and compliance with safety, environmental, and operational requirements.

12.4.4.1 Substation Types

Substations can be broadly categorized based on construction and operational characteristics (Table 12.11).

➢ ***Note on Scope Inclusion***

Substations may fall within the scope of electrical infrastructure works or within building internal systems, depending on project-specific definitions, company operational practices, contractual frameworks, and national regulatory structures.

In some jurisdictions, primary substations (including MV switchgear and transformers) are delivered and owned by the public utility and therefore excluded from the project infrastructure scope.

Table 12.11 Types of Substations and Their Applications

Type	Description	Typical Application
Outdoor Open-Air Substation	Components (switchgear, transformers) exposed or housed under shelters	Large industrial sites, utility transmission systems
Indoor Substation	All components enclosed within a building structure	Commercial buildings, urban sites with space constraints
Package Substation (Unit Substation)	Pre-fabricated, factory-assembled MV switchgear, transformer, and LV board in a single enclosure	Small sites, modular installations, rapid deployment
Hybrid Substation	Combination of indoor MV switchgear with outdoor transformers	Semi-urban campuses, mixed-use developments

In other cases, particularly for private developments and industrial sites, substations are part of the electrical infrastructure works and must be designed, installed, and maintained by the project developer or operator.

It is essential to clearly define substation ownership, design responsibility, and operational jurisdiction during the early design and contractual stages to prevent scope conflicts and ensure regulatory compliance

12.4.4.2 Substation Location Selection Criteria

Substation location must be carefully determined early in the infrastructure planning phase. Key factors influencing location include:

- **Proximity to Load Centers:**
  - Minimize cable lengths and voltage drops between substations and primary load areas.
- **Access for Installation and Maintenance:**
  - Provide clear vehicular access for delivery, crane operations, maintenance personnel, and emergency response.
- **Environmental Conditions:**
  - Avoid flood-prone areas or provide flood mitigation measures (e.g., elevated plinths).
  - Ensure adequate natural ventilation or mechanical cooling if required.
- **Security Considerations:**
  - Locate substations within secured zones or provide fencing, CCTV surveillance, and access controls.
- **Civil Coordination:**
  - Integrate substation locations with site masterplans, avoiding conflicts with roads, drainage paths, or critical landscaping features.
- **Utility Company Interface Requirements:**
  - Meet minimum distance requirements from public roads, utility access points, and property boundaries as mandated by utility providers.

12.4.4.3 Substation Layout and Internal Zoning

Efficient internal substation layout contributes significantly to safe operation, ease of maintenance, and future expansion (Table 12.12).

Design Best Practices:

- Maintain clear segregation between MV, LV, and auxiliary systems.
- Provide sufficient working clearances as per IEC 61936-1 (IEC, 2021) and local codes.
- Allow logical cable routing pathways (e.g., separated cable trenches or risers for MV and LV).

Table 12.12 Substation Zoning

Zone	Components Typically Located
MV Equipment Zone	Incoming MV feeders, MV switchgear, metering panels
Transformer Zone	Dry-type or oil-immersed transformers
LV Equipment Zone	Main LV distribution boards, outgoing feeders
Auxiliary Services Zone	Battery chargers, fire suppression panels, communication panels

12.4.4.4 Environmental, Safety, and Regulatory Considerations

Substation designs must incorporate environmental protection, personnel safety, and regulatory compliance measures:

- **Fire Safety:**
  - Install fire-rated walls and doors between MV, transformer, and LV rooms where required.
  - Include fire detection and suppression systems as applicable.
- **Oil Containment:**
  - For oil-immersed transformers, design oil catchment pits and bunds to contain leaks.
- **Ventilation:**
  - Provide natural or mechanical ventilation to dissipate heat from transformers and switchgear.
- **Acoustic Treatment:**
  - Apply acoustic panels or barriers if substations are close to noise-sensitive areas.
- **Lightning and Surge Protection:**
  - Install surge protection devices (SPDs) at MV incomers and sensitive LV equipment.
- **Regulatory Compliance:**
  - Design substations in accordance with IEC 61936-1 (IEC, 2021) (Power Installations Exceeding 1 kV AC) and local utility requirements.

12.4.4.5 Future Expansion Planning

Infrastructure value is maximized by embedding flexibility into substation designs.

Best practice futureproofing strategies include:

- Reserving space for additional MV or LV switchgear bays.
- Providing spare duct entries and cable trenches.
- Installing oversized busbars and incoming feeders where foreseeable load growth justifies it.
- Designing building structures to support later mechanical expansions (e.g., roof load allowances for future ventilation or solar panels).

12.4.5 LV Systems

LV systems are responsible for distributing electrical power from the secondary side of transformers or main distribution boards (MDBs) to local distribution points and final loads across a site.

In infrastructure works, LV systems bridge the MV distribution with operational systems inside buildings, external lighting, site services, and special facilities.

VE for LV systems focuses on optimizing network topology, sizing for present and future needs, rationalizing distribution points, and ensuring system resilience without unnecessary overdesign (Table 12.13).

12.4.5.1 LV System Topology and Configuration

Selecting an optimal LV topology is critical for balancing reliability, cost, and maintainability (Table 12.14).

Topology Selection Factors:

- Load criticality and operational continuity requirements.
- Site size and geographical spread.
- Capital cost versus downtime risk trade-offs.

Table 12.13 Key Components of LV Systems

Component	Function
Main Distribution Board (MDB)	Receives stepped-down LV power from transformers; primary LV source
Sub-Distribution Boards (SDBs)	Localized boards distributing to specific building areas or external systems
LV Feeder Cables	Link MDBs to SDBs or external loads; major current-carrying elements
Outdoor Feeder Pillars	External LV distribution points feeding site lighting, small buildings, or external equipment
Final Connection Assemblies	Outdoor-rated enclosures, field panels, or junction boxes feeding final loads

Table 12.14 LV System Topology and Uses

Topology Type	Description	Typical Use Case
Radial Distribution	Each load fed from a single source; no alternate path	Simple sites, low-criticality loads
Ring Distribution	Loads connected in a loop with two feeding paths	Higher resilience for critical sites
Parallel Redundant Feeders	Loads fed by independent feeders from different sources	Mission-critical loads needing high availability

12.4.5.2 LV Cable Design and Routing

Proper LV cable design is fundamental to efficient and safe power distribution (Table 12.15).

Routing Must Be Coordinated to Avoid:

- Excessive bends (which increase pulling difficulty and mechanical stress).
- Crossings with high voltage, telecom, or mechanical utility services.
- Placement in areas prone to flooding unless appropriate protection is installed.

Table 12.15 LV Cable Design Considerations

Design Factor	Consideration
Conductor Sizing	Based on load current, voltage drop limits, permissible cable grouping factors
Voltage Drop Criteria	For public LV systems: 3% for lighting circuits and 5% for other uses; For private LV systems: 6% for lighting circuits and 8% for other uses (IEC 60364-5-52, 2009)
Installation Method	Direct burial, ducted, or in covered trenches
Thermal Derating	Apply factors for cables installed in groups, high ambient temperatures, or buried depths
Cable Material	Copper preferred for high current, mechanical robustness; aluminum considered for large, cost-sensitive installations

12.4.5.3 LV Distribution Board Specification

LV MDBs and SDBs must be selected and specified to match project operational and future flexibility needs.

Key Specification Points:

- **Form of Separation (Form 2, 3, or 4):**
  - Form selection depends on protection requirements, operational needs, and maintenance accessibility. Form 4 (particularly Form 4b) is typically used for applications requiring the highest level of segregation, such as hospitals or critical industrial processes (IEC 61439-2, 2020).
- **Ingress Protection (IP) Ratings:**
  - Outdoor panels: Minimum IP54 or IP65 depending on environment (IEC 60529, 2020).
  - Indoor panels: IP42 minimum, higher for dusty or damp locations (IEC 60529, 2020).
- **Short-Circuit Withstand Ratings:**
  - Match to calculated fault levels based on transformer ratings and impedance.
- **Expansion Provision:**
  - Spare ways, oversized busbars, and modular construction to accommodate future loads.
- **Metering and Monitoring:**
  - Integrated energy meters, power quality analyzers, and communication capabilities (e.g., Modbus RTU, TCP/IP).

12.4.5.4 Integration with External Systems

LV systems must interface with:

- **External Lighting Systems:**
  - Streetlights, area lights, parking lots, bollards.
- **Sitewide Small Power Loads:**
  - Security cabins, gatehouses, external equipment enclosures.
- **Emergency Power Systems:**
  - External ATS panels, generator-fed distribution boards where required.

Interface design must ensure appropriate segregation of normal and emergency supplies, compliance with switching and interlocking regulations, and maintainability.

12.4.6 External Earthing Systems

External earthing systems are vital for ensuring electrical safety, equipment protection, and system resilience in infrastructure projects. They provide low-impedance paths for fault currents, lightning strikes, and static discharge, thereby protecting both human life and electrical assets.

VE in earthing system design focuses on optimizing material use, adapting to site-specific soil conditions, ensuring regulatory compliance, and providing sufficient futureproofing for site expansions or changes in use.

➢ **Scope Clarification:** This section addresses earthing systems associated with external infrastructure elements located outside the primary building footprint, including substations, external distribution equipment, external lighting systems, and metallic site structures. Earthing systems internal to building facilities are covered separately under building services design.

Components of External Earthing Systems (Table 12.16)

Table 12.16 Components of External Earthing Systems

Component	Function
Earth Electrodes	Rods, mats, or grids inserted into the ground to disperse fault or lightning currents
Grounding Conductors	Copper or copper-bonded conductors interconnecting earth electrodes, substations, transformers, and external equipment
Earth Pits	Inspection points to enable testing of earth resistance
Equipotential Bonding	Interconnection of metallic systems (pipes, fences, structures) to the earthing system to eliminate voltage differences

12.4.6.1 Design Considerations

Site-Specific Soil Resistivity:

- Conduct a soil resistivity survey early in the design phase.
- Adapt electrode design (depth, number, material) based on soil characteristics (Table 12.17).

Earth Electrode Configuration:

- Single deep-driven rods for small installations.
- Multiple rod arrays, ground rings, or buried grids for larger installations or poor soil conditions.

Material Selection:

- Copper or copper-clad steel for electrodes and conductors.
- Corrosion protection measures (e.g., tinned copper, PVC sleeving) in aggressive soil environments.

12.4.6.2 Integration with Infrastructure Systems

External Earthing Systems Must Integrate With:

- Substation grounding grids.
- Transformer neutral earthing connections.
- External lighting columns and feeder pillars.

Table 12.17 Target Earth Resistance Values

System Type	Typical Target Resistance
LV Systems	<25 ohms (IEEE 142, 2022)
MV Systems	1–5 ohms for industrial installations (IEEE 80, 2020)
Lightning Protection Systems	≤10 ohms common practice IEC advises lowest possible (IEC 62305-3, 2010)
Data Centers/Sensitive Facilities	<1 ohm desirable for low-noise earthing (ANSI/BICSI 002, 2019)

- External metallic structures (e.g., fencing, gates).
- Lightning protection down conductors.

Equipotential Bonding Best Practices:

- Bond all accessible metallic structures within a defined radius of electrical infrastructure (specific distances vary by local codes and application requirements).
- Ensure continuity between earthing systems across different utility services (e.g., telecoms, security systems).

12.4.6.3 Testing and Commissioning Requirements

- Perform earth resistance tests using the Fall-of-Potential method or equivalent recognized techniques (IEEE, 2019).
- Verify continuity of bonding conductors and connections.
- Document all test results and as-built locations of earth pits for future maintenance.

Periodic maintenance testing is recommended every 3–5 years, especially in corrosive soil environments.

12.4.7 Exterior Lighting Systems

Exterior lighting systems are critical components of site infrastructure, providing illumination for safety, security, operational continuity, and aesthetic enhancement.

Properly designed exterior lighting enhances site usability after dark, supports surveillance systems, and contributes to the visual identity of developments.

VE for exterior lighting systems focuses on optimizing lighting levels, energy efficiency, control integration, fixture selection, and long-term maintenance strategies.

Categories of Exterior Lighting (Table 12.18)

Table 12.18 Categories of Exterior Lighting

Lighting Type	Typical Applications
Street and Roadway Lighting	Site roads, driveways, access roads
Area and Site Lighting	Open spaces, plazas, industrial yards, campus areas
Parking Lot Lighting	Parking areas for vehicles, loading bays
Landscape and Architectural Lighting	Trees, water features, building facades, monuments
Security Lighting	Perimeter fences, sensitive external equipment

Each category must be designed according to its specific functional, environmental, and security requirements.

- ***Design Standards and Requirements***

Exterior lighting designs must comply with relevant local, national, and international standards (Table 12.19).

Table 12.19 Exterior Lighting Design Standards

Standard	Scope
EN 13201 Series	Road lighting design and performance requirements (CEN, 2015)
CIE 115	Lighting for roads for motor and pedestrian traffic (CIE, 2010)
IESNA RP-8	Roadway and area lighting design practices
AS/NZS 1158	Lighting for roads and public spaces (Australia/New Zealand)

Key Design Criteria:

- **Illuminance Levels:**
  - Defined by application type (e.g., higher for vehicle areas, lower for pedestrian walkways).
  - Expressed in lux (lx) values based on standard recommendations.
- **Uniformity Ratios:**
  - Ensure consistent lighting without dark patches or excessive glare.
  - Typically defined as minimum-to-average illuminance ratios (CIE, 2010).
- **Glare Control:**
  - Select luminaires and installation angles to limit disability glare and discomfort glare.
- **Energy Efficiency:**
  - Prefer LED technology with high luminous efficacy (>120 lm/W typical), with high-performance LEDs achieving 150–200 lm/W depending on color temperature and application (DOE, 2024).
  - Apply smart control strategies (dimming, scheduling, motion detection) to reduce consumption.
- **Environmental Impact:**
  - Minimize light spill, upward light, and intrusion onto adjacent properties (Dark Sky compliance where applicable).

■ ***Luminaire Selection and Specification***

Choosing the correct luminaires is essential for achieving performance, efficiency, and maintainability targets.

Key Selection Factors:

- **Ingress Protection (IP) Rating:**
  - Minimum IP65 for outdoor environments (commonly required for outdoor street lights to withstand heavy rain and strong winds); higher in dusty or corrosive environments (IEC 60529, 2020).
- **Impact Resistance (IK) Rating:**
  - IK08 or higher recommended for public areas subject to potential vandalism or mechanical impacts (IEC 62262, 2002).

- **Corrosion Resistance:**
  - Materials suitable for site conditions (e.g., coastal areas require marine-grade aluminum or stainless steel).
- **Optical Distribution:**
  - Select beam patterns (Type II, III, IV, V) appropriate to roadway layouts, pedestrian paths, or wide-open areas.
- **Mounting Heights and Pole Spacing:**
  - Optimize to achieve required lux levels and uniformity with minimum number of poles.

■ *Cabling and Power Supply Considerations*

- **Circuit Design:**
  - Typically radial circuits from local LV feeder pillars or distribution boards.
  - Fuse protection within feeder pillars or in pole bases.
- **Cable Routing:**
  - Underground cabling preferred for aesthetics and protection.
  - Use direct burial rated cables or install within conduits in trenches.
- **Voltage Drop Management:**
  - Ensure final pole receives adequate voltage to maintain luminaire performance.
- **Surge Protection:**
  - Install luminaire-level or feeder-level SPDs to guard against transient overvoltages (especially in lightning-prone regions).

■ *Maintenance and Operation Strategies*

Exterior lighting designs must consider future O&M needs:

- **Access for Lamp Replacement and Repairs:**
  - Install lowering mechanisms, access doors, or tilt-down poles where feasible.
- **Spare Parts Standardization:**
  - Limit luminaire model variety across the site to ease maintenance logistics.
- **Smart Monitoring Systems:**
  - Integrate into BMS or dedicated lighting management systems for fault detection and energy reporting.

12.4.8 Lighting Control Systems

Lighting control systems are a fundamental part of exterior lighting infrastructure, enabling automated operation, energy efficiency, reduced maintenance, and enhanced safety. These systems govern when and how exterior luminaires operate based on time, daylight conditions, motion detection, or programmed schedules.

VE in this domain focuses on optimizing control architecture, integration strategy, scalability, and reliability—ensuring lighting systems operate only when needed and in a way that is operationally efficient over the life of the facility.

12.4.8.1 Control Strategies and Modes (Table 12.20)

Table 12.20 External Lighting Control Strategies

Control Mode	Description	Applications
Photocell Control	Lighting activated based on ambient light levels (e.g., dusk/dawn sensors)	Street lighting, general outdoor areas
Time-Based Scheduling	Lighting activated via programmable timers or BMS integration	Commercial plazas, landscape lighting
Motion or Presence Detection	Lighting activated upon detecting movement in defined areas	Security zones, pedestrian walkways
Centralized Control	Remote operation and scheduling via central controller or BMS	Large campuses, smart cities
Hybrid Systems	Combinations of the above	Sites requiring flexibility and efficiency

12.4.8.2 Control Architecture Types (Table 12.21)

Table 12.21 External Lighting Control and Their Use Cases

Architecture	Description	Typical Use Case
Standalone Controls	Each pole or luminaire contains its own photocell or timer	Simple or remote sites
Group-Based Controls	Multiple luminaires connected to a common control point (e.g., in feeder pillar)	Medium-sized sites
Networked Lighting Control (NLC)	Luminaires connected via wired or wireless network to a central controller	Smart infrastructure, energy monitoring needs
BMS-Integrated Lighting Control	Exterior lighting integrated into Building Management System	Mixed-use facilities, commercial estates

12.4.8.3 Control Device Specification

Key Specification Elements:

- **Photocells:**
  - Conform to IEC 63103 for energy performance.
  - Delay-on and delay-off features to prevent nuisance switching due to transient shadows or brief ambient changes.
- **Timers:**
  - Astronomical timers (self-adjusting to sunrise/sunset throughout the year) preferred over fixed-time timers.
  - Backup power memory retention in case of power loss.
- **Motion Sensors:**
  - Select with appropriate detection range, coverage angle, and environmental rating (IP65 minimum).
  - Avoid placing near heat sources or vegetation that could cause false triggers.

- **Controllers and Gateways:**
  - Include secure communication protocols (e.g., DALI-2, Zigbee, LoRaWAN).
  - Remote programmability, diagnostics, and energy monitoring features.

12.4.8.4 VE Opportunities in Lighting Control

- **Optimize Control Zoning:**
  - Group lights based on usage zones (e.g., main road, pedestrian paths, parking) to avoid overlighting.
- **Select Scalable Systems:**
  - Use modular, open-protocol systems that allow future upgrades without full replacement.
- **Balance Capital and Operational Costs:**
  - More advanced control systems (e.g., NLC) may have higher CapEx but deliver energy and maintenance savings over time.
- **Minimize Light Pollution and Nuisance Lighting:**
  - Combine control strategies with proper luminaire optics to reduce light spill and environmental impact.

12.4.8.5 Integration with Sitewide Systems

- **Building Management Systems (BMS):**
  - Exterior lighting can be integrated as a subsystem with scene control, event overrides, and energy dashboards.
- **Security Systems:**
  - Coordinate with CCTV, motion detectors, and perimeter systems for event-driven lighting triggers.
- **Emergency Response Protocols:**
  - Program controls to override into full-on mode during fire alarm, evacuation, or utility power failure scenarios (if generator-backed).

12.4.9 External Generator Systems and Fuel Infrastructure

In many infrastructure projects, standby generators are installed in external yards to supply backup power during grid outages. These systems are critical to operational continuity for essential services and often form part of the project's primary infrastructure. VE in this area focuses on optimizing equipment sizing, yard layout, noise control, fuel logistics, environmental compliance, and integration with electrical and civil systems.

12.4.9.1 Generator Yard Planning Considerations

The siting and layout of generator yards should balance technical performance, accessibility, safety, and regulatory constraints (Table 12.22).

12.4.9.2 Generator Specification and Configuration

Generators must be selected and configured based on critical load requirements, run-time expectations, and integration with the electrical infrastructure.

Table 12.22 Generator Yard Planning Considerations

Planning Factor	Key Considerations
Site Access	Provide 24/7 access for refueling, maintenance, and emergency support
Ventilation and Cooling	Maintain clear airflow paths and separation between generator intakes and exhausts
Acoustic Separation	Minimize noise propagation to occupied buildings and property boundaries; apply sound barriers if required
Security and Fencing	Perimeter fencing, CCTV, and controlled access gates for safety and vandalism prevention
Fire Safety and Clearance	Maintain safe distances from combustible structures; ensure adequate fire-rated enclosures or fire suppression systems if required
Utility Interfaces	Coordinate generator yard access with cable trenches, fuel pipe routing, and utility feeders

Key Specification Areas:

- **Power Rating and Redundancy:**
  - Define critical load classes and align with generator capacity (N, N+1, N+N, etc.).
- **Starting and Transfer Method:**
  - Auto-start via ATS or manual transfer switches (MTS).
  - Black start and soft start capabilities may be required.
- **Emissions Compliance:**
  - Comply with local air quality and environmental regulations. In Europe, generators must meet Euro Stage V standards as defined in EU Regulation 2016/1628, which establishes requirements for gaseous and particulate pollutant emission limits for non-road mobile machinery (European Commission, 2016). In the United States, generators are subject to EPA Tier emission standards, which establish progressively stringent limits for nitrogen oxides and particulate matter emissions from non-road diesel engines (US EPA, 2014).
- **Noise Rating:**
  - Typically 65–75 dB(A) at 1 m for enclosed generators; consider acoustic enclosures or barriers as needed.
- **Fuel Type:**
  - Diesel is standard; natural gas and HVO (hydrotreated vegetable oil) alternatives are increasingly adopted.

12.4.9.3 Fuel Storage and Distribution Infrastructure

Reliable and compliant fuel storage is essential to standby generator systems. Key VE areas include tank sizing, routing efficiency, and environmental mitigation (Table 12.23).

12.4.9.4 Environmental and Regulatory Compliance

External generator systems and fuel storage must comply with applicable national, local, and utility regulations.

Table 12.23 Fuel Storage VE Considerations

Component	VE Considerations
Fuel Tank (Main)	Sized for minimum 6–12 hours of operation; consider bunded above-ground tanks for accessibility and inspection
Day Tank	Smaller tank near generator for short-term operation; simplifies fuel pump control
Fuel Transfer Pumps	Redundancy (dual pumps) if long pumping distances or critical application
Fuel Pipework	Use double-wall piping for environmental protection; minimize pipe length and civil crossings
Spill Containment	Bunds sized for 110% of tank capacity; install leak detection sensors in bunds and pipe trenches
Refill Access	Provide safe and direct truck access for tank refill operations, away from public pathways and air intakes

- **Fire Codes:**
  - Comply with national fire protection standards including NFPA 30: Flammable and Combustible Liquids Code, which provides requirements for the storage, handling, and use of flammable and combustible liquids (NFPA 30, 2024), and NFPA 37: Standard for the Installation and Use of Stationary Combustion Engines and Gas Turbines, which establishes minimum fire-safety requirements for stationary engine installations (NFPA 37, 2021). Local equivalent standards apply in other jurisdictions.
- **Environmental Protection:**
  - Avoid placing fuel tanks in flood zones unless elevated.
  - Include spill detection and automatic shutdown where required.
- **Acoustic and Air Quality:**
  - Conduct environmental impact assessments for noise and emissions if required for permitting.
- **Inspection and Maintenance Provisions:**
  - Allow space and access for periodic fuel quality tests, bund drainage, and generator load bank testing.

12.4.10 EV Charging Infrastructure

The provision of EV charging infrastructure has become a standard requirement in modern developments due to accelerating electrification policies, green building standards, and user expectations.

EV charging systems introduce significant implications for site electrical load, trenching, coordination with civil works, utility interfacing, and long-term scalability—all of which must be addressed early in infrastructure planning and through VE processes.

12.4.10.1 EV Charging Typologies

EV chargers are categorized by output power level, charging speed, and application context (Table 12.24).

The charging modes are defined according to IEC 61851-1, the international standard for EV conductive charging systems, which establishes four distinct charging modes with varying levels of safety features and power capabilities (IEC, 2017).

Selection of charger type and quantity should be based on user profile, site function, power availability, and required turnaround times.

Table 12.24 EV Charger Typologies by Power Level and Application Context *(Classifications based on international standards for electric vehicle charging systems)*

Charger Type	Output Rating (kW)	Typical Application
Mode 2 (AC slow)	≤3.7	Private homes, small fleet areas
Mode 3 (AC fast)	7.4–22	Commercial buildings, parking lots, public spaces
Mode 4 (DC fast)	50–150	Service stations, highways, logistics hubs
High-Power DC	>150	High-turnover commercial or transit hubs

12.4.10.2 Infrastructure Design Considerations

Key VE-related Infrastructure Planning Parameters:

- **Electrical Load Impact:**
  - EV charging can represent significant peak demand. Apply diversity factors or smart load management to avoid costly transformer or switchgear upgrades.
- **Utility Coordination:**
  - Engage early with utility providers to secure additional capacity if required and avoid late-stage delays.
- **Physical Location and Civil Works:**
  - Position chargers close to main cable routes and LV feeder points to minimize trenching and cable runs.
  - Integrate with parking design to avoid rework or conflicts with kerbs, bollards, or drainage features.
- **Conduit and Duct Provisioning:**
  - Install spare ducts and draw pits during early civil works, even if chargers are to be installed later.
- **Access Control and User Management:**
  - Plan for secure access to chargers (e.g., RFID, app-based control) where appropriate.

12.4.10.3 Power Supply and Control Architecture (Table 12.25)

Table 12.25 EV Power Supply Design Considerations

Component	Design Considerations
LV Distribution Board	Dedicated sub-board or panel with surge protection and energy metering
Smart Load Management	Dynamic load balancing between chargers to avoid exceeding site capacity
Integration with BMS or EMS	Enables scheduling, usage monitoring, and energy profiling
Standby Power Coordination	Define whether chargers should be backed by generators or excluded during outages
Lightning and Surge Protection	Required for outdoor installations per established surge protection standards[a]

[a] Lightning and surge protection systems must comply with IEC 61643-11, which establishes requirements and test methods for low-voltage surge protective devices connected to power systems, ensuring protection against lightning and other transient overvoltages (IEC 61643-11, 2011).

12.4.10.4 Futureproofing and Value Engineering Opportunities

- **Install Structural and Conduit Provisions in Advance:**
  - Even if EV chargers are not installed immediately, ducting, draw pits, and feeder boards can be laid during initial civil works at minimal cost.
- **Scalable Distribution Architecture:**
  - Use modular distribution boards and install oversized cable trays or ducts to accommodate future additional chargers.
- **Shared Charger Clusters:**
  - Where turnover is low, cluster chargers to minimize number of powered locations.
- **Precast Foundations and Modular Mounting:**
  - Use standardized mounting bases and bollards to streamline deployment.
- **Renewable Integration Readiness:**
  - Design EV systems with optional future connection to solar PV or on-site battery storage systems.

12.4.10.5 Regulatory and Standard Compliance

EV infrastructure must comply with the data given in Table 12.26.

The IEC 60364-7-722 standard provides specific requirements for low-voltage electrical installations related to EV supply equipment, covering both energy supply to vehicles and feedback of electricity from vehicles to the electrical system. Meanwhile, ISO 15118 establishes the communication interface for vehicle-to-grid applications, enabling advanced functionalities such as plug-and-charge and bidirectional power flow.

It is essential to track evolving regional mandates, as many jurisdictions now require a minimum percentage of parking bays to be EV-enabled or EV-ready.

12.4.11 Sitewide Auxiliary Systems and Enclosures

Infrastructure projects often include a range of auxiliary systems located outside the building envelope. These systems—such as external telecommunications enclosures, security cabins, prefabricated switchgear kiosks, pump control stations, and remote I/O panels—require integration into the electrical distribution and containment network. While typically smaller in scale, these units have a disproportionate impact on trenching, coordination, utility interfacing, and VE outcomes.

Table 12.26 EV Infrastructure Compliance

Standard	Scope
IEC 61851 Series	EV conductive charging system requirements
IEC 60364-7-722	Wiring and installation for EV supply equipment
ISO 15118	Vehicle-to-grid (V2G) communication protocol
National Building and Energy Codes	Varying mandates for EV-ready provisions in new buildings

12.4.11.1 Types of Auxiliary Enclosures and Systems (Table 12.27)

Table 12.27 Auxiliary Enclosures Typical Functions

System Type	Typical Functions
Security or Access Control Cabins	Control gates, cameras, intercoms, badge readers
Telecom Cabinets	Housing for fiber patching, active network devices, ISP demarcation points
Pump Station Kiosks	Control and protection panels for external water/fire pumps or drainage systems
Metering Enclosures	External power metering for tenants, billing, or utility interface
RTU Panels or Remote I/Os	Supervisory control points for site-wide BMS or SCADA integration
Utility Handovers	Prefabricated LV/MV handover kiosks owned or accessed by third parties

These systems are often specified by other disciplines (e.g., security, mechanical, IT) but require careful coordination with electrical works during infrastructure planning.

12.4.11.2 Electrical and Civil Coordination Requirements

- **Power Supply Planning:**
  - Define LV feeder circuits from the nearest SDB or feeder pillar.
  - Apply load diversity where appropriate but ensure critical systems (e.g., fire pump panels) are independently backed or generator-supported.
- **Cable Routing and Access:**
  - Optimize cable trench layouts to serve multiple auxiliary systems with minimal digging.
  - Include draw pits or joint boxes for flexibility in layout and cable pulling.
- **Mounting and Foundation Requirements:**
  - Most outdoor enclosures require concrete plinths or base frames sized per manufacturer installation guides.
  - Ensure raised installation in flood-prone or sloped zones.
- **Environmental and Mechanical Protection:**
  - Specify enclosures with suitable IP ratings (EN 60529, 2000) and IK ratings (IEC 62262, 2002).
  - Add shading or passive ventilation for units containing electronics in exposed locations.

12.4.11.3 VE Opportunities in Auxiliary Systems

- **Enclosure Standardization:**
  - Use consistent enclosure types, form factors, and panel layouts across systems to reduce spares, simplify maintenance, and streamline procurement.
- **Shared Ducting Corridors:**
  - Plan shared containment routes for power, telecom, and control cabling feeding auxiliary systems—especially at perimeters or between security posts.
- **Precast Plinths and Modular Bases:**
  - Specify precast or modular bases for repetitive external systems to reduce on-site fabrication and improve deployment speed.

- **Prefabricated Assembly:**
  - Where feasible, use factory-tested kiosk-type assemblies (e.g., plug-and-play fire pump panel or fiber distribution boxes) to reduce on-site coordination risk.
- **Futureproofing:**
  - Include spare conduits, space for expansion, and additional power breakers for auxiliary growth (e.g., future CCTV or access control additions).

12.4.11.4 Safety, Compliance, and Maintenance

- **Compliance:**
  - Enclosures should comply with IEC 61439-1/-2 for LV assemblies (IEC 61439-1, 2020; IEC 61439-2, 2020) and relevant product-specific directives (e.g., EN 60529 for IP ratings).
- **Access and Maintenance Provisions:**
  - Maintain clearance zones around enclosures (typically 1–1.2 m), and ensure safe access without crossing roadways or landscaped zones.
- **Lightning and Surge Protection:**
  - Bond external metallic enclosures to the site earthing network and install SPDs at control or communication entry points.

12.4.12 Futureproofing Provisions in Infrastructure Design

Infrastructure systems are often installed during the early phases of development, well before the final configuration of buildings, site loads, or technologies is known. For this reason, proactive futureproofing is essential to ensure that infrastructure remains adaptable, scalable, and economical to modify in the future.

VE in this domain focuses on embedding strategic spare capacity, modular elements, and coordinated spatial allowances that reduce the cost and disruption of future changes.

12.4.12.1 Categories of Futureproofing Measures (Table 12.28)

Table 12.28 Futureproofing Measures

Futureproofing Category	Typical Measures
Space Allocation	Reserve additional physical space in manholes, feeder rooms, substations, and ductbanks
Conduit and Cable Tray Provisioning	Install spare conduits and oversized cable trays for future wiring needs
Load Headroom	Design transformers, switchgear, and generators with 10–25% capacity margin (where justifiable)
Panel and Feeder Expansion	Specify panels with spare ways, additional feeder slots, and oversized busbars
System Modularity	Use modular switchgear and scalable UPS or lighting control systems
Monitoring Infrastructure	Include cable pull ropes, smart sensors, or tagged pathways to support digital twin or GIS integration

12.4.12.2 Coordination with Civil and Utility Planning

Many futureproofing strategies depend on early civil and utility planning coordination:

- **Multi-Utility Ductbanks:**
  - Incorporate empty ducts and draw pits reserved for future services (e.g., 5G, security, EV charging expansion).
- **Corridor Preservation:**
  - Define utility reservation zones or sleeves under roadways and pavements to avoid costly re-excavation.
- **Modular Kiosk or Pad Designs:**
  - Design transformer pads, feeder pillar bases, and enclosures with extension capacity or plug-in interfaces.

12.4.12.3 Technology Integration Readiness

Modern infrastructure must accommodate future technologies that may not yet be part of the initial scope:

- EV-Ready Ducting and Load Centers
- Smart Metering or Submetering Capacity
- Integration Points for Renewable Energy (e.g., solar PV busbars, inverters, cable routing)
- Battery Energy Storage System (BESS) Zones
- Edge Data Equipment or Smart City Interfaces

Including structural, electrical, and containment readiness at minimal upfront cost yields major future value.

12.4.12.4 VE Principles for Futureproofing (Table 12.29)

Table 12.29 VE Implication on Futureproofing

Principle	VE Implication
Cost of Change Curve	Intervene early—provisions during infrastructure phase are 5–10x cheaper than retrofit
Standardization	Use repeatable details, modules, and component types to reduce future modification complexity
Risk-Based Expansion Allowance	Apply margin selectively (e.g., high-use or critical zones), not uniformly across all infrastructure
Document and Label Provisions	Ensure spare conduits, pads, and feeders are clearly documented in as-builts for future discovery

12.5 Procurement Considerations

Procurement strategy plays a pivotal role in the successful delivery of electrical infrastructure works. Unlike internal building systems, external infrastructure involves long-lead items, multi-vendor coordination, civil integration, utility approvals, and outdoor environmental performance factors.

A value-engineered procurement approach ensures that infrastructure systems meet technical, regulatory, and functional needs while optimizing total life-cycle cost, lead time, and maintainability.

12.5.1 Infrastructure-Specific Procurement Challenges (Table 12.30)

Table 12.30 Infrastructure-Specific Procurement Challenges

Challenge	Description
Long Lead Times	MV switchgear, packaged substations, custom enclosures, and standby generators often have lead times of 20–36 weeks.
Multi-Stakeholder Dependencies	External works often require approvals or interfaces with utility companies, municipalities, or environmental regulators.
Outdoor Durability Requirements	All procured equipment must comply with outdoor IP ratings (EN 60529, 2000), IK ratings (IEC 62262, 2002), UV resistance, corrosion protection, etc.
Civil Integration Requirements	Equipment must be coordinated with foundations, trenching, and stormwater management works.
Phased Construction Interfaces	Infrastructure systems are often installed in early phases but must remain serviceable across project stages.

12.5.2 Strategic Procurement Approaches for VE (Table 12.31)

Table 12.31 Strategic Procurement Approaches for VE

Strategy	VE Benefit
Early Equipment Specification	Enables pre-approval with utility companies and ensures design-compliant delivery
Framework or Bulk Procurement	Reduces per-unit cost of standard items (e.g., poles, feeder pillars, manholes)
Modular Pre-Assembled Systems	Factory-tested solutions reduce on-site labor and coordination time
Dual-Sourcing Risk Management	Prevents project delays due to single-vendor dependence
Value-Based Prequalification	Evaluates vendors on total value (lead time, warranty, O&M support) rather than lowest price alone

12.5.3 Key Equipment Categories and VE Procurement Guidelines (Table 12.32)

Table 12.32 Key Equipment Categories and VE Procurement Guidelines

Equipment	Procurement Considerations
MV Switchgear	Confirm IEC 62271 compliance, arc classification, future expansion bays (IEC 62271-200, 2021).
Transformers	Validate oil vs dry-type, enclosure rating (IP23+), noise performance (IEC 60076-1, 2011).
Generator Sets	Match critical load, delivery logistics, serviceability access
Feeder Pillars and Panels	Standardize types and dimensions, require factory QA/QC (IEC 61439-1, 2020).

Table 12.32 (Continued)

Equipment	Procurement Considerations
Lighting Columns and Luminaires	Specify impact resistance, material finish, photometric testing certifications (IEC 63103, 2020).
Earthing Components	Require test reports for resistance values, corrosion protection coating data (IEC 61936-1, 2021).
Cables and Ducts	Verify certification to IEC/BS/ASTM standards, ensure availability in required lengths/reels (IEEE 525, 2007)

12.5.4 Procurement Package Structuring

Efficient procurement relies on structuring infrastructure components into clear, manageable packages aligned with contractor and vendor scopes.

Suggested Package Breakdown:

- Package A: MV Switchgear and Substation Equipment
- Package B: Transformers and Associated Civil Works
- Package C: LV Distribution Boards and Feeder Pillars
- Package D: External Lighting Systems and Controls
- Package E: Generator Sets and Fuel Infrastructure
- Package F: Cabling, Earthing, and Containment Systems
- Package G: Smart Systems (EV charging, monitoring, control panels)

Each package should include technical specifications, installation guidelines, compliance checklists, and interface requirements (Figure 12.3).

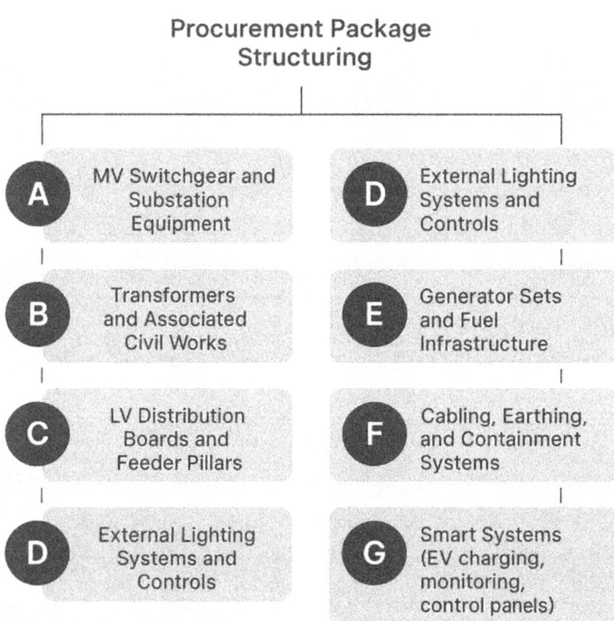

Figure 12.3 Procurement Package Structuring for Electrical Infrastructure Works

Table 12.33 Life-Cycle Cost and Impact on Procurement

Life-Cycle Cost Factor	Procurement Impact
Maintenance Frequency	Select equipment with longer MTBF (mean time between failures) and modular repair options
Energy Efficiency	Consider transformer losses, LED luminaire efficacy, generator fuel consumption
Warranty and Support	Favor vendors offering extended warranties, local servicing, and digital support portals
End-of-Life Handling	Prefer recyclable components and products with known disposal pathways

12.5.5 Life Cycle-Based Procurement Criteria

A value-based procurement process considers not just capital expenditure (CapEx), but also the total cost of ownership (TCO) (Table 12.33).

12.6 Implementation and Site Coordination Considerations

The execution phase of electrical infrastructure works involves complex coordination between civil, electrical, utility, and third-party contractors across various work fronts. Proper site-level implementation planning is essential to realize the value identified during the design and procurement stages. Without it, even well-engineered solutions may suffer from construction delays, access conflicts, interface failures, or compromised system integrity.

VE in implementation focuses on constructability, sequencing, interface clarity, and site-level risk mitigation.

12.6.1 Phasing and Construction Integration

Infrastructure systems are often delivered in parallel with other sitewide works such as roads, drainage, utilities, and structures.

Key VE-oriented phasing considerations are given in Table 12.34.

Table 12.34 Phasing Strategies for Infrastructure Works

Element	Implementation Focus
Ductbanks and Trenches	Sequence prior to road paving or landscaping to avoid rework
Substation Access	Ensure substructure is available early for transformer delivery and cable pulling
External Lighting Foundations	Install bases and conduits before surface finishes are completed
Earthing Network	Install sitewide earth grid concurrently with civil excavation to reduce trenching duplication
Temporary vs Permanent Power	Plan transition from temporary builders' supply to permanent infrastructure-based supply point

Phased handover of infrastructure components must be incorporated into the master construction schedule with sufficient float for approvals and testing (Figure 12.4).

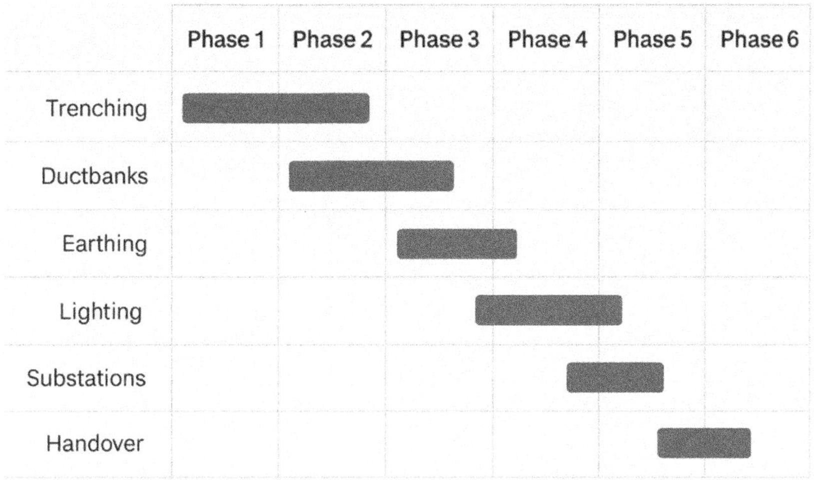

Figure 12.4 Phased Implementation Timeline for Electrical Infrastructure Works. This Gantt-style diagram illustrates the sequential and partially overlapping phases of key infrastructure installation activities, including trenching, ductbank construction, earthing, external lighting, substation deployment, system testing, and final handover. The use of generic phase labels supports adaptation to project-specific timelines.

12.6.2 Civil-Electrical-Utility Interface Management

External infrastructure works demand tight coordination across multiple trades and stakeholders:

- **Civil Works:**
  - Trench alignments, manhole placement, foundations, cable protection, and rebar clearances must be confirmed before excavation.
- **Electrical Contractor:**
  - Must validate cable pull paths, trench depth, feeder lengths, and termination space in panels and switchgear.
- **Utility Providers:**
  - Coordinate inspections, cable jointing, metering, energization, and utility interlock testing.
- **Landscaping and Urban Design Teams:**
  - Validate lighting pole positions, protection of surface finishes, and fixture mounting compatibility.

Use of **coordinated IFC drawings**, combined civil–electrical trench schedules, and regular site coordination meetings are critical to prevent field clashes.

12.6.3 Quality Control and Inspection Planning

Infrastructure components are typically buried, embedded, or installed in permanent exterior conditions—making early inspection and documentation critical (Table 12.35).

Table 12.35 Inception Consideration for Infrastructure Components

Component	Key Inspection Points
Cable Trenches	Confirm depth, bedding, warning tape, backfill compaction
Manholes and Pull Boxes	Verify structural integrity, drainage, cover strength
Earthing Systems	Confirm rod installation, continuity tests, resistance testing prior to backfill
Generator Yards and Fuel Lines	Check bunds, pipe slope, leak detection, access paths
Lighting Columns	Confirm foundation bolts, verticality, cable terminations, corrosion protection
Panel Installations	Confirm mounting, IP ratings, labeling, segregation between circuits

Site inspection checklists, test protocols, and third-party validation may be required based on regulatory or utility body requirements.

12.6.4 Health, Safety, and Environmental Considerations

Infrastructure works often occur outdoors, near roadways, public zones, or environmentally sensitive areas.

Key HSE VE Opportunities:

- **Cable Routing:**
  - Avoid high-traffic crossings where future access is difficult or hazardous.
- **Fuel and Generator Areas:**
  - Provide controlled spill containment, signage, and emergency access paths.
- **Public Interface:**
  - Fence off open pits, provide temporary lighting, and manage works near pedestrian areas.
- **Environmental Protection:**
  - Prevent trench water runoff, manage dust during excavation, and protect tree roots near ducting works.

12.6.5 Documentation and Handover Strategy

Effective project closeout for infrastructure includes:

- **As-built Drawings and Schematics:**
  - Accurately reflect underground cable paths, joint boxes, spare conduits, and earth rods.
- **Test Certificates and Commissioning Reports:**
  - Include insulation resistance, earth resistance, load tests, and protection relay settings.
- **Spare Parts and Asset Register:**
  - Document installed equipment and spares with serial numbers and warranty data.
- **O&M Manuals:**
  - Provide detailed guidance for external systems, including fuel system service intervals, lighting control resets, and enclosure IP protection maintenance.

12.7 Testing, Commissioning, and Handover

Testing and commissioning form the final stage in the delivery of electrical infrastructure systems, validating that all installations meet their design intent, safety standards, and operational requirements.

A structured handover process ensures all stakeholders—owners, operators, utilities, and regulatory authorities—receive verified, compliant systems supported by complete documentation.

VE at this stage focuses on early test planning, interface risk mitigation, documentation standardization, and ensuring long-term maintainability through effective transition of knowledge.

12.7.1 Testing Objectives and Scope

Testing serves both **verification** and **validation** purposes (Table 12.36).

Core Testing Areas Include:

- Cable insulation and continuity
- Earth resistance and bonding checks
- Switchgear operation and protection relay settings
- Transformer turns ratio and winding resistance
- Generator system load testing (with or without load banks)
- Lighting and lighting control functional checks
- Surge protection system integrity
- Interlock and auto-changeover testing (e.g., ATS operation)

12.7.2 Commissioning Sequencing

Commissioning should follow a logical and traceable sequence that reflects system dependencies.

Suggested Commissioning Order:

1) **Visual Inspection and Pre-Test Checks**
   - Confirm labels, torque settings, terminations, and signage.
2) **Dead Testing**
   - Insulation resistance, continuity, earthing validation.
3) **Live Energization of MV and LV Systems**
   - Performed with utility or generator support and safety protocols.

Table 12.36 Testing Scope

Purpose	Scope
Verification	Confirming each system is installed in accordance with drawings, specifications, and codes
Validation	Ensuring systems operate safely and reliably under expected operating conditions

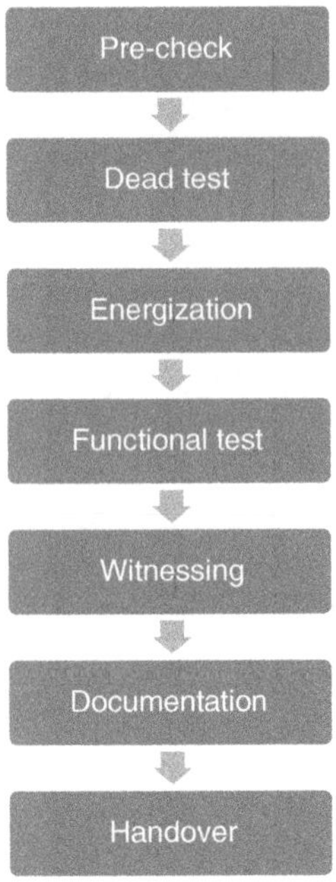

Figure 12.5 Commissioning Sequence Flow Chart

4) **Functional Testing**
   - Control logic, sensor responses, lighting controls, protection device trip settings.
5) **Integrated System Testing (IST)**
   - If applicable, simulate fault or transfer conditions.
6) **Final Validation and Witness Testing**
   - By third parties, utilities, or AHJs as required.

All steps should be supported by a **pre-approved method statement** and **project-specific test checklist**.

Flowchart showing testing and commissioning sequence is shown in Figure 12.5.

12.7.3 Testing Standards and Authority Requirements

All testing activities must align with relevant international and national standards (Table 12.37).

12.7.4 Handover Documentation and Deliverables

Value-based handover planning includes delivery of data given in Table 12.38.

Table 12.37 Testing Standards

Standard	Application
IEC 60364 Series	General electrical installation testing and safety verification
IEC 61439-1/-2	Low-voltage switchgear and controlgear assemblies testing
IEC 60076 Series	Power transformer factory and site testing
IEEE 400 Series	Field testing of power cables
Local utility and AHJ guidelines	Specific documentation and witnessing procedures

Ensure the most current version of each standard is referenced and confirm utility-specific forms or certifications needed for energization approval.

Table 12.38 Handover Deliverables

Deliverable	Content Overview
As-Built Drawings	Cable routes, earth grid layout, panel schedules, equipment elevations
Test Certificates	Validated, signed records for all equipment and systems tested
Commissioning Reports	Test protocols, pass/fail criteria, punch list resolution
Operation & Maintenance Manuals	Manufacturer documentation, system-level procedures, troubleshooting guides
Asset Register	Tagging, serial numbers, warranty durations, maintenance schedule
Training Records	Confirmation of operator and facilities team onboarding and demonstrations

Documentation should be structured and indexed in both print and digital formats, preferably integrated into the project's asset management system.

12.7.5 Common Pitfalls and VE Mitigation Strategies (Table 12.39)

Table 12.39 Common Pitfalls and VE Mitigation Strategies

Common Issue	VE-Oriented Mitigation
Late discovery of testing failures	Early factory testing (FAT), dry run site tests, and pre-functional checklists
Delayed energization due to utility approval	Early engagement, joint inspections, clear documentation of utility interface points
Incomplete documentation	Use of preapproved templates and progressive submission throughout construction
Knowledge loss during handover	Scheduled walkthroughs, user training, and O&M handover checklists

12.8 Summary of VE Strategies for Electrical Infrastructure

Electrical infrastructure external to the building envelope constitutes a significant opportunity domain for VE application across all stages of project development. Owing to its extensive interface with site civil works, regulatory bodies, utilities, and long-term operational requirements, this system layer directly affects both initial construction outcomes and the entire asset life-cycle performance.

The application of structured VE principles to electrical infrastructure achieves both cost and functional optimization while preserving regulatory compliance, safety, and future adaptability. Throughout this chapter, multiple VE domains have been addressed, each contributing to holistic project value:

- **Load Planning and Futureproofing:** Implementation of scalable capacity margins, provision of modular system architectures, allocation of spare conduits, oversized containment, and planned reserve areas for emerging technologies such as EV charging, distributed energy resources, and smart city interfaces.
- **Routing and Civil Integration:** Optimization of trench layouts, ductbank configurations, shared service corridors, and precise spatial coordination with roadways, drainage, utilities, and structural foundations to reduce rework, minimize excavation, and enhance installation efficiency.
- **Equipment Specification and Standard Compliance:** Selection of transformers, switchgear, feeder panels, lighting systems, cables, and protective devices in strict accordance with relevant IEC, IEEE, NFPA, and ISO standards. Specification strategies promote component standardization, factory testing, modular assemblies, and ease of procurement, ensuring both technical performance and procurement efficiency.
- **Procurement Methodologies:** Adoption of value-based procurement approaches incorporating early vendor engagement, framework agreements, dual sourcing, and life-cycle cost evaluation metrics. Modular preassembled equipment and offsite testing methodologies further reduce site coordination complexity and installation risk.

- **Construction and Site Coordination:** Sequenced construction schedules incorporating phased ductbank installation, substation access, lighting base preparations, and integrated civil-electrical trench schedules. Close management of utility approvals, interface milestones, and regulatory inspections is essential to maintain schedule integrity.
- **Testing, Commissioning, and Handover:** Deployment of comprehensive commissioning protocols governed by applicable standards (IEC 60364, IEC 61439, IEEE 400, etc.), systematic verification procedures, documented test results, and structured O&M handover processes. Progressive documentation ensures traceability, regulatory compliance, and knowledge continuity for operational teams.

The structured application of VE across these domains generates measurable outcomes including:

- Reduction of unnecessary overdesign.
- Lower first cost without compromising performance.
- Enhanced constructability and interface control.
- Simplification of future expansions or modifications.
- Optimized long-term maintenance and operational reliability.
- Strengthened resilience against regulatory, environmental, and operational risks.

Electrical infrastructure forms the operational backbone of the entire built environment. VE applied at the infrastructure level produces compounding technical and financial benefits that support both present-day project viability and future system adaptability.

References

ANSI/BICSI. (2019). *002: Data Center Design and Implementation Best Practices.*

Brand, S. (1994). *How Buildings Learn: What Happens After They're Built.* Viking Press.

CEN. (2015). *EN 13201-2:2015 - Road lighting - Part 2: Performance requirements.*

CIE. (2010). *CIE 115:2010 - Lighting of roads for motor and pedestrian traffic.*

Dell'Isola, A. (1997). *Value Engineering: Practical Applications for Design, Construction, Maintenance & Operations.* R.S. Means Company.

European Commission. (2016). *Regulation (EU) 2016/1628 on requirements relating to gaseous and particulate pollutant emission limits and type-approval for internal combustion engines for non-road mobile machinery.* Official Journal of the European Union.

Flanagan, R., Norman, G., Meadows, J., & Robinson, G. (1989). *Life Cycle Costing Theory and Practice.* BSP Professional Books.

IEC 60076-1. (2011). *Power transformers – Part 1: General.*

IEC 60364-6. (2016). *Low-voltage electrical installations – Part 6: Verification.*

IEC 60364-7-722. (2018). *Low-voltage electrical installations – Part 7-722: Requirements for special installations or locations – Supplies for electric vehicles.*

IEC 61439-1. (2020). *Low-voltage switchgear and controlgear assemblies – Part 1: General rules.*

IEC 61439-2. (2020). *Low-voltage switchgear and controlgear assemblies – Part 2: Power switchgear and controlgear assemblies.*

IEC 61643-11. (2011). *Low-voltage surge protective devices – Part 11: Surge protective devices connected to low-voltage power systems – Requirements and test methods.*

IEC 61851-1. (2019). *Electric vehicle conductive charging system – Part 1: General requirements.*

IEC 61936-1. (2021). *Power installations exceeding 1 kV AC – Part 1: Common rules.*

IEC 62262. (2002). *Degrees of protection provided by enclosures for electrical equipment against external mechanical impacts (IK code).*

IEC 62271-200. (2021). *High-voltage switchgear and controlgear – Part 200: AC metal-enclosed switchgear and controlgear for rated voltages above 1 kV and up to and including 52 kV.*

DOE. (2024). *'Energy conservation standards for general service lamps: Final rule', Federal Register, 89(75).* Department of Energy: U.S. Available at: https://www.energy.gov/articles/doe-finalizes-efficiency-standards-lightbulbs

IEC 60529. (2020). *IEC 60529:2020 Degrees of protection provided by enclosures (IP Code).* Geneva: International Electrotechnical Commission.

IEC 61439-2. (2020). *IEC 61439-2:2020 Low-voltage switchgear and controlgear assemblies – Part 2: Power switchgear and controlgear assemblies.* Geneva: International Electrotechnical Commission.

IEC 60364-5-52. (2009). *IEC 60364-5-52:2009 Low-voltage electrical installations – Part 5-52: Selection and erection of electrical equipment – Wiring systems.* Geneva: International Electrotechnical Commission.

IEC. (2017). *IEC 61851-1:2017 Electric vehicle conductive charging system – Part 1: General requirements.* Geneva: International Electrotechnical Commission.

ANSI/BICSI 002. (2019). *ANSI/BICSI 002-2019 Data center design and implementation best practices.* Tampa, FL: BICSI.

IEEE 80. (2020). *IEEE Std 80-2020 IEEE guide for safety in AC substation grounding.* New York: Institute of Electrical and Electronics Engineers.

IEEE 142. (2022). *IEEE Std 142-2022 IEEE recommended practice for grounding of industrial and commercial power systems (IEEE Green Book).* New York: Institute of Electrical and Electronics Engineers.

IEC 62305-3. (2010). *IEC 62305-3:2010 Protection against lightning – Part 3: Physical damage to structures and life hazard.* Geneva: International Electrotechnical Commission.

IEC 63103. (2020). *IEC 63103:2020 Lighting control devices – Photosensors for lighting control – Energy performance.* Geneva: International Electrotechnical Commission.

US EPA. (2014). *Control of emissions from nonroad diesel engines. Code of Federal Regulations, Title 40, Part 1039.* Washington, DC: United States Environmental Protection Agency. Available at: https://www.epa.gov/regulations-emissions-vehicles-and-engines/final-rule-control-emissions-air-pollution-nonroad-diesel

IEEE 400. (2012). *Guide for Field Testing and Evaluation of the Insulation of Shielded Power Cable Systems.*

IEEE 400.2. (2013). *Guide for Field Testing of Shielded Power Cable Systems Using Very Low Frequency (VLF).*

IEEE 525. (2007). *Guide for the Design and Installation of Cable Systems in Substations.*

Institution of Engineering and Technology. (2024). *Electrical Installation Design Guide: Responsibilities for those in an electrical installation design role.*

ISO 15118-1. (2019). *Road vehicles — Vehicle to grid communication interface — Part 1: General information and use-case definition.*

ISO 15118-20. (2022). *Road vehicles — Vehicle to grid communication interface — Part 20: 2nd generation network layer and application layer requirements.*

ISO 15686-5. (2017). *Buildings and Constructed Assets - Service Life Planning - Part 5: Life-cycle Costing.*

ISO 31000. (2018). *Risk Management - Guidelines.*

Miles, L. D. (1972). *Techniques of Value Analysis and Engineering.* McGraw-Hill.

NFPA 30. (2024). *Flammable and Combustible Liquids Code.*

NFPA 37. (2021). *Standard for the Installation and Use of Stationary Combustion Engines and Gas Turbines.*

13

Financial Strategies and Economic Impact of Value Engineering on Electrical Systems

13.1 Introduction

Value engineering (VE) in electrical systems represents a structured, function-oriented methodology for achieving cost-effective outcomes without compromising performance, reliability, or compliance. This section introduces the financial dimensions of VE, situating them within the broader context of infrastructure and building electrical system design. The financial strategies presented in this chapter build upon the technical optimization principles established in earlier sections of this guide, forming an integrated approach to design, execution, and operational decision-making.

While the financial evaluation methods presented in this chapter—such as life-cycle costing, return on investment (ROI), and risk-based modeling—are applicable across all building systems, the focus here is on their application within the electrical engineering domain. This includes power distribution systems, lighting, emergency power infrastructure, earthing and protection, as well as dependent systems such as ICT, BMS, and fire safety where these fall within the electrical scope of works.

13.1.1 Financial Relevance in Electrical VE

Electrical engineers are increasingly required to engage in financial evaluation processes traditionally dominated by project managers and quantity surveyors. The ability to link technical decisions to financial outcomes has become essential to justify capital expenditures (CAPEX), support life-cycle performance strategies, and enhance value delivery. VE enables informed trade-offs between initial capital investment and long-term operational expenditure, underpinned by structured financial analysis tools.

The financial implications of VE interventions can be broadly classified into the following categories:

- **CAPEX**: Initial cost associated with procurement, installation, and commissioning of electrical systems.
- **Operational Expenditure (OPEX)**: Recurring costs including energy consumption, routine maintenance, system monitoring, and parts replacement.
- **Deferred Cost Risk**: Avoidable future costs due to premature obsolescence, inadequate capacity planning, or regulatory noncompliance.

Value Engineering for Electrical Systems in Construction, First Edition. Hazem Hussein.

Companion website: www.wiley.com/go/hazem

Effective VE seeks to **minimize life-cycle cost** (LCC), which can be mathematically defined as:

$$LCC = C_0 + \sum_{t=1}^{n} \frac{C_{e,t} + C_{m,t} - S_t}{(1+r)^t}$$

Where:

LCC: Life-cycle cost
C_0: Initial capital cost
$C_{e,t}$: Energy cost in year t
$C_{m,t}$: Maintenance cost in year t
S_t: Salvage or residual value in year t
r: Discount rate (expressed as a decimal)
t: Analysis year, from 1 to n
n: Project or system life in years

This equation and life-cycle approach is adapted from National Institute of Standards and Technology's (NIST) Life-Cycle Costing Manual, which is widely adopted in infrastructure projects (NIST, 2020).

13.1.2 Financial Objectives of VE in Electrical Systems

Strategic financial planning within electrical VE aims to achieve the following objectives:

1) Minimize upfront investment without compromising compliance or safety
   - **Smart Capital Allocation:** Prioritize expenditure on critical system components while identifying areas where cost efficiencies can be realized.
   - **Value-Based Material Selection:** Evaluate materials and equipment based on price-to-performance ratios rather than lowest initial cost.
   - **Phased Implementation Strategies:** Deploy systems in strategic phases to distribute CAPEX over time while maintaining regulatory compliance.
   - **Regulatory Requirement Optimization:** Design systems that meet all safety and compliance standards without costly overengineering.
2) Maximize long-term operational returns via efficiency and reliability gains
   - **Energy Efficiency Focus:** Select components and system designs that minimize power consumption and reduce utility costs.
   - **Maintenance Requirement Reduction:** Implement designs that lower the frequency and complexity of required maintenance interventions.
   - **Reliability Enhancement:** Invest strategically in system redundancy and quality components where failure would result in significant operational costs.
   - **Performance Monitoring Integration:** Incorporate monitoring systems that enable proactive maintenance and prevent costly emergency repairs.
3) Optimize life-cycle expenditure for capital assets
   - **Total Cost of Ownership (TCO) Analysis:** Evaluate acquisition, operation, maintenance, and eventual replacement costs throughout the asset life cycle.
   - **Depreciation and Replacement Planning:** Schedule system upgrades to align with financial depreciation schedules and technological advancement cycles.
   - **Scalability Considerations:** Design systems with future expansion capabilities to avoid premature replacement costs.

- **End-of-Life Value Recovery:** Plan for recycling, repurposing, or residual value capture of components at retirement.

4) Identify alternative solutions that maintain or enhance system functionality
   - **Technological Alternative Assessment:** Evaluate emerging technologies against traditional approaches based on both performance and financial metrics.
   - **System Architecture Optimization:** Consider multiple design configurations to identify the most cost-effective approach to meeting functional requirements.
   - **Multi-Purpose Component Utilization:** Select equipment that can serve multiple functions to reduce overall system complexity and cost.
   - **Cross-System Integration:** Identify opportunities where electrical systems can be integrated with other building systems for enhanced efficiency.
5) Quantify both tangible and intangible financial benefits
   - **Direct Cost Savings Calculation:** Establish clear metrics for measuring reductions in energy consumption, maintenance expenses, and operational costs.
   - **Risk Mitigation Valuation:** Assign financial value to reduced system failure probabilities and enhanced resilience.
   - **Productivity Enhancement Measurement:** Quantify benefits from improved system performance on operational productivity and output.
   - **Environmental Impact Assessment:** Calculate financial advantages from reduced carbon footprint, including potential carbon credit benefits and regulatory compliance.

These priorities reflect standard best practices in cost-conscious engineering and are foundational to value-based infrastructure design. Financial objectives of VE in electrical systems are illustrated in Figure 13.1.

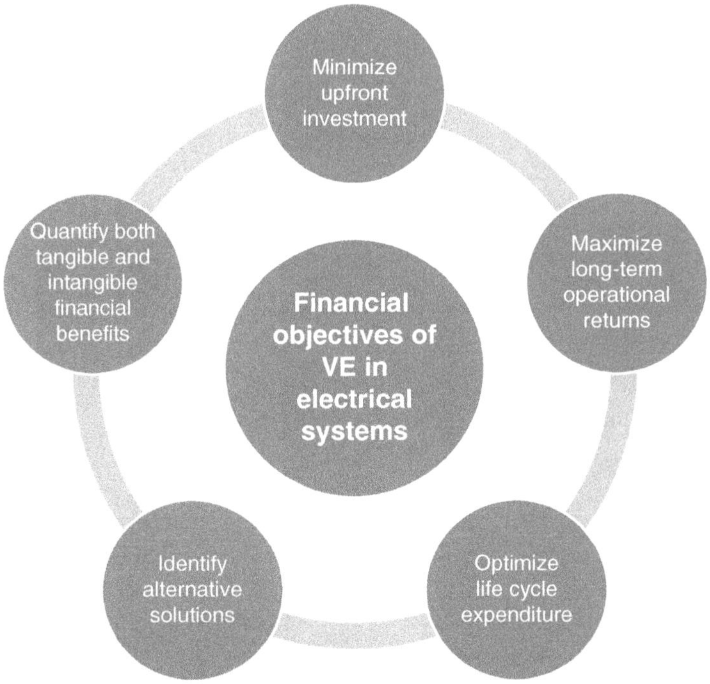

Figure 13.1 Financial Objectives of VE in Electrical Systems

13.1.3 Economic Benefits of VE Adoption

Integrating VE into electrical system planning has both microeconomic and macroeconomic implications. At the project level, the following benefits are typically realized:

- **Direct Cost Savings**: Reductions in quantities, materials, and labor
- **Operational Efficiencies**: Lower utility bills and optimized maintenance schedules
- **Increased Asset Value**: Enhanced marketability of energy-optimized properties
- **Compliance Advantages**: Easier attainment of sustainability certifications (e.g., LEED, BREEAM)
- **Risk Mitigation**: Avoidance of penalties and retrofit costs

These benefits are well documented across case studies in commercial, healthcare, and data center sectors and can serve as a key justification for early-stage VE integration.

13.1.4 Role of Engineers in Financial Planning

Successful implementation of VE requires electrical engineers to adopt a financially literate mindset. This involves:

- Interpreting economic data (e.g., payback period, net present value [NPV], ROI)
- Participating in cost-benefit discussions with commercial stakeholders
- Evaluating technical trade-offs not solely by performance metrics, but also by TCO

Life cycle-based financial decision-making is now a core competency in multidisciplinary design teams and is essential for demonstrating the business case of high-performance electrical systems.

13.2 Strategic Financial Planning in VE

Strategic financial planning forms the cornerstone of successful VE in electrical systems. While technical decisions typically revolve around performance, safety, and compliance, financial planning aligns these decisions with project affordability, long-term viability, and investment optimization. This section outlines key financial planning principles essential to VE implementation, including cost structuring, life-cycle modeling, stakeholder alignment, and adaptability to market dynamics.

13.2.1 Cost Structuring and Life-Cycle Thinking

A well-structured financial plan begins with an accurate understanding of TCO. In VE applications, this requires categorizing costs into the following layers:

- **Capital Costs (CAPEX):** Initial procurement, installation, and commissioning of electrical infrastructure, including switchgear, uninterruptible power supply (UPS), distribution cabling, and controls.
- **Operational Costs (OPEX):** Ongoing expenditures such as energy consumption, periodic maintenance, software updates for controls, and performance testing.
- **End-of-Life and Residual Costs:** Disposal, replacement, or salvage values associated with major equipment.

- **Contingency and Risk Allocation:** Budget allowances for design uncertainties, technology obsolescence, or scope changes.

As introduced in Section 13.1, the LCC model forms the basis for comparing TCO across alternative solutions. In financial planning, this model is expanded and integrated with project-specific parameters as follows:

$$LCC = C_0 + \sum_{t=1}^{n} \frac{C_{e,t} + C_{m,t} - S_t}{(1 + r)^t}$$

Where:

LCC: Life-cycle cost
$\boldsymbol{C_0}$: Initial capital cost
$\boldsymbol{C_{e,t}}$: Energy cost in year t
$\boldsymbol{C_{m,t}}$: Maintenance cost in year t
$\boldsymbol{S_t}$: Salvage or residual value in year t
$\boldsymbol{r}$: Discount rate (expressed as a decimal)
$\boldsymbol{t}$: Analysis year, from 1 to n
$\boldsymbol{n}$: Project or system life in years

This expression enables fair comparison between higher-efficiency, higher-cost solutions and their lower-cost, higher-OPEX counterparts, forming a key decision metric in VE (NIST, 2020).

13.2.2 Setting Financial Evaluation Metrics

In addition to LCC, multiple financial indicators are used to assess the viability of VE options:

- **Simple Payback Period (SPP):**

$$SPP = \frac{C_0}{\text{Annual net savings}}$$

Where:

SPP: Simple Payback Period
$\boldsymbol{C_0}$: Initial capital investment
Annual net savings: Total annual reduction in operational and energy costs resulting from the VE measure

Indicates the time required to recover the initial investment. Short paybacks (e.g., under 5 years) are typically preferred for systems with rapid efficiency gains, such as lighting upgrades.

- **Return on Investment (ROI):**

$$ROI = \left(\frac{\text{Total net benefits}}{C_0}\right) \times 100\%$$

Where:

ROI: Return on Investment (expressed as a percentage)
Total net benefits: Cumulative cost savings (including energy, operations and maintenance, and rebates) minus any recurring operational costs over the analysis period

$\boldsymbol{C_0}$: Initial investment cost
100%: Multiplies the ratio to express ROI as a percentage

- **Net Present Value (NPV):**

$$NPV = \sum_{t=1}^{n} \frac{B_t - C_t}{(1+r)^t}$$

Where:

NPV: Net present value
$\boldsymbol{B_t}$: Financial benefit (e.g., energy savings, rebates) in year t
$\boldsymbol{C_t}$: Total cost (e.g., maintenance, operations) in year t
$\boldsymbol{r}$: Discount rate
$\boldsymbol{t}$: Year index (from 1 to n)
$\boldsymbol{n}$: Analysis period in years

A positive NPV indicates economic favorability.

- **Internal Rate of Return (IRR)**: The discount rate at which NPV becomes zero. This is used to compare against internal

$$\sum_{t=0}^{n} \frac{C_t}{(1+IRR)^t} = 0$$

Where:

IRR: Internal rate of return (the discount rate that sets the NPV of all cash flows to zero).
$\boldsymbol{C_t}$: Net cash flow at time t (can be positive or negative; initial investment is usually negative).
$\boldsymbol{t}$: Time period (starting from 0 for the initial investment).
$\boldsymbol{n}$: Total number of time periods

Using these tools during the VE process ensures not only that alternative solutions perform technically, but that they deliver measurable financial advantage.

13.2.3 Aligning VE with Project Financial Constraints

Projects are frequently constrained by fixed capital budgets, especially in public or institutional sectors. In such environments, VE serves as a cost-redistribution mechanism, shifting funds from overdesigned or noncritical systems into more valuable infrastructure components.

This requires close collaboration between:

- Electrical engineers (assessing functional requirements and technical viability)
- Cost consultants (validating pricing and escalation)
- Owners and sponsors (defining risk appetite and budget priorities)

VE planning must also consider:

- **Cash flow limitations** across multi-phase project schedules
- **Debt service obligations** if financed through loans or bonds
- **Incentive windows** for rebates or tax deductions with expiry dates

A financially strategic VE approach may recommend modular implementation, deferral of non-critical features, or leasing options to reduce upfront capital load.

13.2.4 Market Sensitivity and Flexibility

Electrical system pricing is increasingly sensitive to:

- **Raw material volatility** (e.g., copper, aluminum)
- **Technology obsolescence** (e.g., older PLCs or control interfaces)
- **Regulatory shifts** (e.g., mandatory energy efficiency levels or carbon caps)

Strategic planning must accommodate such risks through:

- **Scenario-based costing** (best/likely/worst)
- **Flexible procurement packages** with design alternates
- **Escalation allowances** and early procurement for long-lead items

A well-calibrated VE plan will use tools such as **sensitivity analysis** to understand the effect of price variation or performance deviation on project outcomes, as covered further in Section 13.4.

13.2.5 Stakeholder Engagement and Decision Governance

Financial strategy in VE is only effective when stakeholders:

- Understand the trade-offs presented
- Have access to quantifiable financial indicators
- Are empowered to make timely decisions

Tools such as **VE matrices**, **cost–value dashboards**, and **scenario comparison sheets** allow project teams to make transparent, data-driven decisions.

An example of cost–value dashboards is given in Table 13.1.

Such comparative displays support rational evaluation, particularly when long-term stakeholders (e.g., operators, financiers) are involved.

Table 13.1 Cost–Value Dashboards Example

Option	Initial Cost (€)	Annual OPEX (€)	Payback (years)	NPV (€)
A (Standard)	80,000	14,000	—	—
B (Efficient)	105,000	7200	3.5	28,500

13.3 Financing and Funding Models for Electrical VE Projects

Implementing VE proposals in electrical systems often requires financial flexibility and creativity. While VE reduces cost through smarter design, it may also suggest **higher-performance alternatives** that require **additional capital**. Therefore, aligning the technical advantages of VE with appropriate financing models is essential to achieving project viability and long-term savings. This section outlines financing mechanisms, funding structures, and economic considerations relevant to VE interventions in electrical infrastructure.

13.3.1 Categorizing Electrical VE Costs

Before selecting an appropriate funding model, it is necessary to identify the types of costs associated with electrical VE initiatives.

13.3.1.1 Hard Costs (Direct Capital Costs)

These include physical materials, installation labor, and technology costs such as:

- High-efficiency transformers and switchgear
- Energy-efficient luminaires and lighting controls
- UPS and battery systems with advanced monitoring
- On-site renewable generation systems
- Advanced metering infrastructure (AMI)
- Labor for electrical installation, commissioning, and testing

13.3.1.2 Soft Costs (Indirect Costs)

These relate to design, approvals, administration, and post-installation activities such as:

- Electrical engineering redesign or revalidation
- Energy simulations and cost modeling
- Commissioning agent fees
- Permit revisions or regulatory resubmissions
- Training for operations and maintenance staff
- Project management of VE implementation

A successful financing strategy considers both categories. For example, energy modeling software licenses or training costs are often overlooked but critical in optimizing post-installation outcomes. Examples of electrical cost categories are illustrated in Figure 13.2.

13.3.2 Financing Mechanisms for VE Implementation

A range of financing strategies can be employed to support VE-driven upgrades in electrical systems. Each has specific use cases and financial implications, as shown in Table 13.2.

Each mechanism shifts **cost ownership and risk** differently. For example, ESPCs reduce upfront capital burden and align payment with performance, but require measurement and verification contracts.

13.3.3 Example: Evaluating VE Financing Options

A university considers a €1.2 million campus-wide LED lighting upgrade. Three financing models are analyzed.

Option A—Traditional Loan:

- Five-year term at 4.5% annual interest
- Annual payments: €273,000
- Total cost over 5 years: €1.365 million

Option B—ESPC Agreement:

- Seven-year term with energy savings guarantees
- No upfront capital required
- Contractor reimbursed through 80% of verified savings

Figure 13.2 Electrical Cost Categories Examples

Table 13.2 Financing Mechanisms for VE Implementation in Electrical Systems

Financing Model	Description	Typical Application in Electrical VE
Traditional Bank Loans	Fixed or variable interest loans with periodic repayment	Standard electrical upgrades (e.g., LED retrofits, panel upgrades)
Green Bonds	Debt issued for energy efficiency or sustainable development	Large-scale VE projects tied to ESG objectives
Energy Savings Performance Contracts (ESPCs)	Third-party finances upgrades; paid back via guaranteed energy savings (DOE, 2023)	Comprehensive lighting, HVAC-electrical integration, UPS upgrades
Power Purchase Agreements (PPAs)	Long-term agreements for on-site renewable electricity	VE proposals incorporating solar PV or cogeneration systems
Equipment Leasing	Operational lease for energy-efficient equipment	Controls, smart panels, battery storage units
Utility On-Bill Financing	Utility-financed upgrades repaid via utility bills	Quick-return upgrades with measurable consumption drops

- Estimated total payment: €1.44 million
- No financial risk if savings fall short

Option C—Utility Rebate + Internal Capital:

- €300,000 rebate from the utility
- €900,000 from the university's improvement fund
- Total effective cost: €900,000
- Fastest payback (~3.2 years)

Outcome: Despite the appeal of zero-upfront ESPC, Option C was selected based on the lowest total cost and internal capital availability.

13.3.4 Grants, Rebates, and Incentives

Many jurisdictions provide financial incentives to promote energy-efficient electrical systems. These can significantly reduce the effective cost of VE proposals.

Common Incentive Types:

- **Government Grants**: Offered for specific project types (e.g., EU Horizon 2020 grants for net-zero facilities)
- **Utility Rebates**: Per-unit incentives (e.g., €25 per LED fixture or €100/kW of reduced load)
- **Tax Incentives**: For example, under some EU frameworks or US Section 179D, owners may deduct a portion of electrical efficiency improvements
- **Demand Response Credits**: Financial compensation for facilities that can reduce load during peak periods
- **PACE Financing**: Property-assessed clean energy loans repaid via local property taxes

These incentives must be **modeled into the project pro forma** to accurately represent total cost and ROI.

13.3.5 Selecting the Right Financing Strategy

The choice of financing model depends on the following factors:

- **Project Size and Complexity**: Larger, integrated VE packages may qualify for structured finance.
- **Owner's Balance Sheet Strength**: Institutions with low debt capacity may prefer ESPCs or PPAs.
- **Risk Profile**: If technology performance is uncertain, third-party performance contracts reduce exposure.
- **Cash Flow Sensitivity**: Projects with tight OPEX constraints benefit from deferred capital solutions.
- **Availability of Local Incentives**: Rebates, grants, and tax deductions may favor self-financing.

A simple decision matrix can aid selection (Table 13.3).

Table 13.3 Decision Matrix to the Right Financing Strategy

Criteria	Preferred Financing Model
No capital budget	ESPC, utility on-bill financing
Fastest ROI	Internal capital + rebate
High savings uncertainty	ESPC with guaranteed performance
Renewable integration	PPA or green bond
High soft cost proportion	Government grant or green bond

13.4 Cost-Benefit and Economic Analysis Techniques

VE in electrical systems must demonstrate not only technical soundness but economic justification. While strategic financial planning provides the framework, **quantitative cost-benefit analysis (CBA)** delivers the evidence needed to compare design options, validate savings, and support stakeholder decisions. This section outlines the most commonly applied techniques for evaluating the financial viability of VE measures and illustrates their application in electrical infrastructure projects.

13.4.1 Return on Investment (ROI)

ROI is one of the simplest and most recognizable financial metrics used to assess VE proposals. It quantifies the net gain relative to the total cost of investment:

$$ROI = \left(\frac{\text{Total net benefits}}{C_0}\right) \times 100\backslash\%$$

Where:

ROI: Return on Investment (expressed as a percentage)
Total net benefits: Cumulative cost savings (including energy, operations and maintenance, and rebates) minus any recurring operational costs over the analysis period
C_0: Initial investment cost
100%: Multiplies the ratio to express ROI as a percentage

13.4.1.1 Example Application

A lighting upgrade costs €25,000 and delivers €6000 per year in energy savings. Over a 5-year period:

- Total savings: €30,000

Where:

ROI: Return on Investment (percentage)
30,000: Total cumulative savings (in euros)
25,000: Initial investment (in euros)
100%: Multiplies the ratio to express as a percentage

This shows the investment breaks even within ~4 years, with a 20% cumulative return over 5 years.

13.4.2 Life-Cycle Cost Analysis (LCCA)

LCCA is the foundational method for assessing long-term value in engineering systems. It considers the TCO from installation to disposal, capturing both CAPEX and OPEX. The standard formula is:

$$LCC = C_0 + \sum_{t=1}^{n} \frac{C_{e,t} + C_{m,t} - S_t}{(1+r)^t}$$

Where:

LCC: Life-cycle cost
$\boldsymbol{C_0}$: Initial capital cost
$\boldsymbol{C_{e,t}}$: Energy cost in year t
$\boldsymbol{C_{m,t}}$: Maintenance cost in year t
$\boldsymbol{S_t}$: Salvage or residual value in year t
$\boldsymbol{r}$: Discount rate (expressed as a decimal)
$\boldsymbol{t}$: Analysis year, from 1 to n
$\boldsymbol{n}$: Project or system life in years

LCCA allows for the fair comparison of high-efficiency vs. standard systems by considering cost over time.

Example: Transformer Option Evaluation (Table 13.4)
Using a 3% discount rate and LCCA calculation, the high-efficiency unit offers approximately €12,000 in life-cycle savings, despite its higher upfront cost (NIST, 2020).

13.4.3 Simple Payback Period

The **SPP** measures how quickly a VE investment will recover its costs:

$$SPP = \frac{C_0}{\text{Annual net savings}}$$

Where:

SPP: Simple Payback Period
$\boldsymbol{C_0}$: Initial capital investment

Table 13.4 LCC Example

Metric	Standard Efficiency	High Efficiency
Initial Cost (€)	16,000	22,000
Annual Loss Cost (€)	3200	1300
Maintenance (€/year)	600	450
Lifespan	20 years	20 years
Residual Value (€)	1500	2200

Annual net savings: Total annual reduction in operational and energy costs resulting from the VE measure

It does not account for the time value of money but is often used as a screening tool.

Rule of Thumb:

- **<3 years:** highly favorable
- **3–5 years:** acceptable for core systems
- **7 years:** typically discouraged unless part of broader long-life asset planning

13.4.4 Net Present Value (NPV)

NPV calculates the value of a stream of savings (or revenues) discounted to today's terms:

$$NPV = \sum_{t=1}^{n} \frac{B_t - C_t}{(1 + r)^t}$$

Where:

NPV: Net present value
B_t: Financial benefit (e.g., energy savings, rebates) in year t
C_t: Total cost (e.g., maintenance, operations) in year t
r: Discount rate
t: Year index (from 1 to n)
n: Analysis period in years

NPV > 0 means the project is economically beneficial. NPV is particularly useful when comparing multiple VE proposals or when financing models differ.

13.4.5 Internal Rate of Return (IRR)

IRR is the discount rate at which NPV = 0. It reflects the effective annual return of an investment:

$$\sum_{t=0}^{n} \frac{C_t}{(1 + IRR)^t} = 0$$

Where:

IRR: Internal rate of return (the discount rate that sets the NPV of all cash flows to zero).
C_t: Net cash flow at time t (can be positive or negative; initial investment is usually negative).
t: Time period (starting from 0 for the initial investment).
n: Total number of time periods

- If IRR > cost of capital: the project is viable.
- IRR is often used in multi-option VE comparisons where investment risks vary.

13.4.6 Sensitivity Analysis

Sensitivity analysis tests how the financial outcome of a VE proposal changes in response to uncertainties in key assumptions such as:

- Energy cost volatility

- Equipment performance variation
- Occupancy or usage patterns
- Maintenance cost escalation
- Discount rate changes

13.4.6.1 Example: LED Retrofit Sensitivity
Base case NPV = €42,000.

- If energy prices drop 20%: NPV = €31,000
- If operating hours increase 10%: NPV = €47,800
- If discount rate rises to 6%: NPV = €36,400

This reveals which assumptions most affect the financial attractiveness, allowing risk-informed decisions.

13.5 Broader Economic Impacts of Electrical VE

Beyond individual project performance, the adoption of VE principles in electrical systems contributes to broader economic value across the building life cycle, real estate market, and sustainability frameworks. This section examines how electrical VE can enhance asset valuation, reduce OPEX at scale, and support achievement of green building certifications—all of which influence long-term ROI.

13.5.1 Market Value Enhancement Through VE

Buildings optimized through electrical VE often outperform peers in market performance due to improved functionality, reduced costs, and superior efficiency.

13.5.1.1 Empirical Indicators from Commercial Property Studies
Research indicates that high-performance buildings can command **5–10% higher rents**, achieve **3–8% valuation premiums**, and reduce vacancy rates by **3–5%**, relative to comparable conventional assets (Eichholtz et al., 2013; WGBC, 2013). Certified buildings also tend to lease **15–20% faster**, making them more attractive to both investors and tenants.

These advantages are increasingly aligned with sustainability and ESG investment strategies, particularly in commercial and institutional portfolios.

13.5.1.2 Example: Post-retrofit Value Increase
A Class B commercial office building that implemented electrical VE—through LED retrofits, upgraded controls, and modernized distribution—recorded the following after 18 months:

- **Energy Cost Reduction**: 26%
- **Tenant Satisfaction Increase**: 18%
- **Property Revaluation Gain**: ~6% (Eichholtz et al., 2013)

Table 13.5 Estimated Operational Expenditure (OPEX) Reductions by VE Implementation Level

VE Implementation Level	Typical Energy Reduction (%)	Annual OPEX Savings per 10,000 m²
Basic (Lighting Only)	10–15	€3000–€4500
Moderate (Lighting + Controls)	15–25	€4500–€7000
Comprehensive (Full Electrical)	25–40	€7000–€10,000
Advanced (Integrated Systems)	35–50	€10,000–€13,000

13.5.2 OPEX Reductions

VE implementations result in measurable utility and maintenance savings. Typical benchmarks for energy cost savings by VE intervention level are given in Table 13.5.

These figures are derived from commercial building case studies across Europe and North America, particularly those implementing smart lighting and high-efficiency transformer strategies (IEA, 2021). Sustained OPEX reduction improves net operating income (NOI), asset value, and investment performance.

13.5.3 Contribution to Sustainability Certifications

Electrical VE supports compliance with major sustainability frameworks such as LEED, BREEAM, and Green Star. In LEED v4 for Building Design and Construction, VE-aligned electrical strategies can contribute up to **35 points**, nearly 30% of the total available (Table 13.6).

Table 13.6 LEED v4—Energy and Atmosphere Credits

Electrical Strategy	Relevant Credit	Points Range
High-Efficiency Electrical Systems (Contributes to Whole-Building Energy Modeling)	EA: Optimize Energy Performance	Up to 18*
Energy-Efficient Lighting with Automated Controls	IEQ: Interior Lighting	2
Renewable Energy Integration (e.g., PV)	EA: Renewable Energy Production	Up to 5
Smart Submetering and Monitoring Systems	EA: Advanced Energy Metering	1
Demand-Responsive Control Systems	EA: Grid Harmonization	2
EV Charging Stations	LT: Green Vehicles	1
Electrical Commissioning	EA: Enhanced Commissioning	6

Total potential contribution: Up to 35 points across EA, IEQ, and LT categories.

* *Note:* The 18 points for Optimize Energy Performance represent whole-building energy savings. Electrical systems typically contribute 30–40% of the total energy reduction through efficient distribution, lighting, controls, and equipment selection (USGBC, 2013).

EA = Energy and Atmosphere; IEQ = Indoor Environmental Quality; LT = Location and Transportation.

These points are directly tied to electrical design decisions and align with VE objectives to reduce load, increase resilience, and optimize cost-performance (USGBC, 2022).

LEED v5 Update: In April 2025, USGBC released LEED v5, which maintains the 110-point structure but reorganizes credits around three impact areas: decarbonization (50% of points), quality of life (25%), and ecological conservation (25%). The new version places greater emphasis on carbon reduction and grid interactivity, making electrical VE strategies even more critical to achieving certification, particularly at gold and platinum levels (USGBC, 2025).

13.5.3.1 Certified Buildings: Economic Advantage

Certified green buildings have been shown to:

- Sell at premiums of 5–11% (Miller et al., 2008; Cushman & Wakefield, 2022)
- Attract lower interest rates from ESG-aligned lenders
- Improve tenant retention due to perceived operational excellence (WGBC, 2013)

These outcomes further reinforce the case for electrical VE as a strategy that enhances both project-level and portfolio-level economic returns.

13.6 Financial Risk Management in Electrical VE Projects

Effective financial risk management is critical to the successful implementation of VE strategies in electrical systems. While VE is designed to enhance value and reduce cost, it can introduce risks related to technology performance, market conditions, procurement timing, and system integration. This section outlines structured approaches to identifying, quantifying, and mitigating financial risks associated with VE interventions in electrical design.

13.6.1 Risk Categories and Their Financial Impact

Electrical VE proposals may encounter several risk categories with potential cost or schedule implications (Table 13.7).

Each risk can affect **initial cost**, **ROI**, **cash flow**, and **payback timing**.

13.6.2 Mitigation Strategies

Financial risk can be managed through a layered approach combining **engineering controls**, **contractual mechanisms**, and **financial instruments**.

Table 13.7 Risk Categories and Financial Implications in Electrical VE Projects

Risk Category	Examples of Potential Impact
Technology Risk	Equipment underperforms or fails to meet savings projections
Market Volatility	Material cost increases (e.g., copper), labor shortages, interest rate hikes
Implementation Delay	Delayed commissioning increases soft costs or disrupts dependencies
Integration Risk	New systems are incompatible with existing infrastructure
Regulatory Risk	Changes in codes or grant conditions mid-project

Engineering and Design Controls:

- **Pilot testing** of new equipment or controls before full deployment
- **Redundancy planning** for critical systems
- **Design reviews** to validate integration compatibility
- **Factory acceptance testing (FAT)** for major switchgear, UPS, or BMS components

Contractual and Commercial Protections:

- **Performance guarantees** in supplier contracts (e.g., minimum efficiency or uptime)
- **Liquidated damages clauses** for delays
- **Fixed-price procurement** for high-risk materials
- **Turnkey delivery** to reduce scope ambiguity

Financial Instruments:

- **Contingency Budgeting**: Typically 5–15% depending on risk profile
- **Insurance**: Covering time-delay, technology performance, or integration failure
- **Energy Savings Insurance**: Available in ESPC or PPA contracts

13.6.3 Quantitative Risk Assessment Techniques

To support transparent decision-making, risk impacts should be quantified and incorporated into financial modeling.

13.6.3.1 Expected Monetary Value (EMV)

$$EMV = \sum_{i=1}^{n}(P_i \times C_i)$$

Where:

EMV: Expected monetary value
n: Total number of distinct risk events considered
P_i: Probability of occurrence of risk event i
C_i: Financial impact (cost) of risk event i

This probabilistic method is commonly used in VE-related project planning and should be included in scenario-based cost modeling (PMI, 2017) (Table 13.8).

This indicates that a contingency of €7100 should be added to the base case NPV model.

13.6.3.2 Monte Carlo Simulation

This technique models the combined effect of multiple uncertainties, such as energy pricing, load variation, and discount rate. The output is a distribution curve of potential project outcomes (e.g., NPV range between €25,000 and €55,000), allowing planners to understand likelihood-weighted financial performance (ISO, 2017).

Table 13.8 Example: Risk Analysis for PV System Upgrade

Risk Event	Probability (%)	Cost Impact (€)	EMV (€)
Inverter Underperformance	15	−12,000	−1800
Installation Delay	20	−9000	−1800
Tariff Reduction	25	−8000	−2000
Roof Structure Upgrade Required	10	−15,000	−1500
Total Risk EMV			**−7100**

13.6.4 Phased Implementation as a Risk Reduction Strategy

Staging VE implementation across phases allows organizations:

- **Test performance** of VE interventions under controlled conditions
- **Defer capital** until early-stage savings are verified
- **Incorporate lessons learned** into subsequent deployments

An example of phased implementation for hospital lighting control upgrades is summarized in Table 13.9.

This approach provided confidence to stakeholders, improved sequencing with clinical operations, and secured early wins to fund later phases.

Table 13.9 Example: Hospital Lighting Control Upgrade

Phase	Area	Cost (€)	Annual Savings (€)	ROI (%)	Risk Level
Phase 1	Admin & corridors (20%)	60,000	18,000	30	Low
Phase 2	Noncritical clinical spaces	110,000	33,000	30	Medium
Phase 3	Critical care and OR zones	140,000	38,000	27	High

13.7 Conclusion and Key Takeaways

VE in electrical systems offers more than technical refinement—it is a financially strategic practice that enables owners, designers, and stakeholders to make decisions that are both economically and operationally sound. This chapter has demonstrated how financial planning tools, funding mechanisms, risk management strategies, and economic impact analysis form the backbone of successful VE integration in modern electrical infrastructure.

Key financial strategies addressed throughout this chapter include:

Summary of Core Principles

1) **Financial Planning Foundations**
   - LCC provides the anchor for VE comparisons by incorporating both CAPEX and OPEX.
   - Economic indicators such as ROI, NPV, and IRR support rational selection among design alternatives.

2) **Financing Mechanisms**
   - Various financing models (e.g., ESPCs, PPAs, green bonds) allow implementation of VE proposals without exceeding capital budgets.
   - Government rebates, utility incentives, and tax allowances improve project cash flow and accelerate payback.
3) **Economic Evaluation**
   - Quantitative assessment techniques (e.g., LCCA, sensitivity analysis) validate the long-term value of electrical upgrades.
   - Tools like payback period and scenario modeling translate engineering choices into boardroom metrics.
4) **Broader Economic Impact**
   - Electrical VE contributes to higher building valuations, rent premiums, and sustainability certification.
   - Operational savings reduce life-cycle costs while strengthening property value and ESG performance.
5) **Risk-Informed Implementation**
   - Financial risks related to market volatility, technology performance, and schedule disruptions must be assessed and mitigated.
   - Risk management strategies—such as phased deployment, EMV modeling, and performance contracts—ensure value realization.

Strategic Implication for Electrical Professionals

The shift toward life-cycle accountability in infrastructure development places electrical engineers at the intersection of design performance and financial decision-making. Mastery of financial analysis tools and strategic planning techniques is no longer optional—it is essential for delivering sustainable value.

Professionals engaged in VE must develop fluency in:

- Translating technical options into economic terms
- Engaging stakeholders with data-backed justifications
- Aligning electrical design choices with broader project financial goals

Author's Note

The field of financial planning and cost analysis in engineering is expansive, spanning detailed capital forecasting, market-based risk modeling, tax optimization, whole-life carbon costing, and beyond. In this chapter of the guidebook, the focus has been deliberately placed on foundational financial strategies that align most directly with the practical application of VE in electrical systems. The aim is to equip engineers and decision-makers with the core methodologies—life-cycle costing, ROI, NPV, and risk evaluation—that support sound value-based decisions. Readers seeking deeper coverage of advanced financial modeling techniques, capital structuring, or cost control in construction economics are encouraged to consult specialized references or collaborate with dedicated cost and finance professionals.

References

Cushman & Wakefield. (2022). *Green is good series: the endurance of green building premiums.* New York: Cushman & Wakefield. Available at: https://www.cushmanwakefield.com/en/united-states/insights/green-is-good-series

Eichholtz, P., Kok, N., & Quigley, J. M. (2013). The economics of green building. *The Review of Economics and Statistics*, 95(1), 50–63.

IEA. (2021). *Energy efficiency 2021: Market report series.* International Energy Agency.

ISO. (2017). *ISO 15686-5:2017 Buildings and constructed assets - Service life planning - Part 5: Life-cycle costing.* Geneva: International Organization for Standardization.

Miller, N., Spivey, J., & Florance, A. (2008). Does green pay off? *Journal of Real Estate Portfolio Management*, 14(4), 385–399.

NIST. (2020). *Life-cycle costing manual for the federal energy management program.* NIST Handbook 135, National Institute of Standards and Technology, U.S. Department of Commerce.

PMI. (2017). *A guide to the project management body of knowledge (PMBOK guide)* (6th ed.). Project Management Institute.

USGBC. (2022). *LEED v4 for building design and construction: Reference guide.* U.S. Green Building Council.

USGBC. (2025). *LEED v5: Building design and construction.* Washington, DC: U.S. Green Building Council. Available at: https://www.usgbc.org/leed/v5

WGBC. (2013). *The business case for green building: A review of the costs and benefits for developers, investors and occupants.* World Green Building Council.

14

Complete Case Study: Value Engineering in a Commercial High-Rise Project

14.1 Project Overview

The case study focuses on a commercial high-rise development located in a **Gulf Cooperation Council** (GCC) country, representing a typical example of a large-scale, mixed-use vertical structure requiring comprehensive electrical infrastructure. The project comprises over 40 floors above ground and multiple basement levels dedicated to parking and services, with a total connected load exceeding 6 MVA.

The **scope of electrical works** encompasses the full spectrum of systems required for a modern high-rise facility, including:

- Medium-voltage (MV) and low-voltage (LV) power distribution
- Emergency and standby power systems (generators, uninterruptible power supply [UPS])
- Interior and exterior lighting systems with controls
- Information and communication technology (ICT) and extra-low voltage (ELV) systems (structured cabling, building management system [BMS] interface, fire alarm)
- Earthing and lightning protection
- Cable containment and routing infrastructure

The project was at the detailed design stage when the value engineering (VE) initiative was formally introduced. By this stage, most of the electrical layouts, equipment schedules, and power distribution schemes had been developed, but sufficient flexibility remained to implement performance-equivalent alternatives without compromising safety or regulatory compliance.

A critical constraint was the limited architectural footprint and service core space, particularly for vertical distribution risers and technical rooms. Moreover, strict adherence to local regulations—including Ministry of Electricity guidelines, civil defense requirements, and local authority clearances—posed a design challenge that shaped the VE proposals. Cost optimization was also a significant driver, with the client mandating life-cycle efficiency, construction feasibility, and streamlined maintenance strategies.

The VE exercise aimed to:

- Rationalize equipment selection and distribution strategies
- Reduce overspecification and unused redundancy
- Improve spatial coordination between disciplines
- Deliver measurable savings in both capital expenditure and life-cycle cost

Value Engineering for Electrical Systems in Construction, First Edition. Hazem Hussein.

Companion website: www.wiley.com/go/hazem

14.2 VE Objectives

The primary objective of the VE exercise was to enhance the overall efficiency and cost-effectiveness of the electrical systems within the constraints of the project's location, regulatory environment, and spatial limitations. The VE process sought to achieve a balance between technical performance, code compliance, capital investment, and long-term operational sustainability.

The key objectives of the VE study were as follows:

1) **Optimization of Capital Expenditure (CAPEX)**
   - Identify and eliminate cost-intensive components that provided marginal functional benefit.
   - Replace premium or overspecified systems with alternatives of equivalent technical compliance and performance.
   - Reconfigure distribution schemes to reduce the number of boards, feeders, and components.
2) **Reduction of Life-Cycle Costs (LCCs)**
   - Implement solutions that would result in lower operational and maintenance expenses over a 25-year period.
   - Improve maintainability and serviceability of electrical infrastructure by rationalizing layouts and selecting modular, durable components.
3) **Space Efficiency and Spatial Coordination**
   - Address significant challenges in riser and technical room planning by reducing the spatial footprint of distribution equipment and cable containment systems.
   - Integrate value-based design alternatives that optimize available architectural space and enhance cross-disciplinary coordination with HVAC, plumbing, and structural systems.
4) **Regulatory and Safety Compliance**
   - Ensure all revised solutions met or exceeded the local authority requirements (e.g., ME regulations, civil defense codes, and applicable international standards such as IEC 61439, IEC 60364).
   - Preserve system integrity with respect to discrimination, fault levels, and voltage-drop performance.
5) **Functional Integrity and System Resilience**
   - Maintain core functionality and load-serving capability across all systems.
   - Where appropriate, consolidate components or reroute feeders without compromising system reliability, selectivity, or operational redundancy.
6) **Constructability and Installation Simplification**
   - Propose alternatives that reduce installation time, minimize on-site fabrication or supervision, and facilitate prefabrication or modular assembly.
   - Consider contractor feedback and market availability of proposed alternatives to ensure feasibility and local sourcing.

14.3 Methodology

The VE process adopted for this project followed a structured, multi-phase approach aligned with best practices in electrical system design and cost optimization. The methodology integrated practical tools and templates from *Chapter 6—Practical Sheets and Templates for Effective Value*

Engineering of Electrical Works, enabling objective evaluation and traceable documentation of all decisions.

14.3.1 VE Process Phases (Table 14.1)

Table 14.1 VE Process Phases Overview

Phase	Description
1. Initiation	Definition of VE scope, formation of the VE team (consultants, client reps, contractor advisors), and identification of critical areas for value intervention.
2. Data Collection	Extraction of electrical design data, load schedules, cost breakdowns, and spatial drawings. Referenced **Sheet: "DATA COLLECTION SHEET FOR VE EXERCISE."**
3. Functional Analysis	Mapping of system functions and their associated cost/value using **Sheet: "FUNCTION ANALYSIS WORKSHEET,"** identifying components offering low functional contribution relative to their cost.
4. Idea Generation	Brainstorming of alternative solutions with inputs from electrical, architectural, and procurement stakeholders. Logged in **Sheet: "VE IDEA GENERATION LOG."**
5. Evaluation and Shortlisting	Scoring of alternatives based on technical feasibility, cost savings, risks, and compliance using **Sheet: "VE IDEA EVALUATION MATRIX."**
6. Cost and Impact Assessment	LCCA conducted for shortlisted proposals to evaluate long-term cost impacts, using the **ELECTRICAL SYSTEM COST BREAKDOWN** and **ENERGY EFFICIENCY IMPACT ASSESSMENT** sheets.
7. Implementation Planning	Approved ideas were documented in the **VE Implementation Tracking Log**, and incorporated into revised drawings and equipment schedules.
8. Post-Evaluation and Knowledge Capture	Lessons learned were summarized in the **VE Lessons Learned Register** for future reference and knowledge transfer.

14.3.2 Tools and Templates Used

The following templates were actively employed:

- VE Project Initiation Template
- Function Analysis Worksheet
- VE Idea Evaluation Matrix
- Load and Capacity Impact Analysis
- Cost Breakdown and LCCA Worksheets
- VE Summary Report

14.3.3 Stakeholder Involvement

- **Client/Developer:** Defined project priorities, approved VE changes, and commissioned the engagement of independent VE consultants.
- **External VE Consultants:** Appointed by the client to lead an objective VE assessment. Worked in coordination with the internal design team to review specifications, develop proposals, and validate technical and cost outcomes.

- **Electrical Consultant:** Led technical simulations, load assessments, and proposed design adaptations for feasibility.
- **Architectural and Mechanical, Electrical, and Plumbing (MEP) Coordinators:** Validated space implications and ensured cross-discipline alignment with structural and mechanical systems.
- **Contractor/Procurement Advisor:** Assessed constructability, procurement risks, and installation logistics based on market knowledge and implementation constraints.

14.4 Applied VE Strategies and Optimization Outcomes

This section outlines the practical application of VE interventions implemented during the detailed design stage. Each subsection corresponds to a key optimization area where measurable savings and system enhancements were achieved.

14.4.1 Power Distribution Optimization

In the original electrical design for the commercial high-rise, power distribution relied heavily on busduct systems, a dense network of sub-main distribution boards (SMDBs), and vertically stacked risers with minimal spatial efficiency. The VE team identified several opportunities to reduce system complexity, improve maintainability, and lower capital and LCCs.

The revised design focused on four strategic value-engineering interventions, documented under new identifiers **VE-E1** to **VE-E4**.

14.4.1.1 VE-E1: Replace Busducts with XLPE Cables in Noncritical Risers

The original design specified cast-resin busducts for all vertical risers, including areas with moderate or low criticality such as basement and mechanical rooms. The VE team proposed replacing busducts in these zones with XLPE power cables on ladder trays (Table 14.2).

Table 14.2 VE-E1: Replace Busducts with XLPE Cables in Noncritical Risers

Category	Original Design	Revised Design
Design Strategy	Cast-resin busducts in all risers	XLPE cables (4C × 240 mm^2 CU) on ladder trays
Riser Height	45 m	Same
Cable Tray Width	—	600 mm (multi-tier trays with segregation)
Capital Cost (USD)	\$120,000	\$84,500
LCC (25 years)	\$128,500	\$91,800
Estimated Savings	—	**\$36,700**
Performance Benefits		• Riser width reduced by 300 mm • Simplified routing • less lead time/faster delivery and install

14.4.1.2 VE-E2: Consolidate SMDBs in Low-Density Floors

SMDBs were originally specified for every floor, including areas with minimal load requirements. This VE measure rationalizes SMDB placement by removing redundant panels and consolidating downstream distribution (Table 14.3).

14.4.1.3 VE-E3: Replace Rooftop Chiller Busducts with Internal Riser Cable Feeders

The original concept routed rooftop chiller feeders using a dedicated vertical busduct rising through the electrical shaft. While functionally sound, this increased riser congestion, introduced complex support requirements, and involved proprietary materials and supervision. The VE alternative routes armored XLPE power cables through the same shaft space, using multi-tier cable trays, while retaining internal routing within the temperature-controlled building envelope (Table 14.4).

Table 14.3 VE-E2: Consolidate SMDBs in Low-Density Floors

Category	Original Design	Revised Design
Design Strategy	SMDBs on every floor (underutilized in many zones)	Fewer SMDBs; consolidated feeders across floors
Panel Quantity Affected	6 units	Removed
New Specification	—	400 A, Form 3b, IP42 (upgraded units)
Capital Cost (USD)	$42,000	$27,000
LCC (25 years)	$45,500	$29,400
Estimated Savings	—	$16,100
Performance Benefits		• 30% reduction in wall-mounted gear • ~20 m^2 space freed • Better riser coordination

Table 14.4 VE-E3: Replace Rooftop Chiller Busducts with Internal Riser Cable Feeders

Category	Original Design	Revised Design
Design Strategy	Internal busduct to rooftop chillers	Internal cable tray riser with XLPE SWA cables
Feeder Route	~55 m vertical busduct	~60 m XLPE cable on riser tray (within building shaft)
Cable Type	N/A	3 runs of 4C × 185 mm^2 XLPE SWA
Chiller Load	450 kVA × 2	Same
Capital Cost (USD)	$65,000	$48,500
LCC (25 years)	$70,800	$52,000
Estimated Savings	—	**$18,800**
Performance Benefits		• Reduced busduct supervision • Lower containment footprint • Complies with thermal design codes

14.4.1.4 VE-E4: Optimize Busbar Layout by Reducing Tap-Off Count

The original design assigned individual tap-offs on the vertical busbar trunking to every distribution board (DB), even when those DBs were physically adjacent or serving a common zone. This created unnecessary joints, extended busbar lengths, and increased containment complexity. The revised strategy involved **grouping multiple DBs under a shared SMDB**, fed from a single higher-rated tap-off. While the capital cost remains approximately neutral, the revised layout improves riser coordination, reduces the number of mechanical joints, and simplifies long-term maintenance (Table 14.5).

➢ **Note on Value beyond Cost**

This VE item illustrates an important principle in VE: **value is not defined by cost reduction alone**. Although VE-E4 does not yield measurable capital or life-cycle savings, it delivers significant improvements in system clarity, riser coordination, long-term maintainability, and safety through simplification. In this context, the proposal was retained not as a cost-saving measure but as a **performance-driven design optimization** that enhances overall project value (Table 14.6).

Table 14.5 VE-E4: Optimize Busbar Layout by Reducing Tap-Off Count

Category	Original Design	Revised Design
Design Strategy	Eight separate tap-offs on busbar trunking	One tap-off feeding a consolidated 400 A SMDB
Tap-Offs Eliminated	Eight units	Replaced with one larger tap-off
New SMDB Added	—	1 × 400 A, Form 3b, IP42 SMDB
Busbar Length Saved	—	~25 m
Busbar Rating	800 A continuous	Same
Capital Cost (USD)	$186,000	~$186,000 *(cost-neutral after equipment swap)*
LCC (25 years)	$195,200	~$195,100 *(no measurable LCC delta)*
Estimated Savings	—	**$0** *(VE retained for operational benefits only)*
Operational Benefits	—	• Fewer joints and brackets • Lower support effort • Easier inspection and simplified riser layout

Table 14.6 Summary of VE-E1 to VE-E4 Cost Impact

VE Ref	Description	Capital Savings (USD)	LCC Savings (USD)
VE-E1	Replace busducts with XLPE cables	$35,500	$36,700
VE-E2	SMDB consolidation	$15,000	$16,100
VE-E3	Chiller feeder reconfiguration	$16,500	$18,800
VE-E4	Tap-off and busbar merging	—	—
Total	—	**$67,000**	**$71,000**

14.4.2 Main Board Optimization

Main low-voltage switchboards (MLVBs) serve as the backbone of electrical power distribution in high-rise commercial projects. In the initial design, all main boards were specified as high-specification, form 4b, IP54-rated assemblies with oversized enclosures, regardless of their actual load environment or segregation requirement.

Through a detailed review of application environments and board configurations, the VE team proposed an optimization strategy to reduce cost, footprint, and enclosure complexity without compromising reliability or compliance.

14.4.2.1 VE-E5: Tailored Form Class and Enclosure Rating of Main LV Boards

The initial design applied a uniform specification of form 4b, IP54-rated main LV boards across all electrical rooms, regardless of their criticality. Through VE review, it was determined that this full segregation was only necessary for emergency and life safety systems. The remaining boards were reclassified to a lower form and IP rating suitable for general distribution use, resulting in significant cost and dimensional optimization without compromising safety or compliance (Table 14.7).

Implementation Considerations:

- All revised boards meet IEC 61439-1/2 and local utility compliance standards.
- No changes in protection settings were required; outgoing feeders and breakers remained unchanged.
- Revised specifications were approved by the local authority following a brief coordination session.

Table 14.7 VE-E5: Tailored Form Class and Enclosure Rating of Main LV Boards

Category	Original Design	Revised Design
Design Strategy	5 × Form 4b, IP54 MLVBs	3 × Form 3b IP42 + 2 × Form 4b IP54
Board Quantity	5	5
Load Coverage	1000–1600 A	Same
Installation Zones	Electrical Rooms (3 per tower)	Same
Unit Cost (avg)	\$84,000	\$69,000
Total Capital Cost	\$420,000	\$345,000
Life-Cycle Cost (25 years)	\$445,000	\$368,000
Estimated Savings	—	\$77,000
Performance Benefits	—	• Reduces equipment depth and width by 20–30% for form 3b boards, easing installation in constrained electrical rooms. • Allows wider vendor competition by avoiding bespoke specifications for general-use boards. • Improves maintenance accessibility and spare parts compatibility. • Retains full arc segregation for critical systems (emergency/life safety).

14.4.3 Lighting System Rationalization

In the original design, a fully addressable DALI (Digital Addressable Lighting Interface) lighting control system was proposed for the entire project, including basements, back-of-house (BOH) areas, and mechanical rooms. While DALI offers advanced programmability and monitoring, applying it universally across all areas was found to be unnecessarily complex and cost-intensive, particularly in zones with static lighting requirements.

14.4.3.1 VE-E6: Limit DALI Control to Occupied and Public Zones

The original design extended DALI lighting control throughout all areas of the building, including technical and low-occupancy zones. While functionally comprehensive, this approach increased material cost, wiring complexity, commissioning effort, and BMS integration requirements. The VE strategy introduced a hybrid system: retaining DALI in front-of-house areas, while replacing it with local switching and standalone PIR sensors in BOH and service zones (Table 14.8).

Implementation Notes:

- Lighting plans were updated with mixed-symbol legends to distinguish DALI-controlled and locally switched zones.
- Conduiting and wiring routes were simplified, reducing containment requirements by 20% in affected areas.
- The hybrid system was designed to remain fully compliant with **IEC 60364-8-1** and regional energy efficiency codes.

Table 14.8 VE-E6: Limit DALI Control to Occupied and Public Zones

Category	Original Design	Revised Design
Design Strategy	Full DALI control via central BMS	DALI in key areas + PIR/local switching in BOH zones
Total DALI Points	~2000	~1200
Local Sensor Points	N/A	~300 PIRs (standalone)
System Topology	Centralized DALI	Hybrid (DALI + conventional)
Affected Area	Entire building	30% of area reverted to local control
Capital Cost (USD)	$92,000	$61,500
Life-Cycle Cost (25 years)	$108,000	$71,200
Estimated Savings	—	**$36,800**
Performance Benefits		• Reduces procurement of DALI drivers, gateways, and central controllers. • Lowers commissioning duration and BMS configuration complexity. • Improves reliability in standalone zones by isolating fault impact. • Retains energy efficiency through occupancy-based control in technical spaces

14.4.4 Cable Infrastructure Sizing

The electrical containment system originally designed for the project was based on conservative derating assumptions, requiring generous physical spacing between power circuits. This resulted in oversized cable trays, excessive riser widths, and inefficient use of shaft space. In high-rise buildings where vertical core areas are premium real estate, this approach introduced avoidable spatial and cost penalties.

14.4.4.1 VE-E7: Optimize Cable Tray Sizing Using 1D Spacing Method

The original design applied a conservative 2D horizontal spacing rule between multicore power cables when calculating cable tray widths, which led to oversized containment systems and unnecessary structural load. The revised strategy applied validated 1D spacing (equal to one cable diameter) in accordance with IEC 60364-5-52, along with appropriate derating assessments, resulting in reduced tray widths and simplified infrastructure in key routing areas (Table 14.9).

Implementation Notes:

- Thermal derating calculations were prepared in accordance with **IEC 60364-5-52**, Section 523, and reviewed by the supervising engineer.
- No change in cable type, current rating, or route length was required.
- Updated routing diagrams were issued with new tray width schedules and adjusted support intervals.

Table 14.9 VE-E7: Optimize Cable Tray Sizing Using 1D Spacing Method

Category	Original Design	Revised Design
Design Strategy	2D spacing (double cable diameter)	1D spacing (single diameter) with derating checks
Max Tray Width (typical)	900 mm	600 mm
Total Containment Length	~1200 m	Same
Redundancy Factor	30% spare capacity	20% spare capacity (maintained)
Capital Cost (USD)	$112,000	$104,000
Life-Cycle Cost (25 years)	$120,500	$110,800
Estimated Savings	—	**$9700**
Performance Benefits		• Reduces cable tray width by up to 300 mm in key zones, freeing riser and ceiling space. • Lowers support system complexity and weight loads on slab anchors. • Decreases waste and handling complexity during installation. • Supports fire-safe cable routing by reducing over-congestion

14.5 Results and LCC Impact

The implementation of the VE measures proposed during the detailed design stage yielded significant improvements in cost efficiency, spatial coordination, and overall system simplicity. The VE initiatives maintained full compliance with local regulations and international standards while addressing project-specific constraints such as limited riser space and overspecified components.

14.5.1 Capital and LCC Savings Summary (Table 14.10)

Table 14.10 Summary of Capital and Life-Cycle Cost Savings (VE-E1 to E7)

VE Ref	Description	Capital Cost Savings (USD)	Life-Cycle Cost Savings (USD)
VE-E1	Replace busducts with XLPE cables	$35,500	$36,700
VE-E2	SMDB consolidation	$15,000	$16,100
VE-E3	Chiller feeder reconfiguration	$16,500	$18,800
VE-E4	Tap-off and busbar merging	0	0
VE-E5	Main board reclassification (form/IP)	$75,000	$77,000
VE-E6	DALI rationalization	$30,500	$36,800
VE-E7	Cable tray sizing (1D spacing)	$8000	$9700
Total	—	**$180,500**	**$195,500**

14.5.2 System Performance Impact

All revised designs preserved or improved:

- **Functional performance** (load capacity, selectivity, protection)
- **Safety** (arc containment, ingress protection, thermal limits)
- **Maintainability** (modular panels, accessible layouts)
- **Coordination** (smaller risers, simplified containment routes)
- **Commissioning** (reduced system complexity, easier verification)

14.5.3 Spatial and Operational Efficiency

- Reduced total SMDB count by approximately **30%** in noncritical zones.
- Reclaimed an estimated **12–15 m^2** of riser and technical room space.
- Reduced total cable tray widths across risers and corridors by up to **35%**.
- Eliminated more than **100 m** of busbar trunking and over **10 tap-offs**.
- Simplified lighting control architecture in **>30%** of project area.

14.5.4 Stakeholder Alignment and Approval

All VE items were reviewed by:

- **Lead MEP Consultant** for technical validation and standards compliance.
- **Client Representative** for life-cycle impact and operations and maintenance (O&M) implications.

- **Architectural and Structural Teams** for spatial coordination checks.
- **Contractor Advisory Team** for buildability and procurement logistics.

Approved items were formally documented in the *VE Implementation Tracking Log* and *VE Summary Report* using templates from **Chapter 6** of this guidebook.

14.6 Lessons Learned and Recommendations

The case study reinforces the critical role of proactive, technically grounded VE in the design and delivery of electrical systems for complex, high-density buildings. By integrating VE during the detailed design stage—while retaining coordination flexibility—substantial benefits were realized across cost, space, and system manageability.

14.6.1 Key Lessons Learned

1) **VE Is Most Effective When Initiated Early**
   - Although introduced during detailed design, many proposed changes (e.g., riser consolidation, lighting control rationalization) would have yielded even greater benefits if initiated at the schematic stage.
   - Early coordination with the architectural team is essential, especially for space-driven decisions like panel elimination or tray resizing.
2) **Fit-for-Purpose Specification Prevents Overdesign**
   - Excessive use of high-specification components (e.g., form 4b boards, full DALI control) does not always translate to added functional value.
   - Design teams should regularly revalidate equipment classification and protection levels based on actual operational risk.
3) **Life-Cycle Thinking Drives More Meaningful Savings**
   - Decisions based solely on capital cost often overlook O&M complexity, vendor lock-in, and unnecessary supervision requirements.
   - LCC metrics—supported by templates from Chapter 6—enabled a more defensible, long-term cost-benefit argument to clients.
4) **Simple Solutions Are Often the Most Effective**
   - Replacing busducts with cables or rationalizing SMDBs are technically straightforward decisions but can deliver disproportionate savings.
   - Implementation of VE does not always require complex re-engineering—clarity, logic, and compliance are sufficient.
5) **Documented Templates Streamline Decision-Making**
   - Use of standardized forms (VE idea logs, evaluation matrices, LCC sheets) ensured transparency, reduced decision turnaround times, and improved cross-discipline communication.
   - Lessons learned were efficiently captured using the VE Lessons Learned Register from Chapter 6.

14.6.2 Recommendations for Future Projects

1) **Embed VE into Stage 2/3 Design Reviews**: Make VE reviews a formal checkpoint during concept and scheme stages to influence layout, equipment zoning, and containment strategies early.

2) **Use Functional Cost Mapping for Prioritization**: Apply function analysis worksheets to allocate cost against performance contribution. Prioritize systems or components where cost significantly outweighs functional value.
3) **Maintain a "VE Readiness Checklist"**: Prepare a project-specific VE readiness checklist—including design freeze dates, coordination deliverables, and authority submission requirements—to ensure VE changes are implementable.
4) **Tailor System Design by Area Function**: Avoid uniform system specifications across the building. Adjust lighting controls, distribution strategies, and panel grades based on area occupancy, criticality, and maintenance frequency.
5) **Monitor Post-Implementation Outcomes**: Establish a feedback loop with facilities teams after handover to assess the performance of VE choices in actual operation and update the organizational knowledge base accordingly.

14.7 Supporting Appendices

The successful implementation of VE in this project was supported by a suite of standardized templates and analysis tools, all drawn from the practical framework established in Chapter 6 of this guidebook. These tools enabled structured decision-making, consistent documentation, and traceable design changes throughout the VE life cycle.

Sheet Title (from Chapter 6)	Application in Case Study Section
VE Project Initiation Sheet	Used to define scope, team roles, and VE goals (Section 14.2)
Data Collection Sheet for VE Exercise	Populated with load schedules, panel lists, and system specifications (Section 14.3)
Function Analysis Worksheet	Applied to assess cost vs. functional value of main boards, risers, and lighting systems (Sections 14.3, 14.4.2, 14.4.3)
VE Idea Generation Log	Used to record initial proposals VE-E1 to VE-E7 (Section 14.3)
VE Idea Evaluation Matrix	Used to shortlist viable VE proposals based on scoring criteria (Section 14.3)
Electrical System Cost Breakdown Sheet	Used to standardize CAPEX inputs and validate savings calculations (Sections 14.4.1–14.4.4)
Electrical Load and Capacity Analysis	Referenced for panel optimization and feeder resizing (VE-E2, VE-E5)
Energy Efficiency Impact Assessment	Used to validate reduced DALI coverage with occupancy sensors (VE-E6)
Value Engineering Summary Report Template	Consolidated documentation of all approved VE items (Section 14.5)
VE Implementation Tracking Log	Maintained live status tracking of accepted/rejected VE ideas (Sections 14.5 and 14.6)
VE Lessons Learned Register	Populated with key insights and improvement strategies (Section 14.6)

14.7.1 VE Project Initiation Form

Project Identification:

- **Project Name:** Commercial Tower VE Study
- **Project Location:** GCC Country

- **Project Phase:** ☑ Detailed Design (RIBA Stage 4)
- **Contract Number:** [To be filled]
- **Date of VE Study:** April 15–24, 2022

14.7.1.1 Electrical Scope Overview

Brief Description: The electrical scope includes full MV and LV power distribution systems, main switchgear, sub-main panels, lighting, emergency power systems, ICT and ELV infrastructure, lightning protection, and containment. The VE study targets key cost centers and overspecified components, focusing on distribution strategy, spatial optimization, and life-cycle performance.

VE Study Parameters:

- **VE Study Duration:** 10 days
- **Workshop Dates:** April 15–24, 2022
- **Location:** Design Office + Remote Stakeholder Coordination

14.7.1.2 VE Team Composition

Role	Name	Company	Contact Info	Expertise
VE Team Leader	[To be filled]	xxxxxx	—	Electrical Systems/LCC
Electrical Engineer	[To be filled]	xxxxx	—	Distribution Systems
Cost Estimator	[To be filled]	xxxxxx	—	CAPEX & LCC Analysis
Project Manager	[To be filled]	xxxxx	—	Project Coordination
Contractor Rep	[To be filled]	xxxxx	—	Site Installation Constraints
Owner Representative	[To be filled]	xxxx	—	Life-Cycle Strategy

Target Electrical Systems for Analysis:

☑ **Power Systems:**

- ☑ Service Entrance and Main Distribution
- ☑ Emergency/Standby Power
- ☑ UPS
- ☑ Branch Circuit Distribution
- ☑ Power Monitoring/Metering
- ☑ Motors and Controls
- ☐ Renewable Energy Systems

☑ **Lighting Systems:**

- ☑ General Lighting
- ☑ Emergency Lighting
- ☑ Lighting Control Systems

☑ **Low Voltage Systems:**

- ☑ Structured Cabling (ICT)
- ☑ Fire Alarm System
- ☐ Security/Access Control
- ☐ Public Address and AV
- ☐ BMS and Controls

☑ **Infrastructure Elements:**

- ☑ Cable Containment
- ☑ Panel Locations and Riser Space
- ☑ Grounding and Lightning Protection

14.7.2 Data Collection and System Inventory

14.7.2.1 Data Collection Sheet for VE Exercise

Project Specifications Overview:

- **Project Name:** Commercial Tower VE Study
- **Project Type:** ☑ Commercial
- **Total Building Area:** Approx. 95,000 m^2
- **Occupancy Classification:** Mixed-use—Office, Retail, Parking
- **Applicable Electrical Codes:**
  - IEC 60364 (Electrical Installations for Buildings)
  - Local Electrical Code (GCC Utility Authority)
  - Civil Defense Fire Safety Code
- **Applicable Standards and Guidelines:**
  - IEC 61439 (Low-Voltage Switchgear and Controlgear Assemblies)
  - IEEE 241 (Recommended Practice for Electric Power Systems in Commercial Buildings)
  - ASHRAE 90.1 (Referenced for energy efficiency where applicable)
- **Special Requirements/Certifications:**
  - Compliance with local civil defense requirements
  - BMS integration for selected systems (lighting, emergency power)
  - Enhanced maintainability and space coordination with structural/MEP

Project Phase:

- ☑ Detailed Design (RIBA Stage 4)
- **Design Completion Percentage:** 85%

14.7.2.2 Electrical System Inventory

Power Distribution Systems

System Component	Current Design Specification	Quantity	Capacity	Reference Document
Utility Service	11 kV RMU (630 A) with Oil immersed transformers (Dyn11, ONAN, N+1 setup)	3 units	2000 kVA each	MV Single Line Diagram
LV Switchgear/ MLVBs	Form 3b (typical), Form 4b (critical), IP42/ IP54 as needed	5 units	1000–1600 A	LV Panel Schedule
Transformers	Dry-type 11/0.4 kV, Dyn11, ANAF	3 units	2000 kVA each	Substation Layout
Distribution Panels	Sub-Main Distribution Boards (SMDBs), type-tested metal-enclosed panels	24 units	250–400 A	Panel SLDs
Branch Circuit Panels	Final DBs, flush/surface mounted	~70	63–160 A	Lighting/ Power Layouts

System Component	Current Design Specification	Quantity	Capacity	Reference Document
Emergency Power System	Diesel generators with ATS and synchronization capability	2 units	800 kVA + 400 kVA	Emergency Power Layout
UPS System	Modular UPS, rack-mount with internal VRLA batteries	2 units	60 kVA each (N+1)	UPS Riser Diagram
Power Monitoring	CT-based smart metering (Modbus protocol) integrated into MLVBs and SMDBs	40+ pts	—	Power Monitoring Spec

Lighting Systems

System Component	Current Design Specification	Quantity	Capacity/ Performance	Reference Document
Interior Fixtures	LED luminaires: troffers, downlights, battens	~4000	8–48 W/unit	Lighting Layout
Exterior Fixtures	LED bollards, floodlights, facade uplights	~350	20–90 W/unit	Landscape Lighting Plan
Lighting Control System	DALI in public/office zones, standalone PIR in BOH areas	Mixed	1 DALI ctrl/ zone	Lighting Control Diagram
Emergency/Exit Lighting	Central battery system with monitored circuits	~300	3–5 W/unit	Emergency Lighting Plan
Specialty Lighting	Scene-based dimming in meeting rooms and feature zones	60 zones	Programmable scenes	Feature Lighting Plan

Extra Low Voltage (ELV) Systems

System Component	Current Design Specification	Quantity	Capacity/ Performance	Reference Document
Fire Alarm System	Addressable FACP with isolator modules	40 loops	~200 devices/ loop	Fire Alarm Layout
Structured Cabling (ICT)	Cat6A with OM3 fiber backbone to IDFs	800 drops	1 Gbps/node	ICT Layout
Building Automation (BMS)	BACnet protocol for HVAC, lighting, and metering control	Central	Integrated with BMS	BMS Architecture

In addition to the quantitative summary provided above, the VE team collected all relevant design documentation, including:

- **Detailed Bill of Quantities (BOQ),**
- **Single-line diagrams (SLDs),**
- **System schematics**, and
- **Equipment specifications** for power distribution, lighting, ELV, and containment.

14.7.3 Function Analysis Worksheet

Project Name: Commercial Tower VE Study:
Project Number: [To be filled]
Date Prepared: April 2022
Prepared By: [To be filled]
Workshop Date: April 15–24, 2022
System Being Analyzed: Power Distribution, Panels, Lighting Control, Emergency Supply

Component/ Subsystem	Primary Function	Secondary Functions	Function Classification	Performance Requirements	Current Cost (USD)	Worth (USD)	Cost/ Worth Ratio	Function Dependencies
1. Busduct (Vertical Risers)	Distribute power	Provide modular tap-off	Basic	800 A, 3P4W, IP55, AL/ Resin body	$120,000	$85,000	1.41	SMDBs, riser layout, ceiling coordination
2. SMDB (Low-Load Areas)	Distribute subcircuits	Isolate floor-level circuits	Secondary	250 A, 12-way MCCB panel	$42,000	$26,000	1.62	DBs, cable tray layout
3. MLVB (Form 4b Standard)	Distribute main power	Protect and monitor loads	Basic	1600 A, IP54, segregated zones	$420,000	$345,000	1.22	Transformers, riser feeders
4. Lighting Control (DALI)	Manage lighting levels	Enable monitoring/ dimming	Support	DALI controller, 64 nodes/ loop	$92,000	$65,000	1.41	Lighting circuits, BMS integration
5. Chiller Busduct Feed	Feed mechanical loads	Provide clean termination	Secondary	630 A, 3P4W, busduct feeder	$65,000	$50,000	1.30	HVAC, structural openings

Notes:

- **Cost/Worth Ratio >1.3** in most cases indicates high potential for VE.
- Items such as the **DALI system** and **underloaded SMDBs** show particularly weak value-to-cost efficiency, justifying their selection as VE priorities.
- Functional dependencies are essential when evaluating impact of changes—especially for risers, panels, and busbar feeds.

14.7.4 VE Idea Generation LOG—Electrical Systems

Project Name: Commercial Tower VE Study:
Project Number: [To be filled]
Brainstorming Date: April 15, 2022
Facilitator: [To be filled]
Focus Area: Power Distribution, Equipment Optimization, Controls
Target Functions: Distribute Power, Control Lighting, Support Critical Loads
Target Cost Reduction: $200,000+

Idea ID	Target System/ Component	Function Addressed	Current Solution	Alternative Proposal	Potential Benefits	Potential Challenges	Feasibility*	Contributor	Notes
VE-E1	Riser busducts	Distribute power	Aluminum busducts in all risers	Replace with XLPE power cables + trays in noncritical risers	Lower cost, easier routing, frees riser space	Cable tray design adjustments, thermal rating review	High	[TBF]	Verify compliance with IEC 60364
VE-E2	Sub-main panels (SMDBs)	Distribute floor loads	SMDBs installed on all levels, some underloaded	Consolidate DBs into adjacent floor SMDBs, remove low-use panels	Fewer panels, reduced cost, smaller closets	Feeder cable sizing and voltage drop	High	[TBF]	Limited to floors with low load demand
VE-E3	Chiller feed busducts	Feed HVAC systems	Busduct routing to rooftop chillers	Replace with riser-mounted XLPE armored cables on internal containment systems	Eliminates proprietary busduct, simplifies routing, reduces riser congestion	Thermal derating in shaft, coordination with riser layout and supports	Medium	[TBF]	Requires shaft design verification and load testing
VE-E4	Vertical busbar tap-offs	Distribute power to DBs	Each DB has a separate tap-off	Merge DBs to shared tap-offs, extend busbar horizontally	Fewer terminations, reduced busbar length, easier support	Revised SLD and load balancing	High	[TBF]	Load grouping study required
VE-E5	Main LV switchboards (MLVBs)	Distribute bulk power	All boards specified as Form 4b, IP54	Use Form 3b IP42 where full segregation is not required	Smaller boards, easier maintenance, cost savings	Client/authority approval needed for classification change	High	[TBF]	Review with authority during VE workshop
VE-E6	Lighting control in BOH areas	Manage lighting control	Full DALI control system everywhere	Use conventional switching and PIR in basements/ plantrooms	Reduced complexity, faster install, lower capital cost	Requires separation of control types	High	[TBF]	DALI retained in front-of-house zones
VE-E7	Cable tray sizing	Support cable routing	Designed using 2D cable spacing assumption	Use 1D spacing with revised derating, reduce tray widths	Reduced tray size and weight, easier coordination	Requires recalculation and verification			

* *Feasibility*: Preliminary assessment based on technical, financial, and compliance considerations. Ratings are indicative and subject to detailed design review.

Additional VE Ideas (Not Pursued)

Idea ID	Target System/ Component	Function Addressed	Current Solution	Alternative Proposal	Potential Benefits	Potential Challenges	Feasibility*	Contributor	Notes
VE-X1	MV transformer cooling	Heat dissipation	Oil-immersed ONAN cooling	Replace with dry-type cast resin transformers	Reduced maintenance, environmental benefit	Higher cost, larger footprint	Low	[TBF]	Rejected due to cost-performance mismatch
VE-X2	Emergency generator rating	Supply critical loads	Two generators (800 + 400 kVA)	Use single 1200 kVA generator with load shedding	Simplifies setup, lower install cost	No redundancy, load shedding complexity	Medium	[TBF]	Rejected due to resilience requirements
VE-X3	Lighting fixture mounting	Illuminate areas	Surface-mounted fixtures throughout	Replace with suspended tracks for general office areas	Flexibility in space planning, aesthetic upgrade	Marginal cost impact, coordination required	Medium	[TBF]	Deferred for architectural coordination
VE-X4	Cable type—copper vs. aluminum	Conduct power	All power cables copper	Use aluminum for selected feeders >100 m	Cost reduction on large runs	Requires larger size, stricter terminations	Low	[TBF]	Rejected based on installation constraints
VE-X5	Central battery for all lighting	Emergency lighting supply	Mix of self-contained + central battery	Shift to fully central battery system with zone monitoring	Simplified maintenance, better monitoring	Single point of failure risk	Medium	[TBF]	Partially accepted for back-of-house zones

* *Feasibility*: Preliminary assessment based on technical, financial, and compliance considerations. Ratings are indicative and subject to detailed design review.

Together with VE-E1 to VE-E7, these ideas show that:

- **Twelve total VE ideas were considered**
- **Seven were selected** based on value, feasibility, and coordination alignment
- **Five were deferred or rejected** for valid technical or cost reasons

14.7.5 VE Idea Evaluation Matrix

14.7.5.1 SCORES (Summary Table)

Idea ID	Initial Cost	LCC Impact	Energy Eff.	Performance	Maintain.	Schedule	Construct.	Total Score
VE-E1	5	4	3	4	5	4	5	**30**
VE-E2	4	4	2	3	4	4	5	**26**
VE-E3	4	4	2	4	4	3	3	**24**
VE-E4	3	3	2	4	3	4	4	**23**
VE-E5	5	5	3	4	4	5	4	**30**
VE-E6	4	4	5	4	5	5	4	**31**
VE-E7	3	3	2	3	3	4	5	**23**

Evaluation Outcome:

- **Highest Scores:** VE-E5, VE-E6, VE-E1
- **Middle Range:** VE-E2, VE-E3
- **Lower Priority (but still viable):** VE-E4, VE-E7
- **Rejected/Deferred Ideas:** Not evaluated in this matrix—documented in the idea log only

14.7.5.2 VE Idea Evaluation—Narrative Summary

Following the structured brainstorming session, each shortlisted VE proposal (VE-E1 through VE-E7) was evaluated against seven criteria: initial capital cost savings, LCC impact, energy efficiency, system performance, maintainability, schedule impact, and constructability. A weighted scoring system using a 1–5 scale was applied by a multidisciplinary team of engineers, cost planners, and client representatives.

The evaluation revealed the following insights:

- **VE-E6 (Lighting Control Rationalization)** received the highest score due to its clear impact on cost, simplicity in implementation, and positive energy efficiency outcome. It provided a strong balance between savings and operational practicality.
- **VE-E5 (MLVB Specification Optimization)** and **VE-E1 (Riser Busduct Replacement)** also ranked highly, offering significant savings while preserving or improving space and maintainability.
- **VE-E2 (SMDB Consolidation)** and **VE-E3 (Chiller Feeder Cabling)** achieved moderate scores, supported by their straightforward implementation and measurable spatial improvements.
- **VE-E4 (Tap-off Merging)** and **VE-E7 (Cable Tray Sizing)** scored lower but were still considered viable due to their minimal disruption risk.
- **Deferred ideas (VE-X1 to VE-X5)** were documented for traceability but excluded from scoring due to technical or strategic limitations.

The matrix provided a defensible basis for selecting which proposals to carry forward into detailed VE implementation, ensuring both engineering integrity and commercial viability.

14.7.6 Electrical Cost Breakdown (Pre- and Post-VE) (Table 14.11)

- **Total Electrical Budget:** $6,400,000
- **VE Scope:** Items VE-E1 to VE-E7 applied
- **Category Weights:** Based on actual BOQ proportions you provided (Table 14.12)

Table 14.11 Overall Electrical Budget Summary (Pre- and Post-VE)

Item	Value
Total Project Construction Budget	$40,000,000 (Core & Shell + MEP)
Total Electrical Budget (Pre-VE)	$6,400,000
Electrical % of Overall Budget	16%
Electrical Cost per Square Meter	$67/m^2 (based on 95,000 m^2 GFA)
Post-VE Electrical Budget	$5,892,000
Net VE Savings	$482,000
Savings as % of Electrical Budget	7.54%
Industry Benchmark (for Similar Projects)	$65–$75/m^2

Table 14.12 Overall Electrical Budget Summary (Pre- and Post-VE)

System Component	% Allocation	Original Cost (USD)	Post-VE Cost (USD)	VE Savings (USD)	Notes (VE Ref)
Main Distribution Boards (MDBs)	21%	$1,344,000	$1,210,000	$134,000	VE-E5: Form class/ IP opt
Sub-Main Panels (SMDBs)	11%	$704,000	$609,000	$95,000	VE-E2: Consolidated
Power Cabling and Termination	25%	$1,600,000	$1,420,000	$180,000	VE-E1, VE-E3
Lighting Fixtures	15%	$960,000	$960,000	—	No VE applied
Lighting Control Systems	5%	$320,000	$245,000	$75,000	VE-E6: Hybrid approach
Emergency Power Systems (Gen)	8%	$512,000	$512,000	—	No change (scope retained)
UPS Systems	4%	$256,000	$256,000	—	No VE applied
Cable Containment Systems	5%	$320,000	$296,000	$24,000	VE-E7: Tray sizing opt
Earthing & Lightning Protection	3%	$192,000	$192,000	—	No VE applied
ELV/ICT Systems	3%	$192,000	$192,000	—	No VE applied
TOTAL	**100%**	**$6,400,000**	**$5,917,400**	**$482,600**	**7.54% savings realized**

Summary:

- **Total VE Savings:** \$482,600
- **Post-VE Electrical Budget:** \$5.92M
- **Savings Realized From:**
  - Equipment reclassification (VE-E5)
  - System consolidation (VE-E2)
  - Smart replacement strategies (VE-E1, VE-E3)
  - Simplified controls (VE-E6)
  - Containment optimization (VE-E7)

Clarifying Notes:

- The electrical scope covers all systems from MV intake to final lighting and ELV, excluding tenant-specific fit-out.
- The total project cost reflects core and shell plus complete MEP systems, consistent with a mid-rise commercial tower in the GCC.
- The original electrical budget was benchmarked using both internal cost models and verified against an external BOQ from a larger tower project.
- All VE-related adjustments were applied to panel specification, riser strategy, containment sizing, and lighting controls.

14.7.7 Energy Efficiency Impact Assessment (Table 14.13)

Table 14.13 Energy Efficiency Impact Assessment

Field	Description
VE Reference	VE-E6: lighting control rationalization
System Affected	Lighting control system (DALI and conventional)
Original Strategy	Full DALI coverage across all floors, including BOH, parking, and plantrooms
Revised Strategy	DALI in public/office zones only; BOH and technical zones use PIR + switches
Energy Saving Mechanism	Use of motion-activated control (occupancy sensors) in low-use areas
Estimated Controlled Area	~30% of total lighting zones (~28,500 m^2 of BOH and technical areas)
Baseline Lighting Load Density	10 W/m^2 (avg across the project)
Adjusted Load (Controlled)	5–7 W/m^2 in optimized zones via occupancy control
System-Level Efficiency Gain	10–15% reduction in lighting energy for those areas
Operational Benefit	Reduced BMS complexity, faster commissioning, lower maintenance needs
Compliant With	IEC 60364-8-1, ASHRAE 90.1, GCC Energy Code
Energy Simulation Required?	Not needed for code compliance; advisable for LEED modeling
Net Annual Energy Reduction	Estimated 120,000–160,000 kWh/year (BOH zones only)
Life-Cycle CO_2 Impact	Reduction of 80–100 tons CO_2/year (based on 0.65 kg CO_2/kWh)

Conclusion

The revised control scheme offers a **measurable energy benefit** in noncritical zones without compromising user comfort or flexibility. It also simplifies the installation, reduces commissioning time, and aligns better with occupancy patterns—making it a **highly efficient VE move**.

14.7.8 VE Summary Report

Project Information:

- **Project Name:** Gulf Commercial Tower VE Study
- **Project Number:** [To be filled]
- **Report Date:** May 2022
- **Prepared By:** [To be filled]
- **Review Period:** April 15–24, 2022

1) **Executive Summary**
   Study Purpose:
   To identify cost-effective, performance-compliant opportunities for optimizing the electrical systems in a mid-sized GCC commercial high-rise project during the detailed design phase.

   Methodology Overview:
   The VE process included team workshops, function analysis, system cost review, feasibility scoring, and LCC evaluation. Templates from the VE handbook (Chapter 6) were applied to document all proposals and outcomes.

Metric	Result
Total Potential Savings Identified	$690,000
Number of VE Proposals Presented	12
Number of VE Proposals Accepted	7
Total Approved Savings	$482,600
% of Original Electrical Budget	7.54% of $6.4M

2) **VE Study Methodology**

   The VE study was conducted in alignment with standard electrical engineering VE practices and the structured templates provided in Chapter 6 of this guidebook. The methodology included the following phases:

   a) **Team Composition**

      A multidisciplinary VE team was assembled, including:

      - **Electrical designers and cost engineers**
      - **Client-side project representatives**
      - **Contractor and installation advisors**

Each member contributed to evaluating functional needs, constructability, compliance, and operational risks associated with proposed alternatives.

b) **Functional and Cost Analysis**

The team employed the **function analysis worksheet** to categorize components into basic, secondary, or support functions. This allowed focus on systems with the highest cost-to-function ratios, such as:

- Overspecified switchboards
- Redundant sub-main panels
- Full-coverage lighting control systems

c) **Idea Generation and Screening**

A total of **12 ideas** were generated during the brainstorming session. Each was recorded in the **VE idea generation log**, and evaluated using:

- **Initial Cost Savings**
- **LCC Impact**
- **Performance and Reliability**
- **Ease of Implementation**

The **VE evaluation matrix** was then used to assign comparative scores and prioritize high-impact ideas.

d) **Technical and Financial Validation**

For the shortlisted seven proposals:

- CAPEX and LCC data were extracted from a validated BOQ and mapped into the **cost breakdown sheet**.
- One proposal (VE-E6) underwent an **energy efficiency impact assessment**, confirming measurable annual energy and CO_2 reduction.

e) **Documentation and Outcome**

All accepted ideas (VE-E1 to VE-E7) were consolidated in:

- The **VE summary report** (this document),
- The **implementation tracking log**, and
- Supporting calculations in the **LCC and BOQ sheets**.

This structured approach ensured traceability, objectivity, and compliance with local authority standards and international engineering practice.

3) **Summary of Accepted VE Proposals**
This section outlines the seven VE items that were accepted for implementation based on their technical feasibility, cost impact, and alignment with the project's operational and spatial goals (Table 14.14).

Table 14.14 Summary of Accepted VE Proposals

VE Ref	System/ Component	Original Design	VE Proposal	Estimated Savings (USD)	Remarks
VE-E1	Vertical power risers	Aluminum busducts throughout risers	Replace with XLPE cables + trays in noncritical risers	$134,000	Verified via thermal/ VD calculations
VE-E2	Sub-main panels (SMDBs)	SMDBs on every level, some lightly loaded	Consolidate floor DBs into fewer SMDBs	$95,000	Applied in basement and BOH zones
VE-E3	Chiller feed via busducts	55 m busduct to rooftop chillers	Replace with riser-mounted XLPE armored cables	$80,000	Eliminates busduct complexity, simplifies riser coordination
VE-E4	Tap-off configuration	Individual tap-offs per DB	Merge DBs under shared tap-offs	0	Shortened busbar runs, fewer joints
VE-E5	Main LV boards (MLVBs)	All boards form 4b IP54	Use form 3b IP42 where segregation not required	$75,000	Complies with IEC 61439
VE-E6	Lighting control strategy	Full DALI everywhere	Use PIR + switches in BOH and technical zones	$75,000	Reduced energy and install effort
VE-E7	Cable tray sizing	2D spacing basis	Adopt 1D spacing with validated derating	$24,000	Reduced tray width by up to 300 mm
TOTAL	—	—	—	**$482,600**	**7.54% of original electrical cost**

14.7.9 VE Implementation Tracking Log

14.7.9.1 Implementation Status Summary

Field	Value
Project Name	Gulf Commercial Tower VE Study
Project Number	[To be filled]
Report Date	May 2022
Prepared By	[To be filled]
Report Period	April–May 2022
Total Approved VE Proposals	7
Proposals Fully Implemented	3
Proposals In Progress	4
Proposals Delayed/At Risk	0
Estimated Total Savings	$482,600
Actual Confirmed Savings	$190,000 (as of May 2022)

Table 14.15 Detailed Implementation Tracking

VE Ref	VE Item Description	Responsible Party	Implementation Status	Target Completion	Confirmed Savings (USD)	Remarks
VE-E1	Replace riser busducts with XLPE cables	Electrical consultant	Fully implemented	April 2022	$134,000	Design revised and IFC issued
VE-E2	Consolidate SMDBs in low-load zones	Consultant + contractor	In progress	June 2022	—	DB grouping study ongoing
VE-E3	Riser-mounted XLPE cabling for chillers	Contractor	In progress	June 2022	—	Routing revised to internal riser; structural check ongoing
VE-E4	Merge tap-offs on vertical busbars	Electrical consultant	In progress	July 2022	—	Awaiting revised SLD release
VE-E5	Reduce form class for selected MLVBs	Consultant + supplier	Fully implemented	April 2022	$75,000	Revised specs approved
VE-E6	Hybrid lighting control with PIRs	Lighting consultant	Fully implemented	April 2022	$81,000	BOH areas reissued with PIR layouts
VE-E7	Optimize tray sizing via 1D spacing	MEP coordinator	In progress	May 2022	—	Coordination with structural ongoing

Table 14.15 summarizes the implementation status of approved VE proposals, including responsible parties, completion targets, confirmed savings, and remarks.

Summary Status:

- **Total VE Items: 7**
- **Fully Implemented: 3**
- **In Progress: 4**
- **Delayed/At Risk: 0**
- **Confirmed Savings to Date: $190,000**
- **Pending Confirmation: $292,600**

14.7.10 Lessons Learned Register

This section captures key takeaways from the Gulf Commercial Tower VE study to guide future projects. It draws from the analysis, implementation tracking, and stakeholder feedback (Table 14.16).

Table 14.16 VE Lessons Learned Register

Category	Observation/Lesson	Recommendation for Future Projects
Scope Definition	Some electrical subsystems (e.g., lighting control) were over-specified in low-occupancy areas	Conduct early-use zoning analysis to match controls to occupancy patterns
Space Coordination	Riser and electrical room space was tighter than assumed in early phases	Involve MEP coordination early in schematic design; reserve flexible zones
System Specification	Several components used high form-class or IP ratings without functional need	Match form class/IP ratings to actual environment and function during design
Life-Cycle Impact	LCC benefits were only apparent after structured modeling	Introduce LCC analysis tools earlier in the design process
Stakeholder Buy-In	VE-E5 and VE-E6 moved faster due to early consultant/client agreement	Engage key decision-makers during initial VE workshops
Documentation	Structured templates (Chapter 6) improved clarity, approvals, and traceability	Maintain formalized logs and scoring systems throughout VE process
Implementation	Some coordination-heavy items (VE-E2, VE-E4) needed phased rollout	Phase complex VE items to align with design freeze and construction schedule

Notice to Readers

This case study is presented as part of *Chapter 14–Complete Case Study: Value Engineering in a Commercial High-Rise Project* within this publication. It has been prepared and included for the purpose of professional education, technical reference, and the demonstration of VE methodologies applied in electrical systems design.

General Disclaimer:

- This case study is for educational and illustrative purposes only. It is not intended to serve as engineering, legal, financial, or construction advice, and must not be relied upon as a substitute for independent professional consultation.
- Readers are advised to apply professional judgment and consult with licensed experts when making decisions related to real-world design, specification, procurement, or regulatory compliance.

Factual Basis and Adaptation:

- The case study reflects a composite scenario derived from real engineering experience, reference data, and historical project records. All values, system specifications, and implementation details have been modified or anonymized to protect confidentiality and avoid direct attribution to any specific project, client, or entity.
- No proprietary, confidential, or restricted information has been knowingly disclosed. If any similarity to actual projects or organizations is observed, it is purely coincidental and unintended.

Cost Data and Market Conditions:

- All cost figures presented in this case study are illustrative and may not reflect current market conditions, regional variations, or specific project circumstances.
- The costs shown were developed for educational purposes and may vary significantly from actual market prices due to factors including but not limited to:
  - Geographic location and local market conditions
  - Supply chain fluctuations and material availability
  - Currency exchange rates and economic conditions
  - Project-specific requirements and specifications
  - Temporal variations in commodity prices (copper, aluminum, steel)
- Readers should not rely on these cost figures for budgeting or estimation purposes and should obtain current pricing from qualified suppliers and contractors.

Index

d

e

f

t

u